Risk, Systems and Decisions

Health, environment, security, energy, technology are problem areas where man-made and natural systems face increasing demands, giving rise to concerns which touch on a range of firms, industries and government agencies. Although a body of powerful background theory about risk, decision, and systems has been generated over the last several decades, the exploitation of this theory in the service of tackling these systemic problems presents a substantial intellectual challenge. This book series includes works dealing with integrated design and solutions for social, technological, environmental, and economic goals. It features research and innovation in cross-disciplinary and transdisciplinary methods of decision analysis, systems analysis, risk assessment, risk management, risk communication, policy analysis, economic analysis, engineering, and the social sciences. The series explores topics at the intersection of the professional and scholarly communities of risk analysis, systems engineering, and decision analysis. Contributions include methodological developments that are well-suited to application for decision makers and managers.

More information about this series at http://www.springer.com/series/13439

David M. Berube
Editor

Pandemic Communication and Resilience

Springer

Editor
David M. Berube
North Carolina State University
Raleigh, NC, USA

ISSN 2626-6717 ISSN 2626-6725 (electronic)
Risk, Systems and Decisions
ISBN 978-3-030-77343-4 ISBN 978-3-030-77344-1 (eBook)
https://doi.org/10.1007/978-3-030-77344-1

This Springer imprint is published by the registered company Springer Nature Switzerland AG
The registered company address is: Gewerbestrasse 11, 6330 Cham, Switzerland

Introduction

"Just because you have an advanced degree does not mean you are a science communicator." The time this line has been spoken by me and many of my communication scholar friends at government meetings to colleagues in science and engineering over the last three decades is too numerous to count.

> "Scientists who engage in science communication must acknowledge that their area of expertise is deep but narrow, and recognize the limitations in their own knowledge.... It is equally imperative to emphasize that being an expert on a topic doesn't automatically make a scholar qualified to communicate it to a nonscientific audience" (Anderson 2020).

There does exist a discipline in communication that focuses on the claims, arguments, and messages designed and delivered to stakeholders. This discipline has been around for over two millennia in one form or another.

Before 2020, the fields of communication and rhetoric were dominated by rhetoricians who followed classical studies involving foundational works by Plato, Aristotle, Cicero, and Quintillian. They were eventually joined by outstanding Muslim scholars of the oratory arts and homiletics such as Ibn al-Jawzi and al-Harfush and the great Ecclesiastical scholars like Augustine of Hippo and Thomas Aquinas.

In the eighteenth century, there was a with a notable split when epistomologists led by Richard Whatley and George Campbell focused on evidence and proof and distinguished themselves from a second movement known as belles-lettres (style, culture, and art) advocated by Hugh Blair which, in turn, was followed by an elocutionary movement proselytized by Thomas Sheridan.

Also, during the eighteenth and nineteenth centuries, scholars Emmanuel Kant, Fredrich Nietzsche, Giambattista Vico informed the criticism of communication and rhetoric with their scholarship in politics and sociology.

Communication (notice no "s") as an academic discipline seemed to have blossomed in the twentieth century with notables including the literary critic Kenneth Burke, Herbert W. Simons, Stephen Toulmin, and others. Contemporarily, colleges of humanist theorists contributed to the corpus of studies in communication as well: Marsilio Ficino, Ernesto Grassi, Martin Heidegger, and many more. When modernism was challenged by post-modernists, Michel Foucault, Jacques Derrida,

Gilles Deleuze, and others, the field unearthed linkages between communication and power and influence.

In the late twentieth and early twenty-first centuries, communication studies drew heavily from cognitive psychology and behavioral economics: psychometrics, heuristics, and nudging as complementary research approaches in understanding messaging and arguments surfaced and was joined by "big data" investigators as we try to make sense of Internet-mediated messaging.

Unfortunately, everyone is not a great spokesperson. While there have been some outstanding envoys in public science and technology, it is not a natural talent. It is an art and a science. Professionals in communication do much more than teach public speaking. We do not base our choices in building messages and campaigns on intuition. Rather we build messages and campaigns by examining social science and humanities research to work from data rather than from instinct or "educated guesses." And, we pretest what we do lest we make a mistake than might take decades to repair as climatologists have realized when they approached climate change communication as a platform to teach science to the public. The public was not impressed, and now communication scholars are trying to clean up the mess.

Speakers such as Rachel Armstrong, Brian Cox, Neil deGrasse Tyson, Richard Feynman, Michio Kaku, Katherine Hayhoe, Carl Sagan, Anthony Fauci, Bill Nye, David Pogue, and David Suzuki are national treasures as are science writers like Bill Bryson, Rachel Carson, James Gleich, Jane Goodhall, Stephen Jay Gould, Brian Greene, Steven Pinker, Simon Singh, and many others. They are some of my favorite and seem to have developed a talent for communicating with a very broad range of stakeholders.

However, for the remainder of the scientific community it is fortunate that communication studies has strong sub-disciplines in science communication, especially health communication. It is time for all of us to take science communication very seriously so as to preclude communication disasters from the past, including messaging associated with subjects such as vaccine hesitancy and events like weather disaster. The science communication community is here to help and not soliciting their participation in science messaging is unfortunate and needs to be changed.

The chapters selected for inclusion in one of the first edited volumes on the inner workings of designing as well as understanding complex messaging about outbreaks, epidemics, and pandemics draws from a broad team of students and scholars. The goal as editor was to bring on some outstanding doctoral students as well as professors at the assistant level and early in their careers as well as full professors with long histories and outstanding publication records. In addition, there was a very strong effort to include women and people of color as well as strong Western and non-Western scholars and perspectives. Each part of this edited volume will introduce the reader to principal theory associated with communicating pandemics and building resiliency among stakeholders as well as providing some examples of what communication scholars can bring to the health communication challenge from ethnographic, qualitative, and quantitative disciplines.

They represent many different approaches and allow the readership to sample some of the published research found in professional communication journals.

This book is separated into six parts, and each part is divided into individual chapters. The first part introduces us to the biopolitics of communication (Hall) and messaging and trust (Wise) and follows with an examination of the 1918 Influenza and SARS pandemics (Ding & Tang).

In the second part, we are fortunate that a team of scholars were willing to pull a summary of dominant health communication theories and models together (Spradley & Spradley). They are followed by two chapters on important developments in communication theory: one psychometrics and mental models (Tallapragada) and a very recent development in risk theory (Cummings). It concludes with an analysis of digital social media using one of the models explicated by the Spradleys (Powell).

Part three is about four phenomena in pandemics that affect resiliency and involves messaging during COVID-19. News avoidance and how it affects public understanding of messaging (Eng) is followed by a tongue-in-cheek examination of panic-buying during pandemic events (Berube). The role of celebrities and its interplay with variable interactions with resilience is examined by reflecting on "Good News" (Hatfield). This part concludes by exploring social media messaging by the travel and tourism industry (Edwards).

Part four begins with a rhetorical analysis of Trump's war metaphor as part of America's health campaign during COVID-19 (Esfandiary). In a unique examination of communication of a non-verbal nature, we have a review of mask fashion (Wollslager). This part ends with a comprehensive and detailed study of masking and how it was communicated in the USA by examining Tweets (Bogomoletc, Binder & Goodwin).

No volume would be complete without addressing how pandemics affect women and people of color. It starts with a review of media coverage of Zika cases in the USA and how it generated fear (Sperry & Lane). Next, a team from University of Cincinnati College of Medicine and Cincinnati Children's Hospital address multisector situational awareness and provide a regional study of life during COVID-19 (Hartley, Beck, Seid, Cronin, Shuler, Rainey, Zafar, Kahn & Margolis). The part ends with an autoethnography by and about being pregnant as a woman during COVID-19 (Achiopoli).

The book ends in the last part with international case studies. We begin with a review of Ebola in Liberia and village responses (Daugherty). Next, we find quantitative analysis about school re-opening messaging during COVID-19 in Spain, South Africa, and the Netherlands (Nane, van Schalkwyk, Dudek, Torres-Salinas, Costa, and Robinson-Garcia). News consumption in Ibero-America during COVID-19 follows (Capoano & Rodrigues-Costa). We move next to India, Pakistan, and Bangladesh with a chapter on media and resilience (Saha). Then, we seek the first of two closing chapters on fake news about COVID-19: first in Indonesia (Muzykant, Muqsity & Patomo) and another investigating fake news on WhatsApp in Brazil (Klein).

It was impossible to include every chapter that was submitted for review and that was unfortunate since some were outstanding and hopefully will surface in other publications.

Finally, this work would have been impossible without the many libraries and the men and women who find us the most obscure materials imaginable. As academics, there are hundreds to thank from department heads (some contributing chapters), deans, and other college and university personnel who help us do the jobs we do. For some of us, deep thanks go out to special organizations and government agencies who have found financial opportunities for us to fund the work we do and help us to hire graduate students and postdoctoral scholars.

May God bless the families of our brothers and sisters who have suffered from morbidity or mortality from pandemics. Dying alone in a hospital and away from the ones you love must be one of the worst experiences imaginable.

And for the loved ones who were kept beyond the reach of a grandparent, parent, partner, spouse, or child, it will get easier to get through the day, but we cannot bring your loved one back. I hope this book opens a few eyes to the risks from pandemics and what we can be expected to do to turn pandemics into an historical artifact, if that is possible at all.

David M. Berube

Reference

Anderson S (2020) Opinion: Being scientists doesn't make us science communicators. The Scientist. December 17. https://www.the-scientist.com/news-opinion/opinion-being-scientists-doesnt-make-us-science-communicators--68294. Accessed on 10 Mar 2021

Contents

Contributors

Ashley Archiopoli University of Houston-Downtown, Houston, USA

Victor Barabash Peoples' Friendship University of Russia (RUDN University), Moscow, Russia

Andrew F. Beck Cincinnati Children's Hospital, Cincinnati, OH, USA; University of Cincinnati College of Medicine, Cincinnati, OH, USA

David M. Berube North Carolina State University, Raleigh, USA

Andrew R. Binder North Carolina State University, Raleigh, NC, USA

Ekaterina Bogomoletc North Carolina State University, Raleigh, NC, USA

Edson Capoano Communication and Society Research Centre (CECS) of the University of Minho, Minho, Portugal

Pedro Daniel Rodrigues Costa Communication and Society Research Centre (CECS) of the University of Minho, Minho, Portugal

Rodrigo Costas DST-NRF Centre of Excellence in Scientometrics and Science, Technology and Innovation Policy, Centre for Research On Evaluation, Science and Technology, Stellenbosch University, Stellenbosch, South Africa;
Centre for Science and Technology Studies, Leiden University, Leiden, The Netherlands

Susan Cronin Cincinnati Children's Hospital, Cincinnati, OH, USA

Christopher L. Cummings North Carolina State University, Raleigh, NC, USA; Iowa State University, Ames, IA, USA

Crystal D. Daugherty Northern Kentucky University, Highland Heights, KY, USA

Huiling Ding North Carolina State University, Raleigh, NC, USA

Jonathan Dudek Delft Institute of Applied Mathematics, Delft University of Technology, Delft, The Netherlands;

Centre for Science and Technology Studies, Leiden University, Leiden, The Netherlands

Jennifer Edwards Tarleton State University, Stephenville, TX, USA

Nicholas Eng The Pennsylvania State University, State College, USA

Esmaeil Esfandiary Tuskagee University, Tuskegee, AL, USA

Elizabeth Fish Hatfield University of Houston-Downtown, Houston, TX, USA

Jean Goodwin North Carolina State University, Raleigh, NC, USA

Shreya Gopi Nanyang Technological University, Singapore, Singapore

Kevin Hall Philipps-University Marburg, Marburg, Germany

David M. Hartley Cincinnati Children's Hospital, Cincinnati, OH, USA; University of Cincinnati College of Medicine, Cincinnati, OH, USA

Robert Kahn Cincinnati Children's Hospital, Cincinnati, OH, USA; University of Cincinnati College of Medicine, Cincinnati, OH, USA

Eloisa J. C. Klein Federal University of Pampa, São Borja, Brazil

Gina Lane William Jewell College, Liberty, USA

Peter A. Margolis Cincinnati Children's Hospital, Cincinnati, OH, USA; University of Cincinnati College of Medicine, Cincinnati, OH, USA

Munadhil Abdul Muqsith Peoples' Friendship University of Russia (RUDN University), Moscow, Russia

Valerii L. Muzykant Peoples' Friendship University of Russia (RUDN University), Moscow, Russia

Gabriela F. Nane Delft Institute of Applied Mathematics, Delft University of Technology, Delft, The Netherlands

Aisha Powell Howard University, Washington, USA

Risky Ridho Pratomo Pembangunan National Veteran Jakarta University, Depok, Indonesia

Laura Raney Cincinnati Children's Hospital, Cincinnati, OH, USA

Nicolas Robinson-Garcia Delft Institute of Applied Mathematics, Delft University of Technology, Delft, The Netherlands

Sonny Rosenthal Nanyang Technological University, Singapore, Singapore

Ali Saha Monash University, Melbourne, Australia

Christine L. Schuler Cincinnati Children's Hospital, Cincinnati, OH, USA; University of Cincinnati College of Medicine, Cincinnati, OH, USA

Michael Seid Cincinnati Children's Hospital, Cincinnati, OH, USA; University of Cincinnati College of Medicine, Cincinnati, OH, USA

Elizabeth Sperry William Jewell College, Liberty, USA

Elizabeth Spradley Stephen F. Austin State University, Nacogdoches, TX, USA

R. Tyler Spradley Stephen F. Austin State University, Nacogdoches, TX, USA

Meghnaa Tallapragada Department of Advertising & Public Relations, Klein College of Media & Communication, Temple University, Philadelphia, PA, USA

Yingying Tang Auburn University, Auburn, AL, USA

Daniel Torres-Salinas Departamento de Información y Comunicación, Universidad de Granada, Granada, Spain

François van Schalkwyk DST-NRF Centre of Excellence in Scientometrics and Science, Technology and Innovation Policy, Centre for Research On Evaluation, Science and Technology, Stellenbosch University, Stellenbosch, South Africa

Kurt Wise University of West Florida, Pensacola, FL, USA

M. Eilene Wollslager Regis University, Denver, USA

Muhammad Zafar University of Cincinnati College of Medicine, Cincinnati, OH, USA

Part I
Resilience and Communication

Chapter 1
Biocommunicability. The Biopolitics of Pandemic Communication.

Kevin Hall

Abstract Communication, especially during the Covid-19 pandemic, is about governing people. Rather than relying solely on content, communication is also performative. This chapter discusses Charles L. Briggs and Daniel C. Hallin's concept of (bio-)communicability to consider how roles, values and rationalities are co-produced together with the information in the process of communication. (Bio-)communicability describes the performative way by which news coverage projects models about how phenomena come to be, how knowledge about them circulates and who should attend to them and how. The chapter begins by offering some of the critiques directed at health communication and the models it is based upon. Rather than exclusively drawing on health, health behavior is guided by multiple and competing rationalities. The chapter elaborates Briggs and Hallin's concept of communicability and present three models of biocommunicability. The chapter concludes by offering some thoughts on how to utilize these analytical concepts for communicating about pandemics.

Keywords Actor-network theory · Biocommunicability · Biopolitics · Covid-19 · Critical discourse analysis

1.1 Introduction

Health communication and especially coverage of health related news in the media were omnipresent during the COVID-19 pandemic. Messages like "Wear a mask! Wash your hands! Keep your distance!" aim at modifying the behavior of the population in order to reduce infections. During the COVID-19 pandemic these rules (and rigorous contact tracing) became Germany's central strategy to "flatten the curve". At first sight, these messages seem simple enough. However, they do more than offer health advice. They communicate a set of expectations that states (here the German state) hold towards what it means to be a responsible citizen during this

K. Hall (✉)
Philipps-University Marburg, Marburg, Germany
e-mail: kevin.hall@uni-marburg.de

D. M. Berube (ed.), *Pandemic Communication and Resilience*, Risk, Systems and Decisions, https://doi.org/10.1007/978-3-030-77344-1_1

pandemic. Pandemic citizenship (Davis et al. 2011: 917–918) defines how individuals should relate their bodies and selves to one another and to society in order to control the pandemic. In this respect communication has a biopolitical valence as it draws upon what philosopher Michel Foucault (1993: 203) has called the "techniques or technology of the self". Communication is entangled with governing people.

In a broad definition biopolitics refers to practices and politics that aim "at the administration and regulation of life processes on the level of populations" (Lemke 2011: 4) such as susceptibility of populations to diseases, rates of birth and death. In this sense populations are governed not so much by intervening in the life of each and every singular individual. Rather, biopolitical interventions target the biological features of human existence in their aggregated and measured form of the population. In this respect metrics play a fundamental role in the government of the population (Adams 2016). Disciplines such as statistics, demography, epidemiology and public health play an important role in producing the knowledge that is necessary for developing biopolitical strategies and exercising power. Biopolitical interventions aim at raising the likelihood of certain desired, statistically distributed events in the population and lowering the chances for undesired events. Policies regarding birth control, access to health care and healthy alimentation, as well as workplace safety are just examples of typical biopolitical domains. However, biopolitics is not restricted to the purely biological aspects of human life. Within Foucault's framework of governmentality biopolitics is entangled with "subjectivation processes and moral-political forms of existence" (Lemke 2011: 48).

Foucault's concept of governmentality is not restricted to how states acting indirectly on the living conditions of a population change the latter's free conduct. It also encompasses techniques of power that address the individual as moral being in its capacity to regulate its own conduct (Foucault 2000). Here the aforementioned techniques of the self come into play. Techniques of the self permit the individual to get to know its own motives and goals, declare these and work on its personality, and in doing so aligning itself with and ultimately achieving its goals. Government for Foucault takes "into account the points where the technologies of domination of individuals over one another have recourse to processes by which the individual acts upon himself" as well as "the points where the techniques of the self are integrated into structures of coercion or domination" (Foucault 1993: 203). One way to address techniques of the self and modify, for instance, health related behavior are communication campaigns like the one mentioned in the above example from Germany. As I will argue in the second section health communication often subscribes to a singular hierarchy of values with health at its top, rendering irrational or immoral any other rationalities one might apply to choices about health related behavior.

Health communication is thus entangled with inequities of power (regarding for example access to the means of producing, circulating and receiving health information and health services) and the government of populations. It is therefore necessary to communicate in a way that is sensitive to inequalities. However, communication is not only about transmitting information. Communication also conveys models of communication.

Cultural anthropologist Charles L. Briggs and media scholar Daniel C. Hallin have proposed the concept of communicability to take into account how roles, values and rationalities are co-produced together with the "information" in the process of communication. Communicability describes the performative way by which news coverage not only imparts knowledge but also projects models about how phenomena come to be, how knowledge about them circulates and who should attend to them and how. Briggs and Hallin derived this concept from their research on the "proliferation of health stories in newspapers, television, radio, social media, and the Internet" (Briggs 2011: 1038). While the concept originates in analytical and comparative scholarship, my hope is, that it may also help practitioners to become sensitive to the relations of power as well as cultural and structural causes for what they might otherwise perceive as lack of compliance on the side of their audiences.

To this end I will begin with a cautionary note by offering some of the critiques directed at health communication and the models it is based upon. While designing health campaigns it is important to keep in mind that health is not the only rationality guiding choices on health behavior. Multiple and competing rationalities are at play. I will then elaborate Briggs and Hallin's concept of communicability in the third section. The fourth section presents the three models of biocommunicability that Briggs and Hallin have identified so far. I will conclude my presentation of biocommunicability in offering some thoughts on how to utilize this analytical concept for communicating about pandemics.

1.2 Communication Models and Multiple Rationalities

Health communication often relies on simple messages that prompt people to change their behavior for the sake of health. It's like "quit smoking". Once a certain behavior is identified as risk taking, it should be changed rendering risky behaviors like smoking or refusing to wear a mask irrational. According to social scientist Andy Alaszewski a rational actor model is implicit in risk communication. There is only one rational behavior towards a risk and that is avoiding it. Once a certain behavior is identified as risk-taking it should be changed (Alaszewski 2006).

The rational actor model has been criticized thoroughly throughout the years. For instance, cultural studies scholars Tulloch and Lupton (1997: 18) point out: "Rationality is defined and measured in extremely individualistic and narrow terms (those assumed by health promotion workers) in which good health is privileged over all other 'benefits' and pleasures." Within this framing, experts are the only source of risk and health information. Consequently, the audiences are imagined as passive recipients of information rather than active information seekers (Alaszewski 2006: 163–169). Tulloch and Lupton (1997: 4) argue in the context of the HIV/AIDS scare that people who take risks should not be thought of as "simply 'irrational', as people who may be changed (in attitude and behavior) by the appropriate transmission of 'knowledge'. In their own context they take risks as rational actors; rational, that is, in terms of their economic, social, cultural and cognitive schemas." For example, a study by Johansson et al. (1991) linked smoking behavior to certain work conditions. Their study thus highlighted a connection of risk behaviors to coping strategies in work life.

Tulloch and Lupton's argument draws attention to the ways by which certain communication models entail the imagination of the public as deficient recipient of messages (see also Bakir 2010: 10). Historians Bauer et al. (2007) have identified three phases in the characterization of the public as deficient by studies on communication. In a first phase, the public was thought to lack knowledge about scientific facts obstructing the public's compliance with advice. Starting in the 1980s they identified a second phase where a failure to comply was ascribed to the public's lack of positive attitudes towards science and technology. In a third phase the studies attributed the deficit to the public as well as to scientific institutions and experts. While a lay public may lack knowledge, positive attitude or trust towards traditional sources of information, experts "harbor prejudices about an ignorant public" (Bauer et al. 2007: 85). Importantly, the authors maintain that the three deficit models continue to exist alongside each other. Each deficit model entails a certain solution to alleviate the public's deficit. A lack of knowledge is best remedied by education, while a negative attitude requires campaigning to give knowledge of a certain kind a positive image. And finally, a deficit of trust in experts could be counteracted by the public's participation which is mirrored by forms of deliberation on the experts' side. Tulloch and Lupton (1997) draw attention to the ways by which communication models entail these deficit models of their respective audiences. In particular they point out three models in health communication research: the effects model, the knowledge-attitude-behavior model, and the cultural studies model.

The effects model conceptualizes communication as a linear process and failures of transmission as barriers that have to be overcome. The knowledge-attitude-behavior model identifies attitudes towards types of knowledge as crucial to whether health behaviors are changed. Attitudes are changed through knowledge of health problems. The knowledge relayed to the audience in this model is imagined as "a translation from unproblematic scientific 'facts'" that are supposed to re- place lay myths (Tulloch and Lupton 1997: 17). The cultural studies model of communication complicates the relation between 'sender' and 'receiver' as the audience actively mediates health messages through its own "narratives, experiences and social meaning." (Tulloch and Lupton 1997: 19) Thus, it is an 'active audience' that is involved in the production of meaning by interpreting the received information.

From this brief discussion of communication models, it is clear that communication has to be understood as more than just transmitting information from a sender to a receiver. While the above examples focus on moral values and rationalities Briggs and Hallin's concept of communicability also takes into account how communication transmits models of ideal communication to its audiences alongside the information.

1.3 Communicability: More Than Getting Your Message Across

Charles L. Briggs and Daniel C. Hallin developed the concept of communicability from a need to address symmetrically the concepts relayed by communication as well as communication itself. While biomedical knowledge as well as concepts of class, race and gender (to name just a few) are generally perceived as being productive of subjectivities and social relations this had not been the case for communication. Briggs and Hallin argue that concepts of communication are just as productive as the aforementioned concepts (Briggs 2005: 270). Linguistic devices structure the roles of participants of the communicative processes and locate them "in relation to imagined communicative circuits" (Briggs 2005: 274) rendering them more or less vulnerable to, for example, infectious disease. Communicability can thus be also understood as an analytical program in regard to these communicative processes. It calls for an analysis of:

> (a) how structural violence exposes some people to infection and limits their access to health care, (b) how biomedical epistemologies construct populations as irrational, (c) how access to the production and reception of authoritative knowledge about disease is distributed, and (d) how this communicative process is ideologically constructed in such a way as to make some people seem like producers of knowledge, others like translators and disseminators, others like receivers, and some simply out of the game. (Briggs 2005: 274)

Communicability draws special attention to the last two aspects of communication highlighting how communication performs boundary work that regulates membership in social groups and at the same time creates and maintains boundaries of disciplinary, social, topical areas of expertise. Rather than only analyzing content of messages (which it also does), the research agenda focuses on how the production, circulation and reception of messages "shapes identities and social 'groups' and orders them hierarchically" (Briggs 2005: 275). Briggs and Hallin take the idea that communication is involved in the process of social stratification from Pierre Bourdieu's practice theory where communicative competence functions as symbolic capital. As such it is associated with particular institutions with strict access controls (Briggs and Hallin 2007: 45). The word communicability draws attention to the ease with which something is communicated or someone communicates and is understood transparently. At the same time it likens the ability of messages to find audiences to the ability of microbes and viruses to infect and spread along social relations (Briggs 2005: 274).

Briggs and Hallin draw from a range of Foucauldian and Science-Technology-Studies concepts and methodologies in order to address the issues raised by their research agenda. Specifically, communicability draws on biopolitical analysis and Paul Rabinow's concept of biosociality, critical discourse analysis and concepts from actor-network theory which I will briefly introduce in the following paragraphs.

1.3.1 Biopolitics, Subjectification and Biosociality

The communication and circulation of biomedical knowledge shapes how people think of themselves and in relation to others in terms of health. Today, thanks to advanced genetic testing our diagnostic testing capabilities surpass our capabilities to treat the diagnosed health conditions. Just as with COVID-19, in many cases we are confronted with the question of how to live with a certain condition that can range from a risk factor due to a mutation in a gene (e.g., Alzheimer's Disease, Breast Cancer, type 1 Diabetes), an actual genetic disorder (e.g., Cystic fibrosis, Hemophilia, Sickle cell disease) or being chronically infected with viruses (e.g., Hepatitis C virus, HIV, Herpes Simplex I). The diagnosis of these diseases and other risk factors often entails the adoption of new practices such as taking up a specific diet, redoing one's home and environment, or educating oneself about the condition. These practices are directed at mitigating the risks or the effects of the disease itself without removing its cause. Biology, today, is not only understood as fate anymore. It has become a condition on which one can act upon within certain limits (Rose and Novas 2005: 442). The existence of support groups for breast cancer survivors, diabetes or people living with HIV, where those affected by a particular health condition—patients, survivors or their relatives—can exchange their experiences among each other, is testament to the "formation of new group and individual identities and practices arising out of these new truths." (Rabinow 1996: 102) Particularly with respect to HIV/AIDS anthropologist Thurka Sangaramoorthy (2012) shows how surveillance and quantification in HIV/AIDS prevention has given rise to numerical subjectivities, where HIV positive individuals identify with and derive meaning from their viral load and CD4 count. This reconfiguration of the social in biological terms is what anthropologist Paul Rabinow calls biosociality.

The formation of identities around risk factors and disease is not only a grassroots phenomenon. Pharmaceutical companies market certain products to consumers by promoting online platforms where patients can exchange experiences and treatment options with others. In some cases, company sponsored patient organizations have been known to lobby for certain legislation affecting the market for pharmaceutical products. Irrespective of whether a group is a grassroots organization or corporate funded, there are increasingly instances of activist groups that surpass "mere" support and either campaign for access to better treatment or against discrimination, donate a part of their income, blood or tissue to advance research on their condition, organize special care, or participate in clinical trials. Here, life itself has become an integral part in the generation of wealth, the production of health and the formation of social norms and values. Their biological condition plays a direct part in the political process. This is what Rose and Novas (2005) call biological citizenship. The term was first introduced by Adriana Petryna in connection with victims of the Chernobyl nuclear reactor explosion to describe how "the damaged biology of a population has become the grounds for social membership and the basis for staking citizenship claims" (Petryna 2002: 5). The concept of biological citizenship refers to an individual's or a group's claims to access privileges of citizenship based on their biological condition. These claims are based on biomedical knowledge. Production and communication of biomedical knowledge thus play an important role in "regulating and rationalizing access" to the privileges of citizenship (Briggs and Hallin 2010: 150). While production of biomedical knowledge can easily be understood as something not everybody can participate in, communication should allow everybody to understand how to gain access to health care or provide the knowledge to protect oneself from, for example, infectious disease. However, structural inequalities are often reproduced in health communication. Briggs and Hallin develop their methodology out of critical discourse analysis to analyze how power relations and inequalities are reinforced by media reporting.

1.3.2 Critical Discourse Analysis

Critical discourse analysis is a theory driven approach to the analysis of language and the production of meaning (semiosis) within broader social processes (Faiclough 2002: 121). Studies most often focus on the analysis of communicative inequalities (allocation of speaking rights, linguistic-communicative resources), discrimination and power relations in language. Texts and speech acts are concrete instances of discourse. Thus, discourse can be understood as a consolidated set of speech acts and texts emerging from a particular social practice. Semiosis through language and its durable form—text—is not restricted to a "bounded social interaction" between immediate participants. Moreover, the interaction takes place in a "wider sociocultural and political-economic context" (Baumann and Briggs 1990: 61). According to Michael Halliday a relationship exists "between the grammatical system and the social and personal needs that language is required to serve" (Wodak 2002: 8). Language fulfills three interconnected metafunctions: Language structures experience (ideational function), it constitutes relationships (interpersonal function), and as text it produces coherence and cohesion (textual function) (Wodak 2002: 8). Language in this sense is one social practice, among others. However, other practices such as economic, political, cultural, etc. practices draw on language.

In communicating directly with others or through text information is typically not relayed in every possible word combination available. There is a scarcity of enunciations. Certain ways of wording a message are more frequent than others. Discourse analysis analyses patterns and generative rules for the production of enunciations and the networks they are embedded in. For example, Fairclough (2002: 130–133) argues that linguistic features such as passive verb forms, temporal modalities and syntax can divert attention away from agents of political processes such as market liberalization, rendering them as inevitable natural phenomenon. Just as words acquire meaning in relation to practices and other words, whole texts acquire meaning in relation to their production, circulation, distribution and consumption. Texts have a historicity where certain phrasings and arguments can be traced to other texts or events. This is what Norman Fairclough (2002: 124) calls intertextuality: the way a certain text draws on other discourses, genres, styles and utterances, and relates them to each other and its topical area. The way discourses are related to each other is called an order of discourse. This means that social practices constitute a social order and generate meaning. Social order as a network of social practices generates an order of discourse that can have power effects in that "some ways of making meaning are dominant [...] in a particular order of discourse; others are marginal, or oppositional, or 'alternative'." (Fairclough 2002: 124) If one particular social structure of semiosis becomes part of the legitimization of authority it is called hegemonic.

Fairclough proposes a problem-based methodology for the analysis of discourses (2002: 125). The first step is identifying what Michel Foucault calls problematizations, that is identifying events, that have introduced "difficulties in our previous way of understanding, acting, relating" to the object in question. The problematization is "the ensemble of discursive and non-discursive practices that make something enter into the play of true and false and constitute it as an object of thought" (Foucault cited in Rabinow 2005: 43). For Foucault and Fairclough there are always different possible ways to address difficulties. Thus, in a second step the discourse analysis should identify obstacles to solving the problem. This step might involve going outside of the text to take into consideration its context. How are entities in the world, practices, meanings and discourses arranged and related to each other (in the text and its context) to exclude all other solutions. This analysis involves linguistic and semiotic, intertextual and interdiscursive, interactional and structural analyses. In a third step the role of the problem for the social order should be considered. Does the network of practices constituting the social order need the problem. Fairclough proposes to identify and make explicit alternative solutions that have been silenced and excluded by the dominant discourse. In a final step he calls for a critical reflection on the analysis.

Blommaert and Bulcaen (2000: 448) critically remark that Fairclough pays little attention to the "resources and other 'macro' conditions" of "the production and distribution of discourse." Briggs and Hallin address this lacunae by drawing on Bruno Latour and Michel Callon's notion of actor-network theory. This addition allows them to consider the context of production as well as the context of consumption of discourse by audiences.

1.3.3 Actor-Network Theory

Briggs and Hallin draw on concepts from science and technology studies (STS) to explain how biomedical knowledge is produced and travels from laboratory and clinical spaces to news outlets and the public sphere. Journalists, scientists, health workers and lay persons all form part of an actor-network complicit in the way that health news is narrated (Briggs and Hallin 2016: 112). Practices and technologies of communication are central to the way that biomedicalized subjects and objects are represented and constructed. However, while they draw on the actor-network concept they do not explicate it in their work.

Actor-Network Theory, as presented by Bruno Latour, Michel Callon and John Law, has elements that connect it with critical discourse analysis. It employs a material semiotic analysis to investigate how meaning takes shape through social practices and how these practices relate different actors to each other to form social order. Just as words acquire their meaning in relation to social practices and differences from other words, the material semiotic approach argues that entities acquire meaning when they are put in relation to each other in everyday practices. It is what entities do with other entities that confers to them the attributes we use to describe them. This is what Latour, borrowing from semiotics, calls interdefinition (1988: 9–12). Entities are defined as a list of actions and properties they display when acting with other entities (Latour 1987: 87–88). However, the "meaning" Latour and other proponents of this approach refer to is not one of beliefs but rather of ontologies, plural. Since meaning is caused by concrete practices specifically arranged together they have real world effects (Law 2011: 2). Relationships between entities—their social meaning—gains social and cultural stability through performativity (Law 1999: 3–4). Different realities can be performed at the same time. The adjective "material" in material semiotics is important. Even if, for example, individual groups act as if SARS-CoV-2 did not exist, the virus will still interact with them. This means that material semiotics grants agency to nonhuman actors as entities act upon and with each other. This is one meaning of actor-network.

Another related sense of actor-network is the process of translation. The term translation describes the association of actors with other actors. Translation consists of four moments. The moment of 'problematization' describes how one actor defines a problem and a possible solution. As in Fairclough's problem-based approach in critical discourse analysis identifying and analyzing the problem definition is an important first step in the analysis of networks. In order to form an association or an alliance of actors others have to be convinced to align their interests with those of the networking actor. This can be achieved by connecting their problems to those of the networking actor, so that solving the latter's problems will solve theirs as well. "Interessement" refers to this process by which the first actor interests others in the identities, roles and functions they would fulfill in the solution to the problem definition. For example, in health communication scientist have the role to produce knowledge, while family physicians and journalists are often charged with the burden of translating scientific results to a lay public; which in turn is supposed to act upon their advice. "Enrollment" occurs when the actors actually fulfill their designated roles and associate in order to achieve the common agenda. The moment of "mobilization" describes the actual mobilization and exchange of things between associated actors that is necessary to stabilize the network (Callon 1986). Important for understanding the concept of biocommunicability is that similar moments are passed when one actor wants to convey health messages to an audience.

1.4 Models of Biocommunicability

Briggs and Hallin address a gap between health communication and journalism studies. They observe a dominance of an orientation in health communication they call the "linear-reflectionist perspective" (Hallin and Briggs 2015: 86). In this perspective biomedical experts provide authoritative knowledge about medical subjects and objects. Journalist are ascribed the "role of circulating this pre-existing knowledge" (Hallin and Briggs 2015: 86). Research in health communication according to Hallin and Briggs follows two variants of the linear-reflectionist perspective they call the "representational and the instrumental variants" (Hallin and Briggs 2015: 91). The representational variant focuses on the accuracy of information transmission and "factors that 'distort' it." (Hallin and Briggs 2015: 88). One common topic in health communication is the phenomenon called "interreality distortion". Health news are found to cover health conditions disproportionately to their impact on population health. The underlying assumption of the concept of interreality distortion is that frequency and quantity of media coverage of health conditions should represent their statistical occurrence in a population. The instrumental variant focuses on the purpose of media: health news should educate a lay public.

This role can lead to interreality distortions when public health authorities launch a communication campaign, for instance at the beginning of a pandemic, as a means of prevention. Health reporting thus has to be understood as "not simply the representation of 'reality' but social purposes." (Hallin and Briggs 2015: 92) Health and medical journalism are independent social and cultural practices and forms of knowledge production that fulfill a panoply of social purposes such as mediating social groups and discourses, popularizing knowledge, and creating frames and narratives for understanding and circulating knowledge about, for example, health and medicine. Journalists are also involved in what Hallin and Briggs (2015: 96) call "pre-mediatization of health issues". Journalists embedded in health institutions and marketing departments build media logics into the practices of private and public organizations as well as state actors.

The concept of biocommunicability draws attention to this active role or health journalism in communication. It combines critical discourse analysis and insights from actor-network theory to analyze not only the semantic content of health communication but also its performativity in the subjectivation of and the production of discourse by the actor-network spanning scientists, health workers, journalists and lay persons. Biocommunicability thus addresses the gap between biopolitical practices, biosocial subjectivites and discourses. Briggs and Hallin use the concept to describe how "the constitution of social subjects is embedded in ideologies about the "flow" of information and of discourse, about who constitutes biomedical knowledge, who is authorized to evaluate it and to speak about it, and through what channels it is assumed to flow." (Briggs and Hallin 2007: 46) This projection of roles in the production, circulation and consumption of biomedical knowledge is performative in that it is conveyed through the way health news are reported: the way experts are staged as indisputable authority, as partner and facilitator of expert health knowledge, or even as just another voice among equals taking part in a public controversy. Roles in communication are producers, translators, desseminators and receptors of health information (Briggs and Hallin 2007: 58). Briggs and Hallin identify three models of biocommunicability distinguished by the direction of information flow, the role of experts, journalists and the public, whose and which goals and what types of knowledge are given precedence, and how this knowledge is produced. The three models are the biomedical authority model, the patient-consumer model and the public sphere model.

1.4.1 Biomedical Authority Model

Health communication in the biomedical authority model has the goal to educate an ignorant public. It conceives communication as a unidirectional flow of information. Biomedical scientists produce objective representations of reality in a supposedly objective, highly specialized and technical sphere. In contrast, other spheres are projected as being dominated by populist, relativist or democratic communication ideologies. Thus, biomedical authorities fulfill the role of the sole reliable sources of health information. Biomedical knowledge is treated as information that can be decontextualized from the conditions of its production and used across different social contexts without changing its meaning (Hallin and Briggs 2015: 90). Briggs and Hallin (2010: 151) liken this model to the relation between patient and family physician where the patient's "proper role is to trust and obey" their doctor. Here the public figures as patient who might cast what is at stake in their health issue differently from the doctor. However: The doctor knows best! His/her solution and framing of the health issue are given precedent over the patients'. In this relationship the lay public is projected as receiving information passively (Briggs and Hallin 2007: 50). Journalists participate in this model by translating health knowledge into popular discourses. Their projected role is to educate the ignorant public and displace "popular misinformation with correct information" (Hallin and Briggs 2015: 90). While journalists when reporting about other areas would typically highlight controversy, in the biomedical authority model they present health information as facts. Journalists downplay their own role in the production of news articles by staging voices of biomedical authorities so that they directly address the audiences. They call their audiences to trust in and rely on the biomedical authorities (Briggs and Hallin 2007: 50). Importantly, due to the necessary process of translation, which is mainly undertaken by journalists who are not part of the elite community of biomedical professionals, the latter perceive health coverage as a distortion of the actual information (Briggs and Hallin 2016: 25–30).

While the biomedical authority model suggests a minimal role of journalists in the reproduction and transmission of biomedical information Briggs and Hallin (2016: 112) emphasize the process of biomediatization as co-productive and the result of an actor-network. In particular they draw on Sheldon Ungar's work on life cycles of news reporting on outbreaks to point out how journalists frame the reporting on pandemics along prepackaged narratives. According to Ungar (2008) coverage of outbreaks such as bird flu, Ebola, H1N1 or the COVID-19 pandemic develops over time in three stages. In a first stage he calls "sounding the alarm" news stories focus on numbers of victims and detective narratives about the origins of the virus. The second stage contains mixed messages that aim at calming down public panic with reassuring stories on preparedness measures and conditions in other countries that aggravate the outbreak in its country of origin. Finally, in the stage he calls "hot crisis and containment" stories focus on characteristics of the virus that hinder it from developing into a truly dangerous epidemic for the western hemisphere. This last stage, however, might vary depending on the unfolding epidemic situation. Briggs and Hallin sound a cautionary note on how emergency exercises can influence news narratives of the real thing as journalists draw on the resources from scripted scenario exercises they attended in the past (2016: 120–124).

1.4.2 Patient-Consumer Model

According to Briggs and Hallin (2010: 152) the patient-consumer model is the most prevalent model in most areas of health communication. The reason for this being, they argue, that it is "cheap to produce" as "much of it is syndicated material" and advertising and editorial content can be easily integrated into the story (Briggs and Hallin 2007: 54). It aligns well with conceiving of health services as a market with active, information seeking, medically literate patients making their own independent decisions. Information flow is still unidirectional in that evidence based knowledge is produced by science and finds its way to the patient/consumer via mass media, the internet and health professionals. But it is the patient/consumer seeking information that starts the process of health communication. In this configuration the physician takes the role of an informed adviser who works together with her patient/client to find the best solution. As a necessary pre-requisite, the client has to be imagined as endowed with sufficient financial resources (Briggs and Hallin, 2016: 33–39). Like in in the biomedical authority model biomedical professionals are the primary source of medical knowledge. But the patient-consumer model projects the patient as an active seeker of information. The patient's goals are given precedent and biomedical professionals only facilitate the knowledge necessary to achieve them. However, patients themselves decide what constitutes relevant and adequate knowledge. Patients make their choices without their physician. They are the main actors in news stories drawing on the patient-consumer model. Instead of instructing patients to heed their doctors advice journalists instruct patient/consumers to 'go shopping'—to get a second opinion and compare different offers. "The journalist's role is not to help medical authorities communicate to an ignorant public, but to help consumers exploit the range of options apparently open to them." (Briggs and Hallin 2007: 53) The model projects a rational process of acquiring information and making choices based on a transparent market assumption (Briggs and Hallin 2007: 52).

Both, the patient-consumer and the biomedical authority model have in common that they define the function of health information in terms of how "it helps the individuals to regulate their behavior in the interest of their own health." (Briggs and Hallin 2010: 152) Health is given precedent over all other values. However, in the patient-consumer model this is only true if patient/consumers' choices "can be imagined as based on information that comes from objective, disinterested sources." (Briggs and Hallin 2007: 57) This condition is being undermined or at least complicated by pharmaceutical drug companies marketing their products to physicians at advanced training events effectively rendering physicians as consumers of information and calling into question the disinterestedness of their advice. Journalists often find themselves in a position where they have to assess the trustworthiness of biomedical knowledge and fill in gaps arising from its unreliability, scarcity or even excess due to conflicting studies or rumors (Briggs and Hallin 2016: 38).

1.4.3 Public Sphere Model

The third model outlined by Briggs and Hallin is the public sphere model. Here the usefulness of health information lies primarily in helping citizens and policymakers "make collective decisions about the public interest." (Briggs and Hallin 2010: 152) However, the sources of health information are not necessarily health professionals and their knowledge is not given precedence over other knowledge produced, for example, by lay persons. This is especially the case in stories treating social movements demanding biological citizenship rights. Here lay activists may ally with health professionals to shift the register of health issues from biomedicine and the field of science to political and social fields (Briggs and Hallin 2010: 160). In this model the audience is addressed as observer-citizen who has to make a judgment about collective decisions and social values. Rather than a linear flow of information from medical experts to a lay public, controversy is framed as conflict between different stakeholders or harmed citizens (Briggs and Hallin 2016: 39–40). Health professionals appear as one interest group among others and health information is open to debate. This is particularly the case when health information refers to areas "that involve the state, state funding, regulation and policy" (Briggs and Hallin 2010: 153). In stories using the public sphere model journalists construct themselves in three roles crucial to the political process: they have to decide which knowledge should be public; they investigate "information that has been withheld or improperly channeled"; and they have to construct "the boundaries of public discourse about health." (Briggs and Hallin 2010: 157).

Depending on how a certain health issue is drawn into the public sphere Briggs and Hallin distinguish between different sub-genres of the public sphere model. Stories that focus on political scandal due to, for instance, environmental pollution follow a standard political model of reporting where audiences are addressed in their role as citizen/policymaker. Rather than quoting health professionals journalists include the voices of politicians, affected citizens or other stakeholders. Another sub-genre of the public sphere model is the elite public sphere model. In these stories credentialed experts in public institutions such as the US Food and Drug Administration or the Centers for Disease Control and Prevention discuss potentially controversial decisions on the regulation of health related issues. These stories do not project an egalitarian public sphere where every citizen has a voice but the public is granted the right to observe and judge. Briggs and Hallin (2016: 42) find that the social movement model is used in stories where activists politicize health issues. Examples of this kind of reporting include stories on abortion advocated for by women's rights movements or against by faith based organizations, stories on the mobilization around HIV/AIDS or stories on patients' advocacy groups. This sub-genre also problematizes biocommunicable failures of, for example, governmental and biomedical institutions keeping information on the regulation of access to health care or other health related issues secret. Here journalists "collaborate (implicitly) with activists and researchers in providing alternative circuits for disseminating health information" (Briggs and Hallin 2016: 43). Consequently, in these stories, information is produced by activists outside of health institutions and disseminated by journalists to a general public. This mode of biocommunicability, however, coexists in tension with the biomedical authority model and the patient-consumer model and is mostly drawn upon when they fail (Briggs and Hallin 2016: 46).

1.5 Communicating About Pandemics

Failure of biocommunicability is not necessarily the only reason for a shift of health communication to the public sphere model. Biocommunicability presents communication as a process in which participants are recruited and arranged in an actor-network according to roles they assume in information circuits. Drawing on the national influenza pandemic plan German federal government and local state officials adopted a communication strategy in the COVID-19 pandemic called the one-voice policy (Hall and Wolf 2019). This policy prescribes that officials should draw on the same information sources, namely Federal health authorities, for their communication about a pandemic. Officials adopted this guideline in communicating about pandemic response measures with the public. From this point of view, it would seem that the biomedical authority model with a unidirectional flow of information from science to the public would have been adopted. And, indeed the public complied with a first nationwide shutdown in mid-March 2020. There was a clear message as to what it meant to be a pandemic citizen within a biopolitical strategy to "flatten the curve".

Various scientists made an effort to establish trustworthiness by emphasizing scientific knowledge as a process where earlier assumptions have to be refuted or modified due to new data. While this was widely accepted old scientific narratives, now marginal, continued to circulate in the pandemic discourse throughout 2020. One such example is the comparison to influenza. Not only did this comparison at the beginning of the pandemic play down the actual morbidity and case fatality of

COVID-19, but also disputed the usefulness of masks. These narratives were included in the discourse of so called corona-deniers who constructed a conspiracy narrative around the changing scientific facts and recommendations.

Also, the one-voice policy proved to be restricted in its scope. Due to the federal system pandemic response is the responsibility of state governments. While state level minister-presidents consistently drew on information and recommendations of the Robert Koch-Institute, Germany's Federal infections prevention and control agency, their interpretation and implementation of pandemic response measures on the state level varied considerably. Being a pandemic citizen acquired different meanings according to where one lived or traveled to. This situation was compounded by the federal elections looming in 2021. Minister-presidents competed to stage themselves as most efficient pandemic managers multiplying voices debating the best pandemic response. The one-voice policy on sources was undermined by a polyphonic chorus on appropriate responses. This raises the question whether the biomedical authority model is doomed to fail in polities with a federal system. The biomedical authority model presupposes a medical, public health or health science expert as originator of the health message. In public health crises state officials and not scientists are the primary crises managers and communicators. Their role is associated with the political sphere and therefore the public sphere model of biocommunicability. A polyphonic chorus on response further shifts how audiences perceive biocommunicability to the public sphere model. Since the measures that are communicated are in principle up for discussion and an outcome of political choices (not medical choices) controversy is mandated. This means that communication strategies and the implementation of pandemic responses have to be matched closely to the political system. However, communication strategies cannot solve problems arising from inconsistent application of pandemic responses to different sectors of society.

References

Adams V (2016) Metrics of the global sovereign. Numbers and stories in global health. In: Adams V (ed) Metrics. What counts in global health, pp 19–54. Duke University Press, Durham

Andy A (2006) Health and risk. In: Taylor-Gooby P, Zinn JO (eds) Risk in social science. Oxford University Press, Oxford, pp 160–179

Bakir V (2010) Media and risk: old and new research directions. J Risk Res 13(1):5–18

Bauer MW, Allum N, Miller S (2007) What can we learn from 25 years of PUS survey research? Liberating and expanding the agenda. Public Underst Sci 16(1):79–95

Bauman R, Briggs CL (1990) Poetics and performance as critical perspectives on language and social life. Annu Rev Anthropol 19:59–88

Blommaert J, Bulcaen C (2000) Critical discourse analysis. Annual Rev Anthropol 29:447–466

Briggs CL (2005) Communicability, racial discourse, and disease. Annu Rev Anthropol 34:269–291

Briggs CL, Hallin DC (2007) Biocommunicability. The neoliberal subject and its contradictions in news coverage of health issues. Social Text 25(4):43–66

Briggs CL, Hallin DC (2010) Health reporting as political reporting: biocommunicability and the public sphere. Journalism 11(2):149–165

Briggs CL, Hallin DC (2016) Making health public. How news coverage is remaking media, medicine, and contemporary life. London: Routledge
Callon M (1986) Some elements of a sociology of translation: domestication of the scallops and the fishermen of St Brieuc Bay. In: Law J (ed) Power, action and belief: a new sociology of knowledge? Routledge, London, pp 196–223
Davis M, Stephenson N, Flowers P (2011) Compliant, complacent or panicked? Investigating the problematisation of the Australian general public in pandemic influenza control. Soc Sci Med 72(6):912–918
Fairclough N (2002) Critical discourse analysis as a method in social scientific research. In: Wodak R, Meyer M (eds) Methods of critical discourse analysis. Sage, London, pp 121–138
Foucault M (1993) About the beginning of the hermeneutics of the self: tow lectures at Dartmouth. Political Theory 21(2):198–227
Foucault M (2000) "Omnes et Singulatim": toward a critique of political reason. In: Faubion JD (ed) Power: essential works of Foucault, vol 3. New Press, New York, pp 298–325
Hall K, Wolf M (2019) Whose crisis? Pandemic flu, 'communication disasters' and the struggle for hegemony. Health Interdiscipl J Social Study Health Illness Med https://doi.org/10.1177/1363459319886112
Hallin DC, Briggs CL (2015) Transcending the medical/media opposition in research on news coverage of health and medicine. Media Cult Soc 37(1):85–100
Johansson G, Johnson JV, Hall EM (1991) Smoking and sedentary behaviour as related to work organization. Soc Sci Med 32(7):837–846
Latour B (1987) Science in action. How to follow scientists and engineers through society. Harvard University Press, Cambridge, MA
Latour B (1988) The pasteurization of France. Translated by Sheridan A, Law J, Harvard University Press, Cambridge, MA
Law J (1999) After ANT. Complexity, naming and topology. In: Law J, Hassard J (eds) Actor network theory and after. Blackwell Publishing, Oxford and Malden, MA, pp 1–14
Law J (2011) What's wrong with a one-world world. Heterogeneitiesdotnet. Available at: http://heterogeneities.net/publications/Law2011WhatsWrongWithAOneWorldWorld.pdf. Accessed 30 Sept 2019
Lemke T (2011) Biopolitics. An advanced introduction. With a preface by Casper, Moore LJ. New York University Press, New York
Petryna A (2002) Life exposed. Biological citizens after chernobyl. Princeton University Press, Princeton
Rabinow P (1996) Artificiality and enlightenment: from sociobiology to biosociality. In: Essays on the anthropology of reason, pp 91–111. Princeton University Press, Princeton, NJ
Rabinow P (2005) Midst anthropology's problems. In: Ong A, Collier SJ (eds) Global Assemblages. Technology, politics, and ethics as anthropological problems, pp 40–53. Blackwell Publishing, Malden, MA
Rose N, Novas C (2005) Biological citizenship. In: Ong A, Collier S (eds) Global Assemblages. Technology, politics, and ethics as anthropological problems, pp. 439–463. Blackwell Publishing, Malden, MA
Sangaramoorthy T (2012) Treating the numbers: HIV/AIDS surveillance, subjectivity, and risk. Med Anthropol Cross-Cultural Stud Health Illness 31(4):292–309
Tulloch J, Lupton D (1997) Television, AIDS and risk. A cultural studies approach to health communication. Allen & Unwin, St. Leonards
Ungar S (2008) Global bird flu communication. Sci Commun 29(4):472–497
Wodak R (2002) What CDA is about—a summary of its history, important concepts and its developments. In: Wodak R, Meyer M (eds) Methods of critical discourse analysis. Sage, London, pp 1–13

Chapter 2
Developing Trust in Pandemic Messages

Kurt Wise

Abstract Public health communicators facing pandemics must contend with a vexing problem: significant portions of the American population have lost trust in institutions. Despite the fact that technology has afforded those in the health field with myriad avenues to disseminate messages, those messages may be met with skepticism if not hostility. Public relations has much to offer the health field with regard to developing and maintaining trusting relationships with various publics. Public relations scholarship has demonstrated that the development and maintenance of relationships before a pandemic occurs can increase the likelihood that pandemic-related messages will be received and acted upon in a positive manner. This chapter discusses the key public relations concepts to be employed by those in the health field that can help increase the level of trust between health professionals and various publics.

Keywords Public relations · Trust · Relationship cultivation · Relationship management · Public health

2.1 Introduction

Public health professionals face considerable challenges creating strategic plans and messages to deal with pandemics. The very nature of pandemics may cause many members of the public to resist health initiatives designed to contain pandemics. This is because, as Udow-Phillips and Lance (2020) pointed out, pandemic interventions that protect public health can have significant negative effects on the economy as well as the financial and psychological welfare of citizens. Another challenge faced by health professionals when confronting a pandemic has to do with trust—put simply, trust in American institutions has been declining for years. Because of these challenges, those charged with protecting the public health must turn to all available sources for assistance. Fortunately, public relations scholarship and practice have

K. Wise (✉)
University of West Florida, Pensacola, FL, USA
e-mail: kwise@uwf.edu

D. M. Berube (ed.), *Pandemic Communication and Resilience*, Risk, Systems and Decisions, https://doi.org/10.1007/978-3-030-77344-1_2

much to offer communicators involved in delivering pandemic messages to various publics.

In this chapter the topics to be addressed include a brief discussion of the public relations field, the importance of relationships in public relations, the benefits of relationship management, relationship cultivation strategies, measuring organization-public relationships, and the role of public relations in building and maintaining trust in the public health arena.

2.2 Public Relations and Relationship Management

A complete discussion of the history of public relations is beyond the scope of this chapter (For a thorough discussion, see Cutlip 1995). However, suffice to say the term "public relations" is widely misunderstood. Some equate public relations with advertising or marketing. Others may believe that those who practice public relations are "paid to lie." This impression of public relations is not surprising, especially given the type of public relations that was common about a century ago. Lies, half-truths, and wild exaggeration were common tactics employed by those seeking publicity and self-promotion. Although one can still find examples of such behavior, the field has become far more professional over the years, with the founding of the first public relations professional association in 1936 and the Public Relations Society of America establishing an accreditation system in 1964 (Browning 2018). The first Public Relations Code of Ethics was established in 1950 (Fitzpatrick 2002), and there are more than 70 national/international public relations associations (Yang and Taylor 2014), providing training on a wide variety of topics. Along with the professionalization of the field, one of the most important developments in public relations is the idea that public relations is about far more than just "messaging." Public relations is about *relationships.*

The first to call for a focus on relationships in public relations scholarship was Ferguson (1984). However, it took years before relationships would become a focal point of public relations research. Just before the turn of the last century, Hon and Grunig (1999) stated that a "growing number of public relations practitioners and scholars have come to believe that the fundamental goal of public relations is to build and then enhance on-going or long-term relationships with an organization's key constituencies" (p. 2). Similarly, Ledingham (2003) said it "would be difficult to overstate the importance of the relational concept to public relations" (p. 183). More than 150 studies on various aspects of organization-public relationships were published between 1998 and 2016 (Cheng 2018). The centrality of relationships in public relations is evident in many definitions of public relations. A widely-used textbook defines public relations as "the management function that establishes and maintains mutually beneficial relationships between an organization and the publics on whom its success or failure depends" (Broom and Sha 2013, p. 5).

The basic idea of the relationship management perspective is simple: Public relations practitioners act as managers of the relationships between an organization and its publics. Using a university as an example, a public relations professional in a management position would manage the university's relationships with its most important publics such as faculty, students, staff, donors, legislators, the surrounding community, and local media. It is important for any organization to establish and maintain positive relationships with its publics, because, asserted Bruning and Ledingham (1998b), "organizational involvement in, and support of the community in which it operates, can engender loyalty to an organization among key publics when that involvement/support is known by key publics" (p. 63). This loyalty is extremely important in times of a pandemic. If individuals have developed a positive relationship with a health organization such as a health department or a hospital, they are more likely to be loyal to that organization and to heed its messages.

Ledingham (2003) was the first to explicate relationship management as a general theory of public relations. Ledingham noted four developments which spurred the development of the relational perspective in public relations: (1) recognition of the central role of relationships in public relations, (2) reconceptualizing public relations as a management function, (3) identifying the components and types of organization-public relationships, their linkages to public attitudes, perceptions, knowledge, and behavior, and relationship management strategies, and (4) construction of organization-public relationship models. For the purposes of this chapter, the most important of the four developments above is the third because organization-public relationships have a clear linkage to public attitudes, perceptions, knowledge, and behavior. Those who develop and disseminate pandemic-related messages wish publics to perceive their messages as credible, and then adopt appropriate health behaviors. There is considerable evidence that the perception of a relationship between an organization and its publics can influence both actual and intended behaviors. The establishment and maintenance of such relationships is not merely a "feel good" enterprise—positive relationships can lead to positive health behavior changes. The studies outlined below illustrate some of the benefits of establishing and maintaining positive relationships with key publics. Research on the outcomes of positive organization-public relationships has been conducted in both the for-profit and non-profit sectors.

Among the early research in the for-profit sector examining the outcomes of positive relationships with publics was the work of Bruning and Ledingham (1998a, b). The authors' research involving a telecommunications company indicated that the quality of the relationships between the company and its customers had a direct tie to the customers' intentions to stay with the company or seek another provider (Bruning and Ledingham 1998a), and that relationship dimensions are tied to consumer satisfaction (Bruning and Ledingham 1998a). Another study in the for-profit sector involved two well-known companies. In their study of Starbucks and Apple, Hong and Yang (2009) found that when customers perceived the reputation of a company to be positive, they were more likely to engage in word of mouth about the company. Notably, satisfaction had a direct effect on customers' positive word of mouth intentions.

In the non-profit sector, O'Neil's (2008) study of the relationship between a large non-profit organization and its donors indicated that clear communication with donors about how their donations were being used made a significant difference in building greater amounts of trust, commitment, and satisfaction among donors. Waters' (2008) work on the organization-donor relationship involved a large health-care nonprofit. Donors who were most likely to have had a long-term relationship with the organization, such as repeat and major-gift donors, evaluated the organization more positively.

In summary, modern public relations practice from a relational approach has tangible benefits. In order for an organization or a specific public to benefit from a relationship, there must be, of course, a relationship in place. But how are these relationships formed? What are the best methods for organizations and communication professionals to use when seeking to form relationships with key publics? Much work has been conducted on this topic by public relations scholars that can be applied to health organizations responsible for pandemic-related tasks.

2.3 Relationship Cultivation Strategies

Relationship cultivations strategies refers to the methods used to build and sustain high quality relationships with publics. Of course, those responsible for creating and disseminating pandemic-related messages should first determine which publics are the most important to reach. Strategies and tactics aimed at the "general public" are not as effective as efforts targeted to specific publics. In public relations terms, professionals creating and disseminating pandemic messages should first engage in public *segmentation,* or "dividing one large public into subpublics according to their communication needs from the organization" (Berkowitz and Turnmire 1994, p. 107). Such efforts would be different for each health organization (a hospital and a state health department, for example), but the more specific communication plans are with regard to segmentation, the greater the chances for success.

Once decisions have been made regarding which publics are most important, relationship cultivation strategies may then be employed. Relationship cultivation strategies would be familiar to interpersonal communication experts. Public relations work in this area has drawn much from the interpersonal arena (Wood 1995). The suggestion that relationship cultivation strategies used in interpersonal relationships could be used at the organization level was offered by Hon and Grunig (1999). Ki and Hon (2009a), building on Hon and Grunig's (1999) efforts, developed a multiple-item scale for measuring relationship cultivation strategies. Ki and Hon (2009a) outlined six strategies: access, positivity, openness/disclosure, sharing of tasks, networking, and assurances. Definitions of each concept by Ki and Hon (2009a) are below:

- Access—the degree of effort that an organization puts into providing communication channels or media outlets that assist its strategic publics in reaching it
- Positivity—the degree to which members of publics benefit from the organization's efforts to make the relationship more enjoyable for key publics

- Openness/disclosure—an organization's effort to provide information about the nature of the organization and what it is doing
- Sharing of tasks—an organization's efforts to share in working on projects or solving problems of mutual interest between the organization and its publics
- Networking—the degree of an organization's effort to build networks or coalitions with the same groups that their publics do, such as environmentalists, unions, or community groups
- Assurances—any effort by an organization to assure its strategic publics that they and their concerns are attended to (Ki and Hon 2009a, pp. 6–9).

The relationship cultivation strategies outlined above could be used by any health-related organization. One such body that could use such strategies is a local health department. It is not difficult to imagine a local health department employing these relationship cultivation strategies in order to form positive relationships with strategic publics. If these relationships are formed, the likelihood of pandemic messages having the desired impact increases.

2.4 Local Health Departments

Local health departments are, obviously, a key link in the United States public health system. They are key components in carrying out one of the 10 essential public health services: Communicate effectively to inform and educate people about health, factors that influence it, and how to improve it (U. S. Centers for Disease Control and Prevention 2020). Local health departments are critical to the nation's response to COVID-19, partly because they are best equipped to know the needs of their communities (Martin and Kronstadt 2020). In practical terms, examples are offered below of how a local health department might employ relationship cultivation strategies.

There are numerous steps local health departments can take to make their departments easily *accessible* to members of specific publics (or the general public). Those speaking to the media should make certain to include simple information such as "If you have questions, please call or email us" in every media appearance. It may seem as though traditional media has been usurped by new technologies when it comes to communicating health-related information, but to a degree, that is not the case. Park et al. (2019) found that television news was the most important channel for Zika information in 2016. Park et al. (2019) stated that "practitioners should be advised that television news outlets remain critical in the dissemination of health crisis information, with value added by placing medical professionals as sources on television news" (p. 5). Thus, it would be advisable for local health departments to have those with medical backgrounds to appear on traditional media, stressing contact information for members of the public. (Although trust in many sectors of society has been declining for years, as will be discussed later in this chapter, trust in doctors and nurses remains relatively high (Brenan 2018).

Of course, traditional media is hardly the only method local health departments can use to increase accessibility. Social media is used by many local health departments across the country. Lamarca Madden (2019) found that about 66% of local health departments use at least one social media channel. Not unexpectedly, Lamarca Madden found spending per 100,000 population was a significant indicator of whether a local health department used social media platforms, with departments that spent the most money being more likely to use social media for communication purposes.

With regard to *positivity*, local health departments are often involved in various community events that, for lack of a better term, could be characterized as "fun." In Arizona, members of the Maricopa County Department of Public Health, along with other organizations, celebrated National Public Health Week by speaking to youth in foster care about nutrition and exercise, giving the young people a chance to practice Zumba (Krisberg 2014). Local health departments regularly take part in health fairs, and such fairs often feature family activities.

Local health departments should, on a regular basis, provide information on what the organization is doing—in other words, to be *open* and *disclose.* The National Association of City and County Health Officials (NACCHO) is one source where those involved in communication activities can access a "toolbox" that provides guidance to help departments communicate their value to publics before a pandemic occurs. A NACCHO (2001) report found that more than two-thirds of local health departments provide the following core services: adult and childhood immunizations, epidemiology and surveillance, environmental health regulation, tuberculosis testing, and—most importantly for this chapter—communicable disease control. Unfortunately, many constituents living within the jurisdiction of a local health department may not have any knowledge about the value the department is providing on a regular basis, which lessens the likelihood citizens will turn to the department for reliable pandemic information.

Local health departments are steeped in the concepts of *sharing of tasks* and *networking.* There is no question, as Baum et al. (2018), put it, that "Developing new partnerships and strengthening existing collaboration efforts are important components to the prevention and control of emerging infectious diseases" (p. 405). Local health departments regularly work with state health departments (Meit et al. 2012), community groups (Allen et al. 2008), hospitals (Prybil et al. 2014), faith-based organizations (Barnes and Curtis 2009), and academic health departments (Erwin et al. 2019) among others. Public health funding cuts, as noted by Seaton et al. (2018), make interorganizational collaboration of particular importance in the modern world of public health. In their scoping review of factors that facilitate collaborative approaches to health promotion, Seaton et al. (2018) found that several key factors contribute to effective collaboration including open communication and trust. The presence of a properly trained public relations professional can make a positive difference with respect to a local health department's success with respect to networking. In their examination of the characteristics of local health departments in Kentucky that led to the receipt of information by external stakeholders during the H1N1 outbreak of 2009, Howard et al. (2012) found that the absence of a public information officer was associated with decreased communication with external publics. Of the 26 local health departments that had a public information officer, 65% initiated risk mitigation communication with local health professionals. Nearly two-thirds of the departments with public information officers held conference calls or briefings with members of the local health community. Such activities were not reported by departments that had no public information officer employed.

It is not enough for a local health department to listen to the issues raised by individuals or groups. Local health departments should *assure* those with health-related concerns that such matters are attended to. Such assurances do not have to involve grandiose plans—in fact, many efforts involve relatively simple matters. In Contra Costa County, California, for example, Contra Costa Health Services worked with residents in its Healthy Neighborhoods Project to install speed bumps and street lighting to reduce illegal drug activity (Morgan and Lifshay 2006).

Each of the relationship cultivation examples above—access, positivity, openness/disclosure, sharing of tasks, networking, and assurances—can be utilized by any health organization, not just local health departments. It should be noted that there is no question funding may well be an issue when organizational leaders must decide which of the relationship cultivation strategies to employ and at what level. A complete review of local health department funding is beyond the scope of this chapter. However, it should be noted that a lack of funding contributed to a loss of more than 55,000 jobs at local health departments from 2008–2017 (Trust for America's Health 2019). Each local public health department would have to make strategic choices about its ability to carry out the relationship cultivation strategies outlined above.

Once a relationship between an organization and a public has been formed, it is desirable that the organization keep close track of the status of the relationship. In other words, it would behoove the organization to *measure* its relationships with its most important publics in order to know precisely how each public views the organization. The measurement of organization-public relationships has been a consistent topic in the public relations literature for years.

2.5 Measuring Organization-Public Relationships

In order to measure a public's perception of an organization, Hon and Grunig (1999) proposed the following indexes: trust, control mutuality, commitment, satisfaction, communal relationship, and exchange relationship.

- Trust—one party's level of confidence and willingness to open oneself to the other party
- Control mutuality—the degree to which the parties agree on who has the rightful power to influence one another
- Commitment—the extent to which each party believes and feels the relationship is worth spending energy to maintain and promote

- Satisfaction—whether one party feels favorably to the other
- Communal relationship—the extent to which the parties in a relationship give benefits to each other because they are concerned with the other's welfare
- Exchange relationship—the extent to which one party gives benefits to the other because the other gave benefits in the past or expects to do so in the future (Hon and Grunig 1999, p. 3).

Obviously, some of the concepts proposed by Hon and Grunig (1999) above apply more directly to the public health arena than others. For the purposes of public health professionals responsible for delivering pandemic-related messages, the most important of these is trust. If individuals trust an organization or specific health professionals in the health field, those individuals are more likely to pay attention to and act upon pandemic messages.

2.6 Trust

A complete discussion of trust could fill volumes of books. Trust has been examined by scholars of sociology (Meyer and Ward 2013; Williams 2020), economics (Shupeng and Lianjiang 2018), political science (Panizza et al. 2018), disaster management (Peng et al. 2020), relationship marketing (Morgan and Hunt 1994), and family communication (Myers 1998), among others. As noted earlier, many concepts in public relations originated in the interpersonal field. One of the most important works on the topic of trust was Rotter's (1967) more than a half a century ago. Rotter (1967) proposed a scale for the measurement of interpersonal trust. He defined interpersonal trust as "an expectancy held by an individual or group that the word, promise, verbal or written statement of another individual or group can be relied upon" (p. 615). This definition of interpersonal trust has direct applicability to the public relations field. To begin with, trust can be thought of as an expectancy held by an individual or group. Thus, when thinking about the recipients of pandemic-related messages, the recipients could be individuals or groups such as community organizations. The second part of Rotter's (1967) definition concerned the reliability of verbal or written statements of individuals or groups. The recipients of pandemic-related messages are most likely to find pandemic-related messages reliable if they trust the individual (perhaps a primary care physician) or an organization (such as a hospital or public health department).

Rotter's (1967) work remains important, but it would not be hyperbole to suggest that Luhmann (1979) can be thought of as the godfather of trust research. Luhmann (1979) thought of trust as the glue in society that holds everything together. It is widely agreed upon that Luhmann's (1979) central insight regarding trust in society is that trust reduces social complexity (Kroeger 2019). When people make decisions on whether or not to trust, they reduce complexity by pursuing their decisions in a rational manner. Trust, then, is a conscious choice. Writing of the cognitive process of trust, Lewis and Weigert (1985) said the process of trusting "discriminates among persons and institutions that are trustworthy, distrusted, and unknown. In this sense, we cognitively choose whom we will trust in which respects and under what circumstances …" (p. 970).

Luhmann (1979) also recognized two different forms of trust: Institutional trust and interpersonal trust. Interpersonal trust is negotiated between individuals and involves an emotional bond between the individuals. Institutional trust, or system trust, is trust placed in a system or an institution. Devos et al. (2002) argued institutional trust means that people have faith in an institution: "Trusting an institution entails having confidence that the institution is reliable, observes rules and regulations, works well, and serves the general interest" (p. 484). Regarding interpersonal trust, some individuals have a high propensity to trust (Butler 1999), while others have a low propensity for trust (Chatman 1991). One of the factors that can influence trust is familiarity, a topic addressed by Luhmann (1979). Familiarity, in Luhmann's (1979) conceptualization, is a precondition of trust, but not necessarily a positive one. Familiarity can mean either favorable or unfavorable expectations. In personal relationships, familiarity can be a strong predictor of trust in the early stages of a relationships (Alarcon et al. 2016; Levin et al. 2006). This same principle can be applied to organization-public relationships. The more familiar publics are with an organization or its organizational representatives, the more likely they are to believe pandemic-related messages, if the past experiences have been positive.

Let us suppose a storm has occurred, and a homeowner has lost shingles off of a roof. A stranger happens to walk by and tells the homeowner about a roofing company that provided good service. That same day, the homeowner's neighbor also happens by. They have been neighbors for years, and have dined together. The neighbor tells the homeowner about another roofing company that has provided good service. Which person would the homeowner most likely believe? The neighbor, of course. As Li and Betts (2003) put it, "past interactions and the knowledge/familiarity resulting from repeated interactions are essential for interpersonal trust. People don't trust a stranger very often" (p. 104). This thought was echoed by Gena Reed, a science and policy analyst with the Union of Concerned Scientists. Commenting in *The Nation's Health*, Reed said science alone is often not enough to win people over when they are making health-related decisions. In Reed's words, "The decision-making process for all of us involves a lot more than just science; it involves relationships and trust and values .... You win hearts and minds with relationships and trust" (Wahowwiak 2018, p. 20).

Trust can be difficult to build and maintain for any organization whether in the corporate world, government, or the non-profit sector. Kroeger (2015) called trust "problematic across a wide range of societal sectors" (p. 431) pointing out various studies that indicated declining trust in government, business, health systems, and regulatory bodies. As noted by the Pew Research Center (2020), for example, public trust in the federal government has been at near-record lows for years. The Pew Center reported that only 20% of U. S. adults trusted the government in Washington to do the right thing just about always or most of the time. Americans' trust in the mass media in general has been eroding for years (Brenan 2019), with only 41%

reporting a "a great deal" or a "fair amount" of trust in newspapers, television, and radio to report the news "fully, accurately, and fairly." Local news organizations do fare better in this regard compared to national news organizations. The Knight Foundation partnered with the Gallup corporation in a study of trust in local news organizations. The results indicated that people trust local news media more than the national news media, but still, only 45% of Americans trust reporting by local news organizations "a great deal" or "quite a lot" (Knight Foundation 2019).

2.7 Trust and Public Health

Trust is crucial in public health. There has been considerable growth in the number of research efforts centered around the concept of trust. As reported by Schlesinger et al. (2005), there were 764 articles in the medical and health literature between 1980 and 1995 related to trust or trustworthiness. Between 1995 and 2003 that number had risen to 1612 (Schlesinger et al. 2005). As Kowitt et al. (2017) summarized, "A growing body of research has shown associations between trust in government and health-related behaviors and outcomes" (p. 2). Trust in the sources of messages is important, argued Kowitt et al. (2017), in the key public health issues of our time such as tobacco use, food safety, and vaccines. The work of Kowitt et al. (2017) concerned trust in the Centers for Disease Control and Prevention (CDC) and the Food and Drug Administration, but the general idea can be applied to other health-related bodies such as hospitals, state and local health departments, and community health organizations. Without trust in health organizations, public confidence in communication about health issues may be diminished (Kowitt et al. 2017). Cummings (2014) went so far as to say that "With notable exceptions, it can seem that skepticism and mistrust are now the public's default position as soon as a public health problem comes to prominence" (p. 1045). Cummings (2014) noted lack of trust has been a problem across a wide range of public health issues in recent years including the safety of breast implants, the safety of oral contraceptives, and beef consumption because of concerns about bovine spongiform encephalopathy (mad cow disease). The general mistrust in public institutions is certainly a factor, argued Cummings (2014), along with a "perception that the media cynically amplify public health issues for their own needs" (p. 1045). As an example, trust in the CDC declined in 2020. According to a Kaiser Family Foundation poll, public confidence in the CDC dropped 16 points from April to September of 2020 (Florko 2020).

The trust problems outlined above, both in and out of the public health field, cannot be completely eliminated by modern public relations management and practice. Such problems can, however, be lessened.

2.8 Trust and Public Relations

There is no doubt about the importance of the concept of trust among public relations scholars. Perhaps Bekmeier-Feuerhahn and Eichenlaub (2010) put it best when they said, "The creation and maintenance of trust in the public counts as one of the most important tasks of public relations" (p. 337). Ki and Hon (2009b) said that "trust has been found to be ... an essential indictor of measuring relationship quality across diverse disciplines" (p. 249). After reviewing the concept of trust in the public relations literature, Cacciatore et al. (2016) used similar language in coming to a conclusion about trust: "In short, trust represents the essence of quality relationships between an organization and its various publics" (p. 617).

The work of Ki and Hon (2007) is an example of the interest of public relations scholars in the concept of trust. Ki and Hon's study involved a state farm bureau and its members. The authors investigated how relationship cultivation strategies affected members' perceptions of relationship quality outcomes. Perhaps the most important finding for those in the pandemic field was that providing assurances was a key strategy for all relationship quality outcomes, including trust. As pointed out by Ki and Hon, providing assurances has been found to be an essential predictor of trust in several studies in the interpersonal communication field.

Hon and Grunig (1999) outlined three dimensions of trust: (1) integrity, the belief that an organization is fair and just; (2) dependability, the belief that an organization will do what it says it will do; and (3) competence, the belief that an organization has the ability to do what it says it will do. For health-related organizations—whether state health departments, non-profits, hospitals, local health departments, or national health bodies—the importance of these three dimensions of trust are obvious. Health-related organizations that publics may turn to for pandemic information must be viewed as fair to all segments of the populations, dependable enough that publics have faith the organization will stand by its word, and competent enough to carry out its pandemic-related functions (such as vaccinations). Public relations professionals working at health-related organizations can play a key role in developing and maintaining relationships so that key publics' view organizations as fair, dependable, and competent.

2.9 Conclusion

The costs of pandemics are enormous, both in terms of human lives taken and affected as well as economic damage. Public relations professionals have no role to play in scientific endeavors such as the development of pandemic vaccines and treatments. But by forming and maintaining relationships with key publics before a pandemic occurs, public relations professionals can help make it more likely that people listen to the recommendations of those in public health, and more importantly, trust public health professionals enough to act on those recommendations. Such positive health behaviors, then, can have the most important effect possible: saving lives.

References

Alarcon GM, Lyons JB, Christensen JC (2016) The effect of propensity to trust and familiarity on perceptions of trustworthiness over time. Personality Individ Differ 94:309–315

Allen C, Hsu C, Schwartz PM, Pearson D, Greenwald HP, Beery WL, Flores G, Casey MC (2008) Involving local health departments in community health partnerships: evaluation results from the partnership for the public's health initiative. J Urban Health 85:162–178

Barnes PA, Curtis AB (2009) A national examination of partnerships among local health departments and faith communities in the United States. J Public Health Manag Pract 15:253–263

Baum C, Cantu M, Laird E (2018) Lessons in infection control: the role of local health departments in prevention, preparedness, and response. J Public Health Manage Pract 24:404–406

Bekmeier-Feuerhahn S, Eichenlaub A (2010) What makes for trusting relationships in online communication? J Commun Manag 4:337–355

Berkowitz D, Turnmire K (1994) Community relations and issues management. J Public Relat Res 6:105–123

Brenan M (2018) Nurses again outpace other professions for honesty, ethics. Available at: https://news.gallup.com/poll/245597/nurses-again-outpace-professions-honesty-ethics.aspx?g_source=link_NEWSV9&g_medium=NEWSFEED&g_campaign=item_&g_content=Nurses%2520Again%2520Outpace%2520Other%2520Professions%2520for%2520Honesty%2c%2520Ethics

Brenan M (2019) Americans' trust in mass media edges down to 41%. Available at: https://news.gallup.com/poll/267047/americans-trust-mass-media-edges-down.aspx

Broom G, Sha BL (2013) Cutlip & center's effective public relations, 11th edn. Pearson Education, Essex, England

Browning N (2018) Ethics and the profession: the crystallizing of public relations practice from association to accreditation, 1936–1964. Am J Historians Assoc 35:140–170

Bruning SD, Ledingham JA (1998a) Organization-public relationships and consumer satisfaction: the role of relationships in the satisfaction mix. Commun Res Rep 15:198–208

Bruning SD, Ledingham JA (1998b) Relationship management in public relations: dimensions of an organization-public relationship. Public Relat Rev 24:55–65

Butler JK Jr (1999) Trust expectations, information sharing, climate of trust and negotiation effectiveness and efficacy. Group Org Manag 24:217–238

Cacciatore MA, Meng J, Boyd B, Reber BH (2016) Political ideology, media-source preferences, and messaging strategies: a global perspective on trust building. Public Relat Rev 42:616–626

Chatman JA (1991) Matching people and organizations: selection and socialization in public accounting firms. Adm Sci Q 36:459–484

Cheng Y (2018) Looking back, moving forward: a review and reflection of the organization-public relationship (OPR) research. Public Relat Rev 44:120–130

Cummings L (2014) The "trust" heuristic: arguments from authority in public health. Health Commun 29:1043–1056

Cutlip SM (1995) Public relations history: from the 17th to the 20th century. Lawrence Erlbaum Associates, Mahwah, NJ

Devos T, Spini D, Schwartz SH (2002) Conflicts among human values and trust in institutions. Br J Soc Psychol 41:481–494

Erwin PC, Parks RG, Mazzucca S, Alen P, Baker EA, Hu H, Davis-Joyce J, Brownson RC (2019) Evidence-based health provided through local health departments: importance of academic practice partnerships. Am J Public Health 5:739–747

Ferguson MA (1984) Building theory in organizations: interorganizational relationships as a public relations paradigm. Paper presented to the Public Relations Division, Association for Education in Journalism and Mass Communication Annual Convention, Gainesville, FL
Fitzpatrick K (2002) Evolving standards in public relations: a historical examination of PRSA's codes of ethics. J Mass Media Ethics 17:89–110
Florko N (2020) Public trust in CDC, Fauci, and other top health officials is evaporating, poll finds. Kaiser Family Foundation. Available at: https://www.statnews.com/2020/09/10/trust-cdc-fauci-evaporating/
Knight Foundation (2019) State of public trust in local news. Available at: https://knightfoundation.org/wp-content/uploads/2019/11/State_of_Public_Trust_in_Local_Media_final_.pdf
Hon LC, Grunig JE (1999) Guidelines for measuring relationships in public relations. Institute for Public Relations. Retrieved from: https://www.instituteforpr.org/wp-content/uploads/Guidelines_Measuring_Relationships.pdf
Hong SY, Yang S-U (2009) Effects of reputation, relational satisfaction, and customer-company identification on positive word-of-mouth intentions. J Public Relat Res 21:381–403
Howard AF, Bush HM, Shapiro RM, Dearinger A (2012) Characteristics of Kentucky local health departments that influence public health communication during times of crisis: information dissemination associated with H1N1 novel influenza. J Public Health Manag Pract 18:169–174
Ki E-J, Hon LC (2007) Reliability and validity of organization-public relationship measurement and linkages among relationship indicators in a membership organization. Journalism Mass Commun Q 3:419–438
Ki E-J, Hon LC (2009a) Causal linkages between relationship cultivation strategies and relationship quality outcomes. Int J Strateg Commun 3:242–263
Ki E-J, Hon LC (2009b) A measure of relationship cultivation strategies. J Public Relat Res 21:1–24
Kowitt SD, Schmidt AM, Hannan A, Goldstein AO (2017) Awareness and trust of the FDA and CDC: results from a national sample of U.S. adults and adolescents. Plos One 12:1–14
Krisberg K (2014) Communities hold national public health week events: health departments, schools, celebrate national observance. The Nation's Health 44:19–33
Kroeger F (2015) The development, escalation, and collapse of system trust: from the financial crisis to society at large. Eur Manag J 33:431–437
Kroeger F (2019) Unlocking the treasure trove: how can Luhmann's theory of trust enrich trust research? J Trust Res 9:110–124
Lamarca Madden S (2019) Exploring the factors associated with social media use in local health departments (Unpublished doctoral dissertation). Georgia Southern University, Statesboro, GA
Ledingham JA (2003) Explicating relationship management as a general theory of public relations. J Public Relat Res 15:181–198
Levin DZ, Whitener EM, Cross R (2006) Perceived trustworthiness of knowledge sources: the mediating impact of relationship length. J Appl Psychol 91:1163–1171
Lewis JD, Weigert A (1985) Trust as a social reality. Soc Forces 63:967–985
Li F, Betts SC (2003) Trust: what it is and what it is not. Int Bus Econ Res J 2:103–108
Luhmann N (1979) Trust and power. John Wiley & Sons, New York
Martin EG, Kronstadt J (2020) No longer invisible: the critical role of local health departments in responding to COVID-19. Health affairs. Available at: https://www.healthaffairs.org/do/10.1377/hblog20200408.106373/full/
Meit M, Kronstadt J, Brown A (2012) Promising practices in the coordination of state and local public health. NORC: University of Chicago. Available at: https://www.astho.org/Research/Major-Publications/Promising-Practices-in-the-Coordination-of-State-and-Local-Public-Health/
Meyer SB, Ward PR (2013) Differentiating between trust and dependence of patients with coronary heart disease: furthering the sociology of trust. Health Risk Soc 15:279–293
Morgan RM, Hunt SD (1994) The commitment-trust theory of relationship marketing. J Mark 58:20–38

Morgan MA, Lifshay J (2006) Community engagement in public health. Contra Costa Health Services, Martinez, CA. Available at: https://cchealth.org/public-health/pdf/community_engagement_in_ph.pdf
Myers SA (1998) Sibling communication satisfaction as a function of interpersonal solidarity, individualized trust, and self-disclosure. Commun Res Rep 15:309–317
NACCHO (National Association of County and City Health Officials) 2001 Local public health agency infrastructure: a chartbook. NACCHO, Washington, DC
O'Neil J (2008) Linking public relations tactics to long-term success: an investigation of how communications contribute to trust, satisfaction, and commitment in a nonprofit organization. J Promot Manag 14:263–274
Panizza F, Peters BG, Ramos Larraburu CR (2018) Roles, trust, and skills: a typology of patronage appointments. Public Adm 97:147–161
Park S, Boatwright B, Johnson Avery E (2019) Information channel preference in health crisis: exploring the roles of perceived risk, preparedness, knowledge and intent to follow directives. Public Relat Rev 45:1–8
Peng L, Tan J, Deng W, Liu Y (2020) Farmers' participation in community-based disaster management: the role of trust, place attachment, and self-efficacy. Int J Disaster Risk Reduction 51. https://doi.org/10.1016/j.ijdrr.2020.101895
Pew Research Center (2020) American's views of government: low trust, but some positive performance ratings. Available at: https://www.pewresearch.org/politics/2020/09/14/americans-views-of-government-low-trust-but-some-positive-performance-ratings/
Prybil L, Scutchfield F, Douglas K, Rex KA, Mays GP, Carman A, Levey S, Fardo DW (2014) Improving community health through hospital-public health collaboration: insights and lessons learned from successful partnerships. Health management and policy faculty book gallery 2. University of Kentucky, Lexington, Kentucky
Rotter JB (1967) A new scale for the measurement of interpersonal trust. J Pers 35:651–665
Schlesinger M, Quon N, Wynia M, Cummins D, Gray B (2005) Profit-seeking, corporate control, and the trustworthiness of health care organizations: assessments of health plan performance by their affiliated physicians. Health Serv Res 40:605–645
Seaton CL, Holm N, Bottorff JL, Jones-Bricker M, Errey S, Caperchione CM, Lamont S, Johnson ST, Healy T (2018) Factors that impact the success of interorganizational health promotion collaborations: a scoping review. Am J Health Promot 32:1095–1109
Shupeng L, Lianjiang L (2018) The dual effect of economic development on trust in the central government: evidence on two waves of a national survey in China. China Rev 18:65–86
Trust for America's Health (2019) New TFAH report: persistent underfunding of America's public health system makes the nation vulnerable and puts lives at risk. Available at: https://www.tfah.org/article/new-tfah-report-persistent-underfunding-of-americas-public-health-system-makes-the-nation-vulnerable-and-puts-lives-at-risk/
U. S. Centers for Disease Control and Prevention. (2020) 10 essential public health services. https://www.cdc.gov/publichealthgateway/publichealthservices/essentialhealthservices.html
Udow-Phillips M, Lantz PM (2020) Trust in public health is essential during the COVID-19 pandemic. J Hosp Med 15:431–433
Wahowwiak L (2018) Public health workers to fight misinformation through trust, relationships. The Nation's Health 48:1–20
Waters RD (2008) Applying relationship management theory to the fundraising process for individual donors. J Commun Manag 12:73–87
Williams DM (2020) Relationships of horizontalism and hierarchy: exploring divergent forms of sociopolitical trust. Soc Sci Q 101:1150–1164
Wood JT (1995) Relational communication. Wadsworth, Belmont, CA
Yang A, Taylor M (2014) A global perspective on public relations professionalism: mapping the structure of public relations associations' international networks. Journalism Mass Commun Q 9:508–529

Chapter 3
Outbreak Narrative in Pandemics: Resilience Building in Communicating About 1918 Influenza and SARS

Huiling Ding and Yingying Tang

Abstract This chapter examines pandemic communication from traditional media about two emerging pandemics, the 1918 Influenza and SARS in Hong Kong and Toronto, two of the epicenters. It focuses on how authorities, experts, and communities made sense and took action to cope with unknown health risks when little was known about the two pandemics. Focusing on pandemic communication from traditional media, it explores various ways resilience was built by different communities to better cope with the health risks brought by these two emerging pandemics, paying particular attention to expert-public communication, social scapegoating, racial profiling, networked collaboration, and community call for action.

Keywords 1918 influenza · SARS · Expert-public communication · Social scapegoating · Racial profiling · Networked collaboration · Call for action

3.1 Introduction to Resilience

In her study of the rhetoric of epidemics, Wald defines "outbreak narrative" as "a formulaic plot that begins with the identification of an emerging infection, includes a discussion of the global networks throughout which it travels, and chronicles the epidemiological work" that results in disease containment (p. 2). While circulating across media and genres, outbreak narratives "consistently register anxieties about the global village that reflexively imagined the containment of disease in national terms" (p. 63) while "making the act of imagining the community a central feature of its preservation" (p. 52). Treichler (1999) emphasized the need to study epidemics such as AIDS as both material and linguistic entities, or in her words, as "an epidemic of transmissible lethal disease and an epidemic of signification" (p. 12). To contain emerging infections, institutions, communities and individuals have to

H. Ding (✉)
North Carolina State University, Raleigh, NC, USA
e-mail: hding@ncsu.edu

Y. Tang
Auburn University, Auburn, AL, USA

D. M. Berube (ed.), *Pandemic Communication and Resilience*, Risk, Systems and Decisions, https://doi.org/10.1007/978-3-030-77344-1_3

develop resilience in communicating about and mobilizing resources to effectively cope with novel health risks. But how can resilience be developed, and what roles can traditional media play in such resilience-building processes?

The Oxford English Dictionary (OED) defined resilience as "The quality or fact of being able to recover quickly or easily from, or resist being affected by, a misfortune, shock, illness, etc.; robustness; adaptability." At the individual level, (Ungar 2011) proposed a multilayered social and ecological understanding of resilience, which expands the study of resilience from individual traits to social and physical ecology to account for not only individual factors but more importantly environmental factors. Ungar identified four types of ecological interaction, which include microsystem interactions (with family, peers, and teachers) and mesosystem interactions (between microsystems) for analysis of available supportive resources. Exosystem interactions focus on the institutional and communal environments (Boyden et al. 2005; Leadbeater et al. 2005) while macrosystem interactions examine laws, customs, and cultural practices (Dawes and Donald 2000; McCubbin and McCubbin 2005; Seidman and Pedersen 2003). To help better define and operationalize resilience, (Ungar 2011) highlighted four principles: decentrality, i.e., a shift from individuals to processes as well as facilitative social and physical ecologies; complexity, i.e., impacts of numerous uncontrollable contextual and temporal factors; atypicality, i.e., atypical use of developmental resources and adaptive behaviors in adverse environments, and cultural relativity, i.e., culturally and temporally/historically embedded processes. Emphasizing both navigation and negotiation in social and physical ecologies, (Ungar 2008) defined resilience as:

> both the capacity of individuals to navigate their way to the psychological, social, cultural, and physical resources that sustain their wellbeing, and their capacity individually and collectively to negotiate for these resources to be provided and experienced in culturally meaningful ways (p. 225).

Serfilippi and Ramnath (2018) defined resilience as "the capacity of people, communities, or systems to prepare for and to react to stressors and shocks in ways that limit vulnerability and promote sustainability" (p. 647). For them, stressors can be acute or chronic with brief or long-term impacts, which can include earthquakes, outbreaks, unemployment, poverty, and aging. Communities can be defined in terms of shared locations, cultures, or purposes (Tropman et al. 2006). Allmark et al. (2014) defined resilience as "always (i) of something (ii) to something (iii) to an endpoint, as in (i) a rubber ball, (ii) to a blunt force, (iii) to its original shape" (p. 2). Many characteristics attribute to resilient communities based on existing studies, which include a shared sense of belonging, values, and common purpose; social and organizational networks; infrastructure and support services (Buikstra et al. 2010; Seccombe 2002); cultural and contextual factors (Gilligan 2001; Ungar 2008), learning and leadership, "different bundles of resources" and socio-economic stressors (Platts-Fowler and Robinson 2013), and participatory decision making and action (Cadell et al. 2001).

Taking an asset-based approach drawing on and developing positive factors existing or potential in the community (Allmark et al. 2014), researchers identified seven types of static assets, which can be co-existing capital, that communities have control over and can contribute to livelihoods and public health. These capitals are *human capital* (i.e., skills, expertise, and education), *social capital* (i.e., social networks), *physical capital* (i.e., facilities and infrastructure), *natural capital* (i.e., natural resources and climate change variables), *financial capital* (i.e., incomes, saving, credit, and funding), *political capital* (i.e., access to institutions and capacity to exert political influence), and *cultural capital* (i.e., beliefs, values, and lifestyles) (Haines 2009). Therefore, a case of community resilience might be: a neighborhood manages to return to normality in a ravaging flu pandemic through the use of public health and communication expertise (*human capital*), close-knit networks (*social capital*), and surveillance technologies (*both physical capital and economic capital*).

Going beyond capital as the static dimension of resilience, (Serfilippi and Ramnath 2018) called attention to three temporal types of capacities, i.e., the absorptive, adaptive, and transformative as the dynamic dimension of resilience, which helps to analyze the stability, flexibility, and change of a system. While the absorptive capacity refers to the ability to reduce risks (preparedness) and to absorb short-term impacts (mitigation), the adaptive capacities refer to longer-term responses to social, economic, and environmental changes. Transformative capacities, in turn, refer to major changes in the structure and function of the system when the adaptive capacities are overwhelmed by external shocks.

Bergström (2018) wrote about a global push, "through policy and government campaigns, to emphasise a local and decentralised responsibility for societal safety and security" through the use of the notion of resilience (p. 31). Such shifts are driven by the political need to define societal safety and security as "decentralised adaptive capacities and preparedness" that depends on local actors and networks (p. 36). As "a new form of societal safety and security biopolitics," the new focus on resilience shifts increasing responsibilities from governments to citizens who are expected to prepare for, adapt to, and pay for emerging, ambiguous, and dynamic risks. This discursive shift also silences discussion of the ethics and distribution of risk exposure, the acceptability of risk, and the need to care for the vulnerable.

In what follows, we examine the news media coverage of the 1918 influenza and SARS, focusing on episodes of community resilience that helped build multiple capitals and capacities to help bring the outbreaks under control.

3.2 Resilience Building and Traditional Media in the 1918 Flu Pandemic

As one of the deadliest pandemics in human history, the 1918 influenza epidemic it infected about one-third of the world's population (500 million people) and killed 50–100 million people ("1918 Pandemic|CDC" 2020). In the US alone, the pandemic caused more deaths "than in World War I, World War II, and the conflicts in Korea, Vietnam, Afghanistan, and Iraq—combined" (Outka 2019, p. 1). This unusually deadly epidemic shattered the wartime world in multiple ways: occurring mostly at the end of World War I (WWI), the pandemic hit the small towns and villages far away from the battlefield. The once safe and tranquil domestic spheres during wartime were penetrated by mass and sudden death of the familiar disease, influenza. Unexpectedly, while most influenza outbreaks killed more of the oldest and the youngest population, the 1918 influenza pandemic had a mortality rate much higher than expected in young people (Tang 2020). Its "characteristic W-shaped mortality curve over age with a peak around 30 years of age constituted a demographic shock" that caused many families to lose adult parents in prime working age (Rao and Greve 2018, p. 10), leaving only children and the elderly in the war-worn families and communities.

Arriving in the United States in spring 1918, the first and relatively mild wave of the 1918 flu pandemic "largely escaped public notice when American attention was gradually geared towards the First World War" (Schwartz 2018, p. 1455). When it returned in fall 1918, the pandemic quickly swept across the United States "with far greater virulence" (Schwartz 2018, p. 1455). Besides the "multiple, closely spaced pandemic waves," wartime conditions including "poor sanitation, overcrowding, and limited health services" caused "tremendous morbidity and death rates" of the virus (Jester et al. 2018, p. 2596). At the time, "genomics, vaccines, antibiotics, mechanical ventilators, and other features of high-technology medicine" were not available in the 1918 world (Parmet and Rothstein 2018, p. 1435). Doctors could only relieve the symptoms of flu and its complications rather than really prevent and treat them. Moreover, 30% of US physicians were engaged in military service (Jester et al. 2018, p. 2596), causing a further shortage of medical resources that were already insufficient.

3.3 Community Resilience: Escape Communities and Nonpharmaceutical Measures in the 1918 Flu Pandemic

As Jester et al. pointed it out, the unexpected severity of the pandemic combined with the world condition in 1918 urged communities to try what they could to slow and stop the spread of influenza, especially in the absence of "national strategy or support from the federal government" or antivirals or antibiotics (Jester et al. 2018, p. 2597). Communities relied on available human, social, physical, natural, financial, political, and cultural capitals to build community resilience. Lacking a unified national deployment, the resilient measures took by communities greatly varied in terms of "the timing of their initiation, and their duration" (Jester et al. 2018, p. 2597), which suggests the localized and decentralized characteristics of community resilience (Bergström 2018).

According to the research conducted by the Center for the History of Medicine at the University of Michigan, among the 43 cities examined in the study, seven communities successfully escaped the pandemic with extremely low infection rates and no influenza-related deaths by promoting nonpharmaceutical measures such as "school closure, cancellation of public gatherings, and isolation and quarantine" (Markel et al. 2007). These seven communities (Yerba Buena, Gunnison, Princeton Student Naval Training Corps, the Western Pennsylvania Institute for the Blind, Trudeau Sanitorium, Bryn Mawr College, and Fletcher) are therefore termed as "escape communities" for their success in minimizing the impacts of the pandemic (Markel et al. 2007).

These seven escape communities that successfully survived the pandemic best embodied how community resilience was built when catastrophes took place: by taking nonpharmaceutical measures, these communities exerted all the possible and potential "psychological, social, cultural, and physical resources" (Ungar 2008, p. 225) to react to the pandemic "in ways that limit vulnerability and promote sustainability" (Serfilippi and Ramnath 2018, p. 647). Nonpharmaceutical measures are "often framed as short-term measures, minimizing the effects of an epidemic until effective vaccines or therapies can be developed, produced, and distributed" (Schwartz 2018, p. 1455). Markel et al. (2007) study indicated "a strong association between early, sustained, and layered application of nonpharmaceutical interventions and mitigating the consequences of the 1918–1919 influenza pandemic in the United States" (p. 644). Promoting nonpharmaceutical measures such as "isolation of those who are ill, quarantine of those suspected of having contact with those who are ill, school and selected business closure, and public gathering cancellations" (Markel et al. 2007, p. 644), this study eventually "became the basis for the CDC's pandemic response guidelines" ("Nonpharmaceutical Interventions" 2007) and its guidelines are strictly enforced in combating the current Covid-19 pandemic.

3.4 The Role of Traditional Media in Building Community Resilience

In the heat of the 1918 pandemic, communities across the country used all resources and took various measures to delay the effects of the pandemic, reduce the overall and peak attack rate, and reduce the number of deaths (Markel et al. 2007). However, these efforts would not work unless community members had adequate resources, effective guidelines, and willingness to comply with communal requirements. With people having limited access to television and radio, newspapers functioned as both the most important information channel and the key platform for ordinary households and communities for building resilience.

This section examines how the traditional media, namely, newspapers, assisted the resilient efforts of US communities to slow down and stop the spread of influenza during the 1918 pandemic. Based on the archival materials of the library and the Center for the History of Medicine at the University of Michigan, this study mainly looks at the news stories and reports of the seven escape communities and investigates how newspapers worked as the most important platform to convene the different political, physical, social, financial capitals within and beyond their communities. We also identify how newspapers served as the essential information channel for individuals to navigate "to the psychological, social, cultural, and physical resources that sustain their wellbeing" (Ungar 2008, p. 255).

A keyword analysis was conducted on the titles of the news reports related to the 1918 influenza in local newspapers in the seven escape communities. We reviewed 809 titles of pandemic-related news reports from 20 local newspapers of the escape communities. By running the titles of news reports of each community via an online tool, Word Counter,[1] we got seven lists of the most frequent keyphrases of each community's news reports. We then coded all the key phrases into community assets (human capital, social capital, physical capital, natural capital, financial capital, political capital, and cultural capital) to see what resources communities relied on to sustain and improve the livelihood and public health of the community during the pandemic (see Table 3.1).

As the keyword analysis shows, as the most important information source in 1918, the newspaper played a key role in connecting and reinforcing the existing and potential resources of the community to help promote community resilience. Many newspapers "ran lengthy articles on influenza" to provide readers timely, in-depth reports of the pandemic ("1918 Influenza Escape Communities|U-M Center for the History of Medicine" 2007a, b). In the era when there was no instant communication system, newspapers were the best platform for authorities to release the latest influenza-related policies, orders, and plans. Via newspaper, individuals were tightly connected to their community's political capitals such as "flu ban", "influenza ban", "army plan", "ban lifted", "closing order", "school open" to guide their school, work, travel, and other activities during the pandemic. They could use culture capitals such

Table 3.1 Community resilience assets

Types of capital	Examples[a]
Political capital	Flu ban (1%), ban lifted (1%), army plan (2%), under government (2%), influenza ban (3%), closing order (2%), everything closed (1%), school open (2%)
Physical capital	Wear masks (6%), gauze masks (2%), flu mask (2%), state item (2%), mayo vaccine (1%)
Social capital	Health board (9%), red cross (6%), centennial state (2%)
Natural capital	Open air (2%)
Financial capital	Crowd waiting loan (2%), loan booth opens (2%)
Cultural capital	Influenza precautions (1%), influenza toll (2%), statistics (1%)

[a]Percentages of the key phrases were calculated by adding up the frequencies of the same key phrase appearing in all the seven lists to show the total frequency of the specific key phrase in various news reports of seven communities

[1] Word Counter (https://wordcounter.net/) is an online tool can show the top keywords and keyword density of the article users upload to it.

as "influenza precautions", "influenza toll", "statistics" to decide what they should do to protect themselves from influenza. They learned how natural capitals like "open air" could reduce virus transmission. They also learned what physical capitals like "masks", "state item", "mayo vaccine", and financial capitals like "loans" were available for them. Through the newspaper, individuals knew that they could turn to social networks such as the "health board", "red cross", and "centennial state" that were on the front line fighting against the epidemic. As Paton et al. put it:

> Household and community resilience will be a function of the level of people's pandemic knowledge, the resources available to facilitate self-reliance, the development of neighbor and community relationships to provide social support, and the development of the plans required to use the knowledge and resources to adapt to the consequences of a pandemic. (p. 45).

As the platform on which these "pandemic knowledge", "resources", and "plans" were disseminated, newspapers helped individuals to "navigate their way to the psychological, social, cultural, and physical resources that sustain their wellbeing" and further transform such individual endeavors into community resilience building (Ungar 2008, p. 255).

3.5 Global Resilience-Enhancing Outbreak Narratives in SARS

Originated in southern China in November 2002, SARS started to spread globally in February 2003. A total of 8422 people were infected with SARS, which resulted in 916 deaths, a fatality rate of 11 percent. The US reported 192 cases with no deaths. Canada recorded 225 cases in Toronto, with 88 (39%) being medical care workers.

The Ministry of Health in China under-reported the outbreak until April 20, 2003. Pressured by travel advisories issued by the World Health Organization (WHO) against travel in Guangdong and Beijing, President Hu fired the Minister of Health and Beijing Mayor on April 21 and quickly implemented new mechanisms to ensure transparent case reporting and accountability. In late April, China launched a nationwide anti-SARS campaign with mass mobilization while constructing Xiao Tangshan Hospital, which served as the designated SARS hospital for Beijing when local hospitals were overwhelmed by the quickly escalating cases. These efforts helped to quickly contain and eradicate the SARS outbreak in China, with WHO removing all Chinese cities from its travel advisory list on June 23, 2003 (see Ding 2014). Several countries and regions were seriously affected by SARS and made to WHO's travel advisory lists, including Canada, Hong Kong, Singapore, and Taiwan. According to WHO).

Ding (2014) divided the international coverage of SARS in the US media into three stages, with the first one focusing on the underreporting and under-calculation of SARS cases in China and a quick logical jump to the Communist Party's traditional penchant for secrecy. The second stage started after China began its belated-yet-full-scale anti-SARS campaign on April 21. The US media offered little coverage of China's new anti-SARS efforts while focusing on issues such as early inaction, censorship, secrecy, and cover-up of SARS, which remained a recurring theme throughout the SARS pandemic. The third stage focused on China's efforts to contain SARS and questioned the quality and reliability of China's data when WHO announced no systematic underreporting was found in cities such as Shanghai or Beijing in late April. The media repeatedly predicted massive outbreaks in China's economically backward and medically under-equipped hinterland (Bradsher 2003; Eckholm 2003). In early June, SARS was almost brought under control in China. The US media, however, witnessed a surge of questions about the reliability of China's data, the "many imponderables about the SARS epidemic," and "the crackdown on the media to maintain tight control" (Altman 2003, p. A18; Pomfret 2003, p. A16).

In domestic coverage, American and Canadian media witnessed widespread practices of racializing and politicizing SARS as an emerging epidemic originated in China. Hate speech and vicious rumors were employed to demonize immigrants, international students, and ethnic communities based on skin colors, ethnic and cultural backgrounds, or countries of origin (Ding 2014). The widespread use of racial discrimination of Chinatowns and Asian communities during SARS brought unnecessary panic and alienation as well as sharp economic losses to those ethnic communities (Matthews 2003; Rosenwald 2003; Surendran 2003). These racial profiling incidents inserted theories of social contagion and transgression into SARS outbreaks and pathologized ethnic communities with perceived connections with epicenters.

In response to these incidents of racial profiling, health and political authorities took action to help local Chinatowns combat discrimination and business loss. Federal senators, including Senator Hillary Clinton, not only participated in press releases but also dined in the Chinatown in New York City. Mayors of Boston and New York ate lunches in Chinatowns, and the Canadian Premier visited Toronto's Chinatown to show moral support and to promote tourism (Li 2003; Rosenwald 2003; Westfeldt 2003). Ethnic Chinese newspapers highlighted these symbolic gestures of support from prominent officials to dispel rumors of SARS cases in Chinatowns. They also heavily emphasized the economic and cultural contributions made by Chinese and other Asian communities to local areas to build their political capital and cultural capital.

In what follows, we will explore how community resilience was constructed in media coverage of SARS, focusing on two incidents: extra-institutional risk communication about locations of SARS cases in Hong Kong and efforts to reduce cross-infection of nurses in Canada.

3.6 Sosick.org and SARS Location: Participatory Risk Communication in Hong Kong

Starting in early March 2003, SARS cases started to appear in non-hospital settings in Hong Kong. Despite repeated requests for information about locations where infected cases had been found, the Hong Kong government rejected such requests because of its concerns about privacy and mass panic. Such information vacuum resulted in rumor-mongering in Hong Kong, i.e., Hong Kong being officially declared as "an infection port" and SARS cases being "underreported in Hong Kong" (Philips 2003, p. 3). On March 31, a total of 329 atypical pneumonia cases and later 33 deaths were reported in Amoy Gardens, a housing estate in Kowloon, which resulted in the quarantine of all its residents (Pomfret and Weiss 2003). Denied open access to detailed risk information, the public was deeply concerned about the possibility of unknowingly contracting the SARS virus when working or traveling near locations with recently infected cases.

To fill in this information void, four young computer engineers in Hong Kong worked together to compile and publish online information about locations of SARS patients in Hong Kong. With the necessary human capital, i.e., skills and experiences in hand, these engineers put into use their research skills, web development expertise, and engineering background to address an urgent public need. On March 31, they published the first Sosick.org SARS building list, which reported data on buildings and locations where suspected SARS patients had visited or had been found. Its information was in such high demand that in two days, sosick.org attracted over 200,000 accumulated hits and over 100 emails a day. Soon, site visitors started to send the Web site owners data about SARS-affected buildings, which were gathered through online or onsite research from news reports, news releases, or building management announcements about possible cases in communities. With thousands of voluntary distributed collaborators from all over Hong Kong, sosick.org started to leverage cultural capital, namely, the public value of transparent risk information and open communication, while coalescing social capital by serving as the platform to augment public concerns and experiment with collective endeavors to address such concerns using Web 1.0 technologies such as emails and web pages. Through public participation in risk information gathering, complication, and verification, sosick.org developed its absorptive capacity as a flexible community infrastructure that helped the public to reduce possible risks of infection. Meanwhile, through the combined use of "strengths, attributes, and resources available to individual[s] and community," Sosick.org managed to "undertake actions to reduce adverse impacts [and] moderate harm" (IPCC 2012), which is a core feature of adaptive capacity.

The four young men collected data from users who posted information on their Web site, and then checked the credibility of such data before using them to update their list of infected buildings. They also offered all information in both Chinese and English, which enhanced its international visibility and impacts as a bilingual resource. As a grassroots effort to list all the infected housing and commercial estates, the Web site functioned as a major player in the transnational, extra-institutional risk communication about SARS in and out of Hong Kong.

To enhance the credibility and impacts of their risk messages, the sosick.org owners decided to offer authoritative and credible references and external links to all buildings included in their constantly growing list. To ensure the accuracy of any new buildings they added to the Web site, they took rigorous fact-checking measures. Users sharing such data with them were required to send them valid proofs such as "newspaper articles [publicizing] the names of infected buildings, or letters issued by the Department of Health (DH) to the estates' management about people on an estate being infected with SARS" (Lais 2003, p. 6). After receiving the original data from users, the owners took rigorous measures of credibility checks to verify the news before posting the information online for public dissemination (Taylor 2003). In addition, because of the rapid submissions of infected buildings from the visitors, the website greatly facilitated the timely circulation of urgently needed public health information among the public.

Sosick.org attracted so much attention that by April 9, the corresponding email address for the site received over one thousand messages reporting infected locations every day. On April 11, its cumulative hit count reached two million, and on April 12, Sosick.org became one of the top ten most searched words in Yahoo.com.hk, one of Hong Kong's top search engines ("Sosick.org Timeline"). On the same day, the DH's official Web site published for the first time a list of infected buildings where confirmed SARS cases had been found. The public believed that the move was prompted by sosick.org, "which had [been], for more than a week, providing details of SARS hot spots, including the names of buildings where infections had been confirmed" (Lais 2003).

Sosick.org's extra-institutional risk communication was lauded by over twenty transnational media, including CNN and National Public Radio (NPR) in the United States (SARS Web site in Hong Kong 2003). *Lancet*, a leading medical journal, listed it as one of the most prominent Web sources for SARS information along with official websites of WHO, national health ministries, and world-renowned research institutions (Larkin, p. 389). Such transnational media exposure demonstrates the huge impact Sosick.org had on the global risk communication processes and its transformative capacity to help introduce major policy changes through participatory grassroots risk efforts. Focusing on individual and collective wellbeing, sosick.org managed to create an ad hoc coalition that helped introduce individual agency, public deliberation, and eventually policy changes.

3.7 Building Resilience Among Nurses in Canada to Reduce SARS Cross-Infection

In Toronto, out of a population of approximately 3 million, about 30,000 people got quarantined in the SARS outbreak (Naylor 2003; Rothstein et al. 2003), with medical care workers comprising 39.5% of those quarantined (DiGiovanni et al. 2004). Although nurses consisted of 46.4% of all medical care workers in Canada, about half of them were employed as contingent labor and thus had to work on multiple jobs "without benefits or disability income protection". Moreover, in addition to widespread discrimination and social avoidance, nurses were often exposed to SARS patients without appropriate infection control measures or financial compensation arrangements. Last but not least, while most medical care workers were praised for their altruism and bravery, Canada did not offer official compensation to medical care workers contracting SARS from work until late May 2003.

Toronto witnessed two phases of SARS, which we will refer to as SARS I and SARS II below. SARS I started in early March and was believed to be under control in early May when WHO removed Toronto from its list of areas with recent local transmissions on May 14. On May 13, to lobby WHO to remove it from the advisory list, Toronto made a politically driven decision to remove all infection control measures and workplace safety precautions in hospitals. This measure left medical care workers exposed to health risks without adequate protection. SARS II started when a cluster of five SARS cases was identified on May 20, 2003, which later was proved as an ongoing, undetected outbreak simmering at North York General Hospital between April 20 and May 7 before spreading to other hospitals. On May 26, WHO added Toronto back to its list of areas with ongoing SARS transmission and did not remove it until July 2. *The SARS Commission Executive Summary* lamented "the full extent of worker safety failings" and the tragic impacts of official decisions to relax precautions in the middle of a spreading outbreak (Campbell 2006, p. 22).

Leveraging social capital, human capital, and cultural and political capitals, professional organizations such as the Ontario Nurses' Association (ONA), Registered Nurses' Association of Ontario (RNAO), and the Ontario Public Service Employees Union (OPSEU) pushed for official support and compensation for medical care workers while lobbying for a full public inquiry into the SARS outbreaks. These organizations played important roles in building resilience in medical care communities by conducting research using focus groups and interviews and collecting personal narratives from thousands of nurses directly impacted by SARS. In public hearings held in September 2003, both ONA and RNAO presented to the SARS Commission their findings, which called attention to numerous problems faced by nurses: anxiety, isolation, stigmatization, segregation, economic losses, and physical and emotional repercussions brought by SARS.

RNAO made active efforts to lobby politicians for a full public inquiry into the SARS outbreaks. RNAO's President Adeline Falk-Rafael delivered to Premier Eernie eves a written request for an open inquiry. To call for a full public inquiry, RNAO's board of directors also held a press conference, the largest media event in its history, and employed powerful rhetorical strategies to call attention to the issues surrounding the systemic ignorance of nurses' expertise. As a visual reminder of the mistreatment medical care workers went through, RNAO invited nurses "trying to blow the whistle on their workplaces [but] disregard" to sit silently on stage wearing masks that read: "muzzled," "silenced", and "ignored" (Grinspun 2013, p. 6).

This carefully executed rhetorical mobilization of human, cultural, and political capitals generated repercussions in major newspapers across Canada, which published the images of nurse whistleblowers wearing masks at RNAO's press conference (Grinspun 2013, p. 6). Howard Hampton, a politician in Ontario, urged Premier Ernie Eves to consider RNAO's request for a public inquiry into SARS II, saying, "They raised warnings with hospital administrators ... on that SARS was re-emerging in our hospitals, yet their concerns were ignored" (Di Costanzo 2013). Leveraging political capital, these high-profile lobbying efforts led to two full-scale investigations about the ways Toronto dealt with SARS in 2003 (Grinspun 2013, p. 6).

To sum up, these nursing organizations leveraged human capital (individual leadership and expertise in nursing), social capital (the large networks of nursing practitioners), and cultural capital (professional values and beliefs in open communication) to build political capital. Doing so allowed them to gain access to media institutions and to exert political influence with their professional credibility and collective voice and to argue for financial capital, i.e., increasing funding to support medical care workers affected by SARS. Working together, they built both absorptive capacities (short-term risk reduction) and adaptive capacities (longer-term risk responses to emerging outbreaks), which helped to reduce health risks faced by medical care workers, to push for accountability investigations, and to build political and financial support for Medical care workers.

3.8 Discussions: Traditional Media and Resilience Building in Outbreaks

Our analysis of both the 1918 flu and SARS shows that in emerging infections, communities and organizations often relied on their own human capital, social capital, and cultural capital to initiate efforts, exploit existing knowledge to seek innovative solutions, and develop resilience. Traditional media can help strengthen such efforts, however, by functioning as a platform to help these communities to build political capital and seek financial capital, which in turn makes it possible to produce sustainable adaptive capacities and to generate transformative capacities, or major changes in the larger system. It takes intensive rhetorical, intellectual, and emotional labor for leaders of communities and organizations to make the jump from a locally initiated endeavor to one that attracts regional or national attention through media coverage (Ding 2019; Greene 1998). Such media attention often functions as the critical force that helps drive policy changes and thus generates transformative capacities.

When comparing the world in 1918 when people relied heavily on newspaper, to our current age "with its panoply of newspapers, magazines, radio, cable, Internet Web sites, Web logs, discussion groups", and social media, we have made a lot of progress in "the technology, speed, and variety" of "the media's coverage of pandemic events" (Markel et al. 2006, p. 19). However, such changes do not necessarily lead to better pandemic communication. As Markel et al. (2006) argue, widespread media coverage of pandemics "has the power to both inform and misinform" (p. 19). The development of instant messaging technology and self-media enables everyone to broadcast the news around him/her at any time, which greatly increases the amount and speed of news production. While this brings us the real-time epidemic briefings and statistics, travel regulations, and protection requirements, such user-generated content is also full of fake news and misinformation due to the lack of editing and fact-checking.

Our era is marked by "political polarization", "fake news", "tribal politics", and "trust in the media, government officials, and even science is fading" (Parmet and Rothstein 2018, p. 1435). Under such circumstances, "the public's failure to trust the guidance offered by public health officials" can be catastrophic when the next pandemic or another type of crisis arises (Parmet and Rothstein 2018, p. 1435). Therefore, effective, trust-winning pandemic communication is key to the building of community resilience. As Reynolds and Quinn point out (Reynolds and Quinn 2008): "During a crisis, an open and empathetic style of communication that engenders the public's trust is the most effective when officials are attempting to galvanize the population to take a positive action or refrain from a harmful act" (p. 13). Trust building through "empathy and caring, competence and expertise, honesty and openness, and dedication and commitment" (Reynolds and Quinn 2008, p. 13) is the essential part of community resilience.

References

Allmark P, Bhanbhro S, Chrisp T (2014) An argument against the focus on community resilience in public health. BMC Public Health 14(1):62–62. https://doi.org/10.1186/1471-2458-14-62

Altman L. (2003) The SARS enigma: drop in cases encourages the WHO but fears and questions remain. New York Times A18

Bergström J (2018) An archaeology of societal resilience. Saf Sci 110:31–38

Boyden J, Mann G (2005) Children's risk, resilience, and coping in extreme situations. In: Ungar M (ed) Handbookfor working with children and youth: pathways to resilience across cultures and contexts. Sage, Thousand Oaks, CA, pp 3–26

Bradsher K (2003) Relapse by SARS patients probably not from syndrome. New York Times A5

Buikstra E, Ross H, King CA, Baker PG, Hegney D, McLachlan K et al (2010) The components of resilience—perceptions of an Australian rural community. J Community Psychol 38(8):975–991

Cadell S, Karabanow J, Sanchez M (2001) Community, empowerment, and resilience: paths to wellness. Canadian J Community Mental Health = Revue Canadienne De Santé Mentale Communautaire 20(1):21–35

Campbell A (2006) The SARS Commission Executive Summary. http://www.archives.gov.on.ca/en/e_records/sars/report/v1.html. Accessed 10 Nov 2020

Chm.med.umich.edu (2007a) 1918 Influenza Escape Communities|U-M center for the history of medicine. Available at: http://chm.med.umich.edu/research/1918-influenza-escape-communities/. Accessed 17 Dec 2020

Chm.med.umich.edu. (2007b). Nonpharmaceutical interventions implemented by US cities during the 1918–1919 influenza pandemic|U-M center for the history of medicine. Available at: http://chm.med.umich.edu/research/nonpharmaceutical-interventions-implemented-by-us-cities-during-the-1918-1919-influenza-pandemic/. Accessed 26 Dec 2020

Dawes A, Donald D (2000) Improving children's chances: developmental theory and effective interventions in community contexts. In: Donald D, Dawes A, Louw J (eds) Addressing childhood adversity. David Philip, Cape Town, South Africa, pp 1–25
Ding H (2014) Rhetoric of global epidemic: transcultural communication about SARS. Southern Illinois University Press
Ding H (2019) The materialist rhetoric about SARS sequelae in China: networked risk communication, social justice, and immaterial labor. In: Walsh L, Gruber D (eds) Routledge handbook of language & science
Di Costanzo M (2013) Key players respond to questions about RNAO's role during SARS, and how transformative the outbreak was for Ontario. RNAO. Retrieved from http://rnao.ca/resources/rnj/%5Bfield_rnj_pubdate_yyyy%5D/%5Bfield_rnj_pubdate-mm%5F/5%Bfield_rnj_pubdate-dd%5D/key-players-res
DiGiovanni C, Conley J, Chiu D, Zaborski J (2004) Factors influencing compliance with quarantine in Toronto during the 2003 SARS outbreak. Biosecur Bioterror Biodef Strategy Pract Sci 2(4):265–272. https://doi.org/10.1089/bsp.2004.2.265
Eckholm E (2003) Virus badly underreported in Beijing, the WHO team finds. New York Times A13
Greene RW (1998) Another materialist rhetoric. Crit Stud Mass Commun 15:21–40
Gilligan R (2001) Promoting resilience: a resource guide on working with children in the care system. British Agencies for Adoption and Fostering, London
Grinspun D (2013) SARS: a transformative event for all. Registered Nurses J 6
Haines A (2009) Asset-based community development. Phillips R, Pittman RH (eds) An introduction to community development. Routledge, New York
IPCC (2012) 'Summary for policymakers', in managing the risks of extreme events and disasters to advance climate change adaptation. A special report of working groups I and II of the IPCC, Cambridge University Press
Jester B, Uyeki T, Jernigan D (2018) Readiness for responding to a severe pandemic 100 years after 1918. Am J Epidemiol 187(12):2596–2602
Lais C (2003) Lifting the Lid on the Spread of SARS. South China Morning Post, p 6
Li J (2003) Canadian Premier Visited Chinatown. Dajiyuan
Leadbeater B, Dodgen D, Solarz A (2005) The resilience revolution: a paradigm shift for research and policy. In: Peters RD, Leadbeater B, McMahon RJ (eds) Resilience in children, families, and communities: linking context to practice and policy. Kluwer, New York, NY, pp 47–63
Markel H, Stern A, Navarro A, Michalsen J (2006) A historical assessment of nonpharmaceutical disease containment strategies employed by selected US communities during the second wave of the 1918–1920 influenza pandemic
Markel H, Lipman HB, Navarro JA, Sloan A, Michalsen JR, Stern AM, Cetron MS (2007) Nonpharmaceutical interventions implemented by US cities during the 1918–1919 influenza pandemic. JAMA 298(6):644–654
Matthews K (2003) SARS Scare hits New York's Chinatown. Associated Press Worldstream
McCubbin LD, McCubbin HI (2005) Culture and ethnic identity in family resilience: dynamic processes in trauma and transformation of indigenous people. In: Ungar M (ed) Handbook for working with children and youth: pathways to resilience across cultures and contexts. Sage, Thousand Oaks, CA, pp 27–44
Naylor D (2003) Lessons from SARS: renewal of public health in Canada. Public Health Agency of Canada
Parmet WE, Rothstein MA (2018) The 1918 influenza pandemic: lessons learned and not—introduction to the special section
Philips H (2003) The data they didn't want to release; 'it would be difficult to conclude there is a need for such information'. South China Morning Post, p 1
Platts-Fowler D, Robinson D (2013) Neighbourhood resilience in Sheffield: getting by in hard times. Sheffield, Available from: http://www.shu.ac.uk/research/cresr/sites/shu.ac.uk/files/neighbourhood-resilience-sheffield.pdf. Accessed 17 Dec 2020

Pomfret J (2003) China raises disease's death toll: under pressure, officials admit spread of infection to Beijing. The Washington Post A16
Pomfret J, Weiss R (2003) Hong Kong quarantines complex to control spread of epidemic. The Washington Post A02
Rao H, Greve HR (2018) Disasters and community resilience: Spanish flu and the formation of retail cooperatives in Norway. Acad Manag J 61(1):5–25
Reynolds B, Quinn SC (2008) Effective communication during an influenza pandemic: the value of using a crisis and emergency risk communication framework. Health Promot Pract 9(4_suppl):13S–17S
Rosenwald M S (2003) Chinatown fights fears restaurants slump on SARS worries. The Boston Globe B1
Rothstein M (2004) Are traditional public health strategies consistent with contemporary American health values? Temple Law Rev 77:175–192
SARS Web site in Hong Kong (2003) NPR Morning Edition. https://advance-lexis-com.prox.lib.ncsu.edu/api/document?collection=newsandid=urn:contentItem:48KW-3PS0-00KC-W3C8-00000-00andcontext=1516831. Accessed 17 Dec 2020
Schwartz JL (2018) The Spanish flu, epidemics, and the turn to biomedical responses. Am J Public Health 108(11):1455–1458
Seccombe K (2002) "Beating the odds" versus "changing the odds": poverty, resilience, and family policy. J Marriage Fam 64(2):384–394
Seidman E, Pedersen S (2003) Holistic contextual perspectives on risk, protection and competence among low-income urban adolescents. In: Luthar SS (ed) Resilience and vulnerability: adaptation in the context of childhood adversities. Cambridge University Press, Cambridge, England, pp 318–342
Serfilippi E, Ramnath G (2018) Resilience measurement and conceptual frameworks: a review of the literature. Ann Public Coop Econ 89(4):645–664. https://doi.org/10.1111/apce.12202
Surendran A (2003) Chinatown hit by SARS scare; business owners seek to allay fears about the mystery illness. The Philadelphia Inquirer A1
Tang Y (2020) Review of Outka, Elizabeth, viral modernism: the influenza pandemic and interwar literature. H-diplo, H-net reviews. https://www.h-net.org/reviews/showrev.php?id=55267. Accessed 26 Dec 2020
Taylor N (2003) Net-spread panic proves catchier than a killer virus. South China Morning Post, p 2
Treichler PA (1999) How to have theory in an epidemic: cultural chronicles of AIDS. Duke University Press, Durham, NC
Tropman J, Erlich J, Rothman J (2006) Tactics and techniques of community intervention. Wadsworth, Belmont
Ungar M (2008) Resilience across cultures. Br J Soc Work 38:218–235
Ungar M (2011) The social ecology of resilience: addressing contextual and cultural ambiguity of a nascent construct. Am J Orthopsychiatry 81:1–17
Westfeldt A (2003) Survey: chinatown business down since 9/11, SARS. The Associated Press State & Local Wire

Part II
Pandemic Communication Theory

Chapter 4
The Building Blocks of Pandemic Communication Strategy: Models to Enable Resilient Risk and Crisis Communication

R. Tyler Spradley and Elizabeth Spradley

Abstract The use of theory to strategize, construct, disseminate, and evaluate pandemic communication is more likely to achieve positive outcomes. This chapter reviews relevant theoretical models including each theory's basic definition, relevance to pandemics, use, and cases or examples to illustrate its practicality. Theories are reviewed with roots in behavior change, health communication, persuasion, public relations, and risk and crisis communication. Within the chapter, tables serve as quick-reference guides for theory overview and resources. The goal is to help readers gain basic and applied understanding of each theory in relation to pandemic communication.

Keywords Behavior change theories · Health communication theories · Persuasive communication theories · Public relations theories · Risk and crisis communication theories · Theory

4.1 Introduction

Theoretical models enable the enactment of effective risk and crisis communication during pandemics by providing practitioners tested processes and practices and the tools by which to assess them. Simply, good theory is applicable to the practice of communication. The high equivocality of crisis events demands adequate sensemaking models to enact results that meet the complexity of environments (Weick 1995). Empirical models, or theory, help practitioners learn. Risk and crisis communication theories are designed to systematically explain and/or predict social phenomena and their outcomes. Such frameworks in crisis, health, persuasion, and risk communication direct practitioners to evidence-based ways to craft more competent pandemic messages, thus enhancing resiliency in times of flux (Spradley 2017).

R. T. Spradley (✉) · E. Spradley
Stephen F. Austin State University, Nacogdoches, TX, USA
e-mail: rspradley@sfasu.edu

D. M. Berube (ed.), *Pandemic Communication and Resilience*, Risk, Systems and Decisions, https://doi.org/10.1007/978-3-030-77344-1_4

This chapter reviews relevant theoretical models including each theory's basic definition, relevance to pandemics, use, and cases or examples to illustrate its practicality. Within the chapter, tables serve as quick-reference guides for theory overview and resources. The goal is to help readers, like you, gain basic and applied understanding of each framework to enable sensemaking during future pandemics or crises.

4.2 Why Use Theoretical Models to Aid Communication Strategies in a Pandemic?

Effective risk and crisis communication in a pandemic is both expected and desired. While communication practitioners may have been sidelined in public health or emergency response in the past, tides have turned. Communication practitioners are increasingly recognized as experts relying on evidence-based, theoretical approaches to exchange information, persuade audiences, and build relationships (Paek et al. 2010; Schiavo 2007). Thus, it is incumbent upon communication practitioners constructing pandemic communication to fulfill such theory-based expectations and contribute positively to the increased prominence of communication.

That is not say that theoretical models are the miracle pill for managing pandemics. As O'Keefe (2012) argues, there are no clear, generalizable theory-based strategies that communication practitioners can use as their "go to" to ensure effectiveness. There are complexities to consider when selecting which theories to use and how to use them. In other words, there are many considerations to match theoretical models to such complexities as diverse audiences, a multiplicity of communication channels, and specific conditions and stages uniquely related to pandemics. So, while theory-based communication is not an easy fix, it should be a thoughtful, strategic approach toward "fixing" pandemic-related problems.

The use of theory to strategize, construct, disseminate, and evaluate pandemic communication is more likely to achieve positive outcomes (Noar 2006). Consider the anxieties, disruptions, fears, risks, and uncertainties that typify pandemic conditions. Poor pandemic communication may exacerbate these complexities and increase harm rather than achieving its intended purpose. For example, Spradley and Spradley (2020) found that early communication concerning the use of face masks from the White House Coronavirus Task Force and the Centers for Disease Control and Prevention contributed to contradictory face covering recommendations that did not adequately address counter argumentation in public service announcements. Subsequently, the contradictory messages constructed double binds for the public as they experienced peer and societal pressures from the bifurcated practices. Ideally, pandemic communication, even as it evolves to match new information and dynamic conditions, should address individual and collective behavior changes, mitigate risk, prompt efficacious action, and reduce uncertainty. Applying proven theoretical models can enable pandemic communication practitioners achieve those outcomes.

4.3 What Theoretical Models Should Be Considered When Crafting Pandemic Communication Strategies?

Effective pandemic communication strategies rely on the theoretical models used by crisis managers and crisis communication practitioners. Given that different theories feature different assumptions, goals, and strategies about crises and communication, selecting one or a constellation of theories means finding a good match between models, pandemic conditions, and intended communication outcomes. This section of the chapter outlines extant theoretical models relevant to pandemic communication. To be clear, not all theories are exhausted in this chapter. For example, prospect theory did not make the list; yet communication practitioners may find the nuances of loss and gain frames more or less relevant. With that said, the theories selected below represent different disciplines—crisis communication, health communication/public health, persuasion, public relations, and organizational communication—and have demonstrated their heuristic and practical value through empirical research and implementation during crises.

To assist with comparing and contrasting theories, see Table 4.1: Applying Theoretical Models to Pandemic Communication that provides an overview of the theory's primary proposition and its pandemic application, but each theory is further explored within the chapter. Additionally, Table 4.2: Disciplinary Influences of Pandemic Communication Models categorizes the primary disciplinary influences of the theoretical models to help understand the assumptions and applications of each pandemic communication theory.

While many of these theories cross disciplinary boundaries and transcend categorization, there are certain disciplines that the theory either originated within or has been claimed by specific disciplines. In that sense, despite the fact the health communication research heavily draws on persuasion theories from the rhetorical tradition and behavior change theories from social and cognitive psychology, the roots of theories like the Transtheoretical Model or Social Cognitive Theory are not in health communication. Thus, Table 4.2 is not designed to limit the theory's application, but rather, it is designed to help understand prevailing thought and influence on the theory's development. However, given the fluid nature of crises, practitioners and researchers should continually seek to add more multidisciplinary approaches to risk and crisis communication (Seeger 2018).

4.4 Crisis and Emergency Risk Communication (CERC)

Following the 9/11 terrorist and anthrax attacks, the Centers for Disease Control and Prevention (CDC) embarked on a process to develop a way to systematically address these types of crises (Sellnow and Seeger 2013). "The five-stage CERC model assumes that crises will develop in largely predictable and systematic ways: from risk, to eruption, to clean-up and recovery on into evaluation" (Reynolds and Seeger

Table 4.1 Applying theoretical models to pandemic communication

Theoretical model	Primary proposition	Pandemic application
Crisis and emergency risk communication (CERC) model	Stage model of crisis and emergency communication includes: (1) pre-crisis, (2) initial event, (3) maintenance, (4) resolution and (5) evaluation	Apply prescriptions for each stage of a crisis to address the needs of the stage, particularly applicable with *interorganizational* response
Discourse of renewal theory	Crisis is an opportunity for organizations to recreate or repurpose around past or present core values and to correct vulnerabilities and limitations	Learning, ethical communication, prospective vision, and effective rhetoric
Elaboration likelihood model (ELM)	Two routes to persuasion include the central (highly elaborative) and the peripheral route (less elaboration)	Select elaboration based on audience's likelihood to process information centrally or peripherally and develop strong, issue-relevant, tailored argumentation for behavior change
Extended parallel process model (EPPM)	Perceptions of high threat and high efficacy lead to the adaptive response termed danger control; whereas perceptions of high threat but low efficacy led to the maladaptive response termed fear control	Develop messages that advance self, response, and collective efficacy and that demonstrate severity and susceptibility to the threat. These messages should be directed at a danger control behavior
Image repair theory	Crises threaten organizations' and leaders' image, and in response to the crises, organizations and leaders use a variety of image repair strategies to reduce or eliminate threats and repair image. Responses to the pandemic can emerge as threats to image depending on how the response is perceived by the public	Communicate swiftly and accurately about image threats using *corrective action*, when appropriate, given the public's positive response to this strategy when there are perceptions of wrongdoing
Situational crisis communication theory (SCCT)	Drawing on attribution theory, the focus is on stakeholders' attribution of the cause—organization or external factors. Organizations respond with one or more postures: deny, diminish, rebuild, or bolster	Crisis response should fit the crisis type, crisis history, and prior organizational reputation

(continued)

Table 4.1 (continued)

Theoretical model	Primary proposition	Pandemic application
Social cognitive theory	Learned behaviors result from cognitive representations that model the behavior and the feedback that positively or negatively reinforce the behavior when enacted	Develop messages that model desired behaviors, demonstrate positive reinforcement of behaviors, and use messengers that the target audience identifies with or trusts
Stakeholder theory	Effective crisis communication is influenced by the organization's relationship with stakeholders: (1) primary, (2) secondary, (3) the public, and (4) the news media	Develop relationships with stakeholders prior to an event and advocate the inclusion of all stakeholder groups influencing decision making and influenced by decision making before, during, and after crises.
Theory of reasoned action and theory of planned behavior (TRA/TPB)	Behavioral intention is dependent on subjective norms, attitudes, and behavioral control	Develop messages that appeal to normative behaviors, positive and strong attitudes, and ability to control one's behavior in order to adopt the recommended behavior
Transtheoretical model (TTM)	Stage model explains behavior change using: (1) precontemplation, (2) contemplation, (3) preparation, (4) action, and (5) maintenance	Using information about the target's stage of change, use messages to address the stage and processes to persuade and equip the target to transition to the next stage

Table 4.2 Disciplinary influences of pandemic communication models

Disciplinary influence	Pandemic communication models
Behavior change/intervention	Social cognitive theory
	Theory of reasoned action
	Theory of planned behavior
	Transtheoretical model
Crisis and risk communication	CERC model
	Discourse of renewal theory
	Situational crisis communication theory
Health communication	Extended parallel process model
Persuasion	Elaboration likelihood model
Public relations	Image repair
	Stakeholder theory

2005, p. 51). It is noteworthy that the Crisis and Emergency Risk Communication (CERC) model assumes predictability of sequencing, not predictability of crisis. In that regard, the CERC model shares stage model characteristics with models like Turner's Six-Stage Sequence of Failure in Foresight and Fink's Four Stage Cycle in the field of crisis management (Sellnow and Seeger 2013).

Developed by the CDC, the five stages of CERC are: (1) pre-crisis, (2) initial event, (3) maintenance, (4) resolution and (5) evaluation (National Consortium for the Study of Terrorism and Responses to Terrorism 2012). The pre-crisis stage is marked by communication and education campaigns aimed at informing the public of possible health risks. When the crisis occurs, organizations during the initial event stage should quickly disseminate information to reduce uncertainty, facilitate personal responses, empathize and reassure emotional publics, and clarify emergency and medical responses. Next, ongoing uncertainty reduction is essential in the maintenance stage. Organizations must build on the communication efforts begun during the initial event stage, ensure accuracy of information and perceptions, seek to understand background/causal factors and continue to coordinate broad-based efforts. The resolution stage assumes risk is minimal at this point. Communication during this stage is directed toward recovery efforts, open discussion about crisis, causal relationships and possible new risks. Finally, the evaluation stage stresses the importance of communicating lessons learned and improving future effectiveness. CERC, with a communication emphasis, delineates specific guidelines during each stage of a crisis for effective crisis and risk communication (Seeger et al. 2010).

With regard to CERC, the first two stages—pre-crisis and initial event—are significant in risk and crisis mitigation. In the case of a pandemic, the CERC model effectively helped mitigation in the US response to the avian flu and the swine flu, in which there was ample pre-crisis planning and information about pandemic influenza and time to prepare (Seeger et al. 2010). What happens when pandemics are less predictable, thus tightly compressing the pre-crisis and initial event stages? What happens when limited information about the virus impacts pre-crisis and initial event knowledge from which to progress? What happens when recommended health behaviors change as information about the pandemic changes? These are significant questuons for practitioners and researchers to consider in their respective analysis of the COVID-19 response.

4.5 Discourse of Renewal

The Discourse of Renewal (DR) theory is a progressive or transformative theory that frames a crisis as an opportunity for organizations to recreate or repurpose around past or present core values and to correct vulnerabilities and limitations (Sellnow and Seeger 2013). There are four central objectives of DR: (1) learning, (2) ethical communication, (3) prospective vision, and (4) effective rhetoric. First, learning is the commitment to actively find and evaluate vulnerabilities and limitations. Awareness of these potential or real failures is what initiates a renewal process. Openly and honestly discussing what happened and how crises can be prevented in the future demonstrates change is at the heart of learning. Second, the act of communication during a crisis reveals the values of an organization. If not during, after a crisis, the way an organization makes decisions and acts is noticed. When those decisions and actions are unethical, untrustworthy, intentionally ambiguous, or dishonest, a crisis will out those "transgressions." Likewise, if an organization operates openly, honestly, trustworthily, and responsibly, a crisis will out those "noble" or ethical acts. Consequentially, ethical core values breed effective avenues for renewal. Third, a prospective (future oriented) vision is juxtaposed with a retrospective (past oriented) vision. Discourse of Renewal highlights the former by communicating about positive steps forward and hopeful messages versus messages that shift blame and that are divisive. Lastly, a discourse of renewal is effective rhetoric, or effective motivational messages that connect to stakeholders. Using effective argumentation and framing techniques, leaders seek to inspire and empower with a genuine intent to advocate for positive steps forward (Ulmer et al. 2010). Discourse of renewal should be contextual to the crisis (e.g., increasing future prepareness to address failures arising from the pandemic response). Transformational ethics demand leaders and organziations avoid exploiting stakeholder vulnerabilities associated with pandemics, and other crises, to bolster or implement change that is unrelated or unnessecary. The opportunity for renewal is not an opportunity for managerial or political gain.

4.6 Elaboration Likelihood Model (ELM)

Determining how to make messages more persuasive is a monumental task, but the theorizing of Petty and Cacioppo (1986) led to a dual-process model of persuasion entitled the Elaboration Likelihood Model (ELM). At its core, ELM is designed to change attitudes through persuasive messages. Petty and Wengener (1999) clarify the scope and aim of ELM this way:

> ELM was formulated as a theory about how the classic source (e.g., expertise), message (e.g., number of arguments), percipient (e.g., mood), and contextual (e.g., distraction) variables have an impact on attitudes toward various objects, issues, and people. More, generally, though, the theory can be used to understand how any external or internal variable has an impact on some evaluative (e.g., good-bad) or nonevaluative (e.g., likely-unlikely) judgment (p. 42).

ELM posits that there are two elaborative routes to persuasion: (1) central and (2) peripheral. Depending on the route, message elaboration variables, and other variables, there may be an evaluative direction that has positive valence (more persuasive) or negative valence (less persuasive) (O'Keefe 2008). The central route relies on highly elaborative argumentation and is more effective when the audience is motivated to critically process the message, holds pro-attitudinal thoughts about the message, and perceives the message's argumentation as strong (e.g., sound reasoning, evidence-based, clearly argued). Conversely, when the audience is counter attitudinal toward the message or perceives the argumentation as weak because of fallacies in reasoning, lacking evidence, ambiguous argumentation, then the audience is more likely to respond with negative valence. The two variables often tested with relation to persuasion and central route processing are argument quality and source credibility (Stephenson et al. 2001).

Because audiences are not always motivated to critically process elaborative argumentation or audiences may perceive the message as ambiguous or hold neutral attitudes toward the message, the peripheral route to persuasion may be strategically used and accessed by audiences. Peripheral cues consist of persuasive cues that emotionally connect the audience to the message. Using emotional appeals, likable or attractive messengers, attitudinal balance, or other cues like visual storytelling, peripheral cues may grab audience's attention, generate more positive associations between the audience and the message, and possibly, entice the audience to engage through the central route. Overall, Petty and Cacioppo (1986) surmise that audiences engaged through the peripheral route will likely have short-term attitudinal and/or behavior changes, and audiences engaged through the central route, who have favorable rehearsal of the argumentation, will likely have longer-term attitudinal and/or behavior changes. This is quite significant for pandemic message strategies and communication practitioners seeking to engage the public through the central route or move the public from the peripheral to the central route.

There is a set of message strategies related to ELM that are relevant in pandemic conditions—tailored message strategies.

> In classic studies of ELM, central route processing of personally relevant information is indicated by the generation of positive and/or negative thoughts about the advocated position. Thus, an ELM-based evaluation of tailored health communication would suggest that tailored messages may be more effective than nontailored messages because the stimulate greater cognitive activity or elaboration (Kreuter and Wray 2003, p. S229).

Tailoring further specifies targeted messages to increase perceptions of individualized, relevant messages for the recipient and are found to increase cognitive activity in recipients (Kreuter and Wray 2003). Tailoring enables the message strategy to consider how recipients may similarly and differently engage with central route processing and match the message design to fit the recipient's motivation to critically process different types of elaborative argumentation or ways to move the recipient from peripheral to central route processing. Early work in tailoring health behavior change messages used published and completed assessments about recipient preferences to design a set of tailored persuasive messages and subsequent work has made applications to technology (Kreuter et al. 2000; Kreuter and Wray 2003). Tailoring through technology allows the user to answer questions, in which the next set of questions or information is altered by the user's answers.

Overall, for pandemic messaging there a number of practices stemming from ELM that could be applied, especially to persuasive messages aimed at changing the public's behavior to reduce health risks. First, message design should aim to engage central route processing by including quality, strong argumentation from credible sources using credible messengers. Next, selecting messengers that evoke empathy and identification can impact information processing and behavior change

(Stephenson and Palmgreen 2001). Finally, find ways to customize messages to achieve the tailoring effect (Kreuter and Wray 2003).

4.7 Extended Parallel Process Model (EPPM)

At the crux of the Extended Parallel Process Model (EPPM) are fear appeals and the varied responses to them. Pandemic-generated and exacerbated fears may be related to personal contraction of a virus and its physiological impact on the body, or fears may be related to interpersonal relationships, financial security, psychological well-being, or yet another factor. To use fear appeals in pandemic messaging, EPPM provides an empirical framework to guide practice. Kim Witte's (1992, 1994) work extended fear-based persuasive messaging aimed at behavior change to consider the differential responses that audiences may have to fear appeals and the variables that enhance desired behavior change.

The process begins with exposure to the stimuli, fear appeal message. Fear appeal message components should include self-efficacy, response efficacy, susceptibility, and severity if the next stage of appraisal is to prompt the individual to respond (Witte 1992, 1994). The message processing includes two types of appraisals: is efficacy perceived and is threat perceived. To better understand, here is a breakdown of efficacy and threat. Efficacy is comprised of self and response efficacy, and efficacious appraisals mean that the individual believes that they can respond to diminish or eliminate the threat. Perceived efficacy is appraised by (1) self-efficacy, which is the belief that one can enact the healthy behavior or avoid/cease the risky behavior, and (2) response efficacy, which is the belief that the prescribed behavior will have the intended outcome. Threat is comprised of susceptibility and severity, and threatening fear appeals are required for individuals to respond. Perceived threat is appraised by (1) susceptibility, which is the belief that the threat is likely to occur and (2) severity, which is the consequentiality of the threat to incur short-term and/or long-term harm (Roberto et al. 2010).

Three fear appeal outcomes are possible: no response, a maladaptive response, or an adaptive response. First, if no threat is perceived, then no response is made. Second, if a perceived threat is high but perceived efficacy is low, the maladaptive response termed *fear control response* occurs. *Fear control responses* include: (1) defensive avoidance (ignore information or avoid exposure to information sources), (2) denial (choosing to downplay the risk or believe it is real), and (3) reactance (find a reason to reject the risk, like labeling the risk message source as manipulative) (Roberto et al. 2010). Third, if a perceived threat is high and perceived efficacy is also high, the adaptive response termed *danger control* occurs. *Danger Control* is the desired outcome, and practitioners constructing fear appeals should use EPPM to design messages with high efficacy and high threat.

EPPM has been used to analyze, construct, and hypothesize fear appeals to influence behavior in pandemics. In their study on critical infrastructure employees willing to report to work in an influenza pandemic, von Gottberg and colleagues

(2016) found that 20% of public service workers were not willing to report to work due to perceived high risk and perceived low efficacy to prevent infection to self and interpersonal relations. In a timely study of a cross-section of Canadian adults, Lithopoulos and colleagues (2021) surveyed the public with regard to their response to fear appeals to adhere to governmental recommendations to physically distance. Results indicated that older adults, females, and adults with higher education were more likely to perceive high threat and high efficacy resulting in physical distancing (danger control). Whereas, young adults, males, and less educated were more likely to perceive lower threat and/or lower efficacy resulting in less physical distancing (fear control or no response). Recommendations from Lithopoulos and colleagues' (2021) work focus on the need to increase perceptions of self and response efficacy for the fear appeals to result in danger control.

Similarly, Nazione et al. (2021) collected survey data at the onset of COVID-19 outbreak in the US related to participants' media exposure, interpersonal communication, perceived risk, perceived efficacy, and adaptive behaviors. The study supported EPPM's value to pandemic messaging with high efficacy predicting adaptive behaviors. While their findings did not support media consumption affecting perceived risk, interpersonal communication did affect perceived risk suggesting a two-step process of media effects on perceived risk. These studies underscore the need for pandemic messaging to make risk information accessible and relevant to the public and demonstrate efficacious behaviors/information. Additionally, efficacious message construction should add *perceived collective efficacy* as a third component of efficacy messages and appraisals. Bandura (1997) is credited with defining perceived collective efficacy and Roberto et al. (2010) argue that it should be included in EPPM design. Perceived collective efficacy is the group's shared belief that they can enact the healthy behavior or avoid/prevent the risk and that the action will accomplish its intentions. This is of particular interest to pandemic applications of EPPM because it is not just about individual health, it is about public health.

4.8 Image Repair Theory

Benoit (1997) and colleagues work with image repair reminds communication practitioners that pandemics threaten individual and public health, yes, but pandemics also threaten the public trust in governments, public health agencies, organizations, and leaders, which may have a negative impact on the public's willingness to comply with protective actions that they advocate in a pandemic (Low et al. 2011). Image is developed overtime, and as individuals and organizations face threats to image, they respond. Crises threaten image. Threats are attributed as the cause of the crisis and/or the response to the crisis. Image Repair Theory asserts that responses can be categorized as (1) denial, (2) evasion of responsibility, (3) reduction of offense, (4) corrective action, or (5) mortification.

First, *denial* can be simple denial or blame shifting (Benoit 1997, 2018, 2020). *Simple denial* fails to accept responsibility for the act or problem, and *blame shifting* attributes responsibility elsewhere. Second, *evasion of responsibility* includes provocation, defeasibility, accident, and good intentions. *Provocation* evades responsibility by blaming the act or problem on the need to respond to something else, and *defeasibility* evades responsibility by claiming ignorance. Another way to evade responsibility is to attribute the crisis to an *accident*. A final way to evade responsibility shifts attention to well-meaning intentions, which is the *good intentions* strategy. Third, *reducing offensiveness* can be achieved through bolstering, minimization, differentiation, transcendence, attack the accuser, or compensation. *Bolstering* is achieved by focusing on the positive. *Minimization* is a way to de-escalate the seriousness of the crisis by downplaying. *Differentiation* relies on comparison to contrast the crisis with something more serious. *Transcendence* appeals to prevailing values that are framed as more important than the crisis. *Attack accuser* turns the blame game around to question the accuser and their credibility. A final way to reduce offensiveness is to use *compensation* to give victims something of value. Fourth, *corrective action* uses planning and resolution to demonstrate what is being done about the crises. Fifth, *mortification* is an apology (Benoit 1997, 2018, 2020).

Spradley (2020) has used image repair to analyze COVID-19 pandemic communication focusing on the case of Notre Dame's president's rhetoric upon the two-week campus closure in Fall 2020 semester and the president's apology letter for attending an event at the United States White House maskless and contracting the virus. Analysis showed that the image repair response to the two-week closure focused on evasion of responsibility and corrective action, and the response of the university president was mortification and corrective action. Given that the current scholarship supports the use of corrective action as the most effective and appropriate image repair strategy (Benoit and Drew 1997), it is not surprising that corrective action was consistently used by Notre Dame across two different COVID-19 related crises. Being conscious of image repair strategies and their consequences, should improve the appropriateness and effectiveness of responses to image threat. A number of practice-based image repair concerns should be attended to as individuals and organizations communicate in pandemics. For example, Benoit (2018) uses a case study of United Airlines image repair response to a passenger ejection to draw attention to the way social media can compress the time needed between an incident and the response. In the case of United Airlines, their initial response was *differentiation*, but the viral video of the ejection generated outrage resulting in a second round of image repair strategies that relied on *mortification* and *corrective action*.

In a pandemic, communication practitioners should be aware that social media affects rapid dissemination of information. Just as a public health official comments in a press briefing or gives a news interview, there are soundbites and memes going viral. Just as an organization furloughs or lays off workers due to a pandemic's economic impact, there are messages circulating on social media and news reports being announced, and if the organization's upper management continue to enjoy six-figure salaries or accept bonuses from their boards, then the organization will face a barrage of additional image threats on top of the threat of the pandemic. Tarnished images will have more complicated pandemic responses than shining images (imagery borrowed from Brinson and Benoit 1999).

4.9 Situational Crisis Communication Theory (SCCT)

In the same vein as Image Repair Theory, the Situational Crisis Communication Theory (SCCT) has been influenced by and advanced within public relations (Sellnow and Seeger 2013). SCCT answers how stakeholders perceive crisis responsibility and how to respond to their perceptions. The overarching argument of SCCT is to find a good match or fit between the crisis situation and the crisis response (Coombs and Holladay 2002). Drawing on attribution theory, SSCT asserts that stakeholders will attribute crisis cause to either the organization or external factors, and depending on causal attributions, the organization will need to either change stakeholders' perceptions about the crisis or the organization (Coombs 2004). Threats to the organization's reputation are assessed by (1) crisis type, (2) crisis history, and (3) prior reputation National Consortium for the Study of Terrorism and Responses to Terrorism 2012). First, crisis type "is the frame used to define the crisis," which can be organization as victim (minimal crisis responsibility), crisis as accident (low crisis responsibility), or crisis as intentional (strong crisis responsibility) (Coombs 2010, p. 111). The next two threats increase attributions of intentionality and strong responsibility—crisis history and the organization's negative reputation prior to the crisis. For example, if a chain of nail salons is found to be a super spreader in a pandemic, stakeholders, especially those who learn of the chain's responsibility through contact tracing, will attribute greater responsibility if they perceive the chain to have poorly managed the crisis conditions (e.g. do not enforce social distancing, neglected to sanitize stations in between clients …) and perceive the chain's reputation negatively (e.g. news stories exposed fungal infections tied to the chain a year previous to the pandemic). However, if the chain had a positive reputation prior to the crisis, the reputational asset would mitigate the threat posed by the crisis (Coombs 2007).

SCCT identifies four crisis response postures: (1) deny, (2) diminish, (3) rebuild, and (4) bolstering (Coombs 2010). First, denial severs the connection between the crisis and the organization. Attacking the accuser, denying the crisis exists, and identifying a scapegoat are three ways to accomplish the deny posture. Second, the diminish posture minimizes the organization's control over the crisis. Using excuses and justifications, the organization shows that they had limited control over the crisis emerging or responding to it. Third, the rebuild posture seeks to repair organizational image damaged by organizational culpability. To do so, the organization may compensate victims or apologize. Fourth, Coombs (2010) explains that bolstering is supplemental to the other three responses. Bolstering may include complimenting stakeholders to ingratiate the organization toward the stakeholders and reminding stakeholders about the organization's history, accomplishments, or track record. Pandemic complexity should cause organizations to consider how the large-scale crisis may cause a ripple effect of many smaller-scale crises. As in the example of the chain of nail salons, they are not responsible for the pandemic, but their stakeholders may attribute responsibility for a smaller-scale crisis—super spreader. SCCT can help organizations like the chain of nail salons manage their responses to reputational threats, and SCCT can help organizations like hospitals or public health agencies manage reputational threats. Organizations should determine attribution of responsibility and match between the response to the type of attribution.

4.10 Social Cognitive Theory (SCT)

Tenets of Social Cognitive Theory (SCT) have already been explored in relation to the AIDs epidemic (Bandura 1992) and have direct application to pandemic messaging. Stemming from the works of Albert Bandura, SCT asserts that learning a new behavior is dependent on modeling, efficacy, and reinforcement. Describing the role of behavioral modeling, Bandura (2005) writes,

> Cognitive representations conveyed by modeling serve as guides for the production of skilled performances and as standards for making corrective adjustments in the development of behavioral proficiency. Skills are usually perfected by repeated corrective adjustments in conception-matching during behavior production. Monitored enactment with instructive feedback serves as the vehicle for converting conception to proficient performance. The feedback accompanying enactments provides the information for detecting and correcting mismatches between conception and action (p. 12).

Behavioral modeling demonstrates when, where, how, and why to do the intended behavior. In doing so, the demonstrative messages contribute to individual's efficacy—the belief that they can enact the behavior. Then, as the individual enacts the behavior, the individual received positive or negative reinforcement and feedback to identify and correct inadequacies in the behavioral performance (Bandura 1997).

At the core of this theory is the assumption of personal, proxy, and collective agency that enables intentional adaptations and change. The first mode of agency is personal agency, which is individual in nature. Second, proxy agency occurs when "people secure desired outcomes by influencing others to act on their behalf" (Bandura 2002, p. 269). The final type of agency is collective, which alludes to interdependent individuals acting "in concert" to affect a change (p. 269). Considering these modes of agency and the tenets of behavioral modeling there are numerous applications to pandemic communication. Thinking more specifically about Public Service Announcement (PSAs) from public health agencies, behavioral modeling figures prominently in the content. For example, the United States Surgeon General Dr. Jerome Adams served as a messenger for making a homemade face covering. In the PSA, Surgeon General Adams holds up a scarf, a towel, and a T-shirt followed by a verbal script explaining the steps to make a homemade face covering accompanied by a visual script demonstrating how to fold and attach the face covering properly (Centers for Disease Control and Prevention 2020). Increasing personal agency through efficacious messages with behavior modeling, this PSA illustrate the value of SCT to pandemic messages. However, proxy and collective agency also have direct implications. For example, proxy agential messages may focus on influencing interpersonal influencers with a 2-step process and giving those interpersonal influencers rhetorical resources (example messages) to use in conversation.

Targeting collective agency, pandemic message may focus on demonstrating groups of people, communities, or organizational members performing modeled behaviors, which would, in turn, generate a sense of collective capacity to enact the behavior and protect the collective's health in a pandemic.

4.11 Stakeholder Theory (ST)

Stakeholder Theory (ST) expands traditional thinking about groups related to a crisis. Resting on the shoulders of the works of Freeman (1984) and colleagues (Freeman and Gilbert 1987), ST extends the relational scope of an organization beyond its members or of a corporation beyond its shareholders. For example, stakeholders of a local school district would clearly include the staff and students, but additional stakeholders include contractors and suppliers (e.g., bus drivers, mechanics contracted to work on buses, grocery distributer for cafeterias). If a pandemic causes remote education for an extended period, how will the mechanics who maintain buses be impacted when the buses have no wear and tear? How will the grocery distributer be impacted when students are not in school to consume breakfast and lunch?

Effective crisis communication is heavily influenced by the organization's relationship with stakeholders: (1) primary (those directly influenced by the crisis event), (2) secondary (those influenced by their relationship with primary stakeholders or by ripple effects), (3) the public, and (4) the news media. With reference to crises, the literature examining stakeholder relationships recommend that organizations identify, develop, and maintain positive stakeholder relationships pre-crisis (Ulmer et al. 2019). Stakeholders are invested in the organization's success, and trust and cooperation are best built pre-crisis. However, organizations should remember: "Although establishing strong stakeholder relationships will not likely help an organization avert every crisis, it can play an important role in how the organization resolves as crisis it cannot avoid" (Ulmer 2001, p. 593). Additionally, stakeholders can help organizations during a crisis by advocating for them, providing resources, and complimenting their communication strategies.

Specific application of stakeholder relationships in a pandemic are emerging, but initial findings underscore the value of open, accurate, and frequent communication. In response to the COVID-19 pandemic, here are questions that Jong (2021) propose to use with stakeholders. As a practitioner: (1) were stakeholder's needs considered; (2) were communication efforts aligned with stakeholder's communication efforts; (3) were new stakeholders identified and aligned with; (4) was public stakeholder communication assessed as strong, credible, and vocal; and (5) what could be learned with stakeholder communication during this pandemic to shape future pandemic interaction?

4.12 Theory of Reasoned Action (TRA) and Theory of Planned Behavior (TPB)

Similar to ELM's assumption that attitudes will guide behavior, the Theories of Reasoned Action (TRA) and Planned Behavior (TPB) assume that attitudes predict behavior, and thus, if a persuasive message can affect attitudes, it is likely to affect behavior (Ajzen 1991). Attitudes, subjective norms, and perceived behavioral control are antecedents to behavioral intention, which predicts volitional behavior. The Theory of Planned Behavior (TPB) is an extension of Fishbein and Ajzen (1985) work with the Theory of Reasoned Action (TRA) focusing on behavioral control and volition (Madden et al. 1992). As such, this section focuses on the later work of TPB. Steeped in the assumption that attitudes can predict behavioral intention, the Theory of Planned Behavior is a theory that explains a persuasive process and is applied to behavior change messages or interventions. The three predictive components to behavioral intention are (1) *attitudes* comprised of *belief strength* and *belief evaluation,* (2) *subjective norms* comprised of *normative beliefs* and *motivation to conform,* and (3) *perceived behavioral control* comprised of the *control belief* and *perceived power* (Springston et al. 2010).

With regard to the coronavirus pandemic, consider the following illustration of social distancing to reduce risk exposure. A university wants to persuade students to stay six feet apart. Using the TRA and TPB, the university president, chancellor of student affairs, and chief risk and crisis management officer design a series of social distancing messages for students. To address attitudes, they want students to have a positive attitude toward social distancing behaviors (belief evaluation) and increase their confidence in the effectiveness of social distancing to manage risk (belief strength). To address subjective norms, they want students to perceive that other students expect social distancing (normative belief) and that social distancing conformity will be met with strong positive reinforcement by students' peers (motivation to comply). To address perceived behavioral control, they want students to perceive that the university has redesigned space to ensure that social distancing is possible (control belief) and that the students can resist temptations to hug, high five, or stand close to one another (perceived power). The university produces a short 2-min video that shows healthy students positively interacting within social distancing requirements (attitude), students smiling as they back up to maintain the appropriate distance and thanking one another for doing so (subjective norms and perceived behavioral control), and the redesigned campus space to accommodate social distancing with social distancing markers in public spaces like elevators and food court lines (perceived behavioral control). If students' attitudes, subjective norms, and perceived behavioral control are targeted to change their behavioral intention to socially distance, then students' volitional behavior to socially distance is more likely.

4.13 Transtheoretical Model (TTM)

The Transtheoretical Model (TTM) is a stage model that uses stages of change to help predict audience's responses to behavior change messages and adapt message strategies based on the stage. At its core, TTM assumes people intentionally make behavior change decisions in a predictable process and that behavior change decisions are influenced by a constellation of complex variables including personal, societal, and biological factors (Yzer and Nagler 2021). Prochaska and DiClemente (1982) worked in clinical psychology and developed the original iteration of this model to explain smoking cessation by therapy clients and self-changers. Prochaska and DiClemente (1983) argue that people progress forwards, and sometimes backwards, through five stages of behavior change: (1) precontemplation, (2) contemplation, (2) preparation, (4) action, and (5) maintenance, and with a sixth stage added in some applications of the theory, (6) termination (Prochaska and Velicer 1997).

The first stage, *precontemplation*, is marked by individuals having no behavior change intention in the foreseeable future. Precontemplation exists and persists for a variety of reasons such lack of information about the behavior change or consequences of behavior, general apathy toward the behavior change, and feelings of hopelessness because of failed behavior change attempts. Individuals may even avoid information, conversations, or cognitions about the behavior change (Prochaska and Velicer 1997).

Second, the *contemplation* stage is indicative of individuals' intention to act within six months (Prochaska and DiClemente 1983). These individuals seek information about the behavior's consequences and behavior change, discuss the behavior change with interpersonal relations, and consider changing the behavior.

Third, the *preparation* stage includes concrete plans and obtaining required resources and skills to ensure the behavior change is successfully implemented (Prochaska 2008).

Fourth, *action* "is a stage in which the individual has made specific, overt modifications in his or her behavior within the preceding 6 months" (Prochaska 2008, p. 846). Prochaska goes onto explain that not all behavior change, which is observable, is enough change to reach the threshold to count as the action stage. For example, cutting down from one pack of cigarettes to one cigarette per day is an observable behavior change, but Prochaska (2008) argues that "only total abstinence counts" because only total abstinence is "sufficient to reduce the risk of disease" (p. 846). Fifth, the *maintenance* stage is like any type of maintenance; it is the upkeep to ensure that the behavior change is permanent and resistant to relapse. At this juncture, the individual can experience relapse by returning to any of the previous stages of change (Prochaska and Velicer 1997).

Sixth, the *termination* stage is indicative of complete behavioral transformation. In this stage, individuals no longer consider returning to their previous behavior, and they require no maintenance to ensure that the behavior change is maintained. As Prochaska and Velicer (1997) put it, "It is as if they never acquired the habit in the first place" (p. 39). In addition to these five or six stages of change, there are ten processes that describe the relationships between stages: social liberation, consciousness-raising, dramatic relief, environmental re-evaluation, self-reevaluation, self-liberation, contingency management, counterconditioning, stimulus control, and helping relationships (see Prochaska et al. 1996 for more information). Communication practitioners can use both the stages of change and the processes to help identify messages that will address the targets current stage and construct messages that will persuade targets to move to the next stage.

With regard to pandemic communication, consider target groups and the risky health behaviors that are risking individual and public health. If the group is a geographic community spiking in case numbers, what has contact tracing and interviewing revealed about the group's risky health behaviors contributing the case increase? Are they aware of their risky behaviors and how it is affecting their community? Do they know how to change their behaviors? Do they need help identifying a plan and resources to increase their efficacy to act on the behavior change? Were they performing the behavior change and stopped maintenance? What do they need to maintain the change? Of great value to pandemic communication are the processes of contingency management, counterconditioning, stimulus control, and helping relationships that positively affect the behavior change maintenance, especially if a pandemic persists and requires longer-term behavior changes than initially predicted.

What are resources are recommended to prompt resilient pandemic communication strategies?

When using these different empirical frameworks to guide pandemic communication, there are several recommended readings that may help. Table 4.3: Pandemic Communication Theory Resources identifies the theory, the primary scholar attributed with the theory, and one-two recommended readings.

Because of the ubiquitous nature of issues management and crisis response across fields like corporate communication, crisis and risk communication, emergency management, health communication, persuasion, public relations, and organizational communication, to name a few, there are number of academic and practitioner-based studies and resources that are quite useful tools to have on the physical or virtual bookshelf. This list, while not comprehensive, is designed to complement an existing pandemic resource library or assist in the building of one. See Table 4.4: Resources to Build Your Pandemic Communication Theory Library.

4.14 Becoming More Resilient Through the Application of Empirical Frameworks

As this chapter wraps up, return to the buzz term "resilient." One way to conceive of resiliency is to look to Patrice Buzzanell's (2018a, b; 2010) work on organizational resiliency, because after all, organizations are applying theories to address

Table 4.3 Pandemic communication theory resources

Pandemic communication theory	Theorist	Recommended reading
Crisis and emergency risk communication (CERC) model	Centers for disease control and prevention (CDC)	*Crisis & Emergency Risk Communication (CERC).* (2018). Retrieved from https://emergency.cdc.gov/cerc/index.asp
Discourse of renewal	Ulmer, Sellnow, and Seeger	Ulmer, R. R., Sellnow, T. L., & Seeger, M. W. (2019). *Effective crisis communication: Moving from crisis to opportunity* (4th ed.). Los Angeles, CA: Sage
Elaboration likelihood model (ELM)	Petty and Cacioppo	Booth-Butterfield, S. & Welbourne, J. (2002). The Elaboration Likelihood Model: Its impact on persuasion theory and research. In J. P. Dillard & L. Shen [eds.] *The SAGE handbook of persuasion: Developments in theory and practice, 2nd ed.* (pp. 155–172). Thousand Oaks, CA: Sage
Extended parallel process model (EPPM)	Witte	Witte, K. (1992). Putting the fear back into fear appeals: The extended parallel process model. *Communications Monographs, 59*(4), 329–349
Image repair (restoration) theory	Benoit	Benoit, W. L. (2018). Crisis and image repair at United Airlines: Fly the unfriendly skies. *Journal of International Crisis and Risk Communication Research, 1(1),* 11–26. https://doi.org/10.30658/jicrcr.1.1.2
Situational crisis communication theory (SCCT)	Coombs and Holladay	Coombs, W. T., & Holladay, S. J. (2002). Helping crisis managers protect reputational assets: Initial tests of the situational crisis communication theory. *Management Communication Quarterly, 16*(2), 165–186

(continued)

Table 4.3 (continued)

Pandemic communication theory	Theorist	Recommended reading
Social cognitive theory	Alfred Bandura	Bandura, A. (1997). *Self-efficacy: The exercise of control.* New York, NY: Freeman
Stakeholder theory	Freeman and Gilbert	Freeman, R. E. (1984). *Strategic management: A stakeholder approach.* Marshfield, MA: Pitman Publishing
Theory of reasoned action and theory of planned behavior	Fishbein and Ajzen	Ajzen, I. (1991). The theory of planned behavior. *Organizational Behavior and Human Decision Processes, 50,* 179–211
Transtheoretical model (TTM)	Prochaska and DiClemente	Prochaska, J. O. & Velicer, W. F. (1997). The transtheoretical model of health behavior change. *American Journal of Health Promotion, 12(1),* 38–48

Table 4.4 Resources to build your pandemic communication theory library

Resources to build your pandemic communication theory library
Heath, R. L. & O'Hair, H. D. (Eds.). (2010). *Handbook of risk and crisis communication.* New York, NY: Routledge
National Consortium for the Study of Terrorism and Responses to Terrorism. (2012). *Understanding risk communication theory: A guide for emergency managers and communicators.* College Park, MD: Department of Homeland Security Science and Technology Center of Excellence
Sellnow, T. L. & Seeger, M. W. (2013). *Theorizing crisis communication.* Malden, MA: Wiley-Blackwell
Thompson, T. L. & Schulz, J. (Eds.). (2021). *Health communication theory.* Hoboken, NJ: John Wiley & Sons
Ulmer, R. R., Sellnow, T. L., & Seeger, M. W., (2019). *Effective crisis communication: Moving from crisis to opportunity* (4th ed.). Los Angeles, CA: Sage

pandemics, constructing messages and strategies, and reducing risks for their stakeholders, including the general public. Buzzanell (2010, 2018a) theorized resilience as adaptive-transformative processes triggered by loss or disruption and involving five subprocesses: crafting a new normalcy; affirming or anchoring important identities during difficult times; using and/or maintaining salient communication networks;

looking beyond conventional ways of thinking about and doing life by putting alternative logics to work; and foregrounding productive action while backgrounding unproductive behaviors or negative feelings (Buzzanell 2018b, p. 14).

In many ways, Buzzanell's resiliency in the adaptive-transformative process echoes different empirical frameworks reviewed in the chapter. First, the idea of crafting a new normalcy ties into the behavior change/intervention, health communication, and persuasion theories that aim to change individual and collective behaviors to avoid viral spread. The new normal is reflected in behavior changes associated with remote work, social distancing, face coverings, and small-scale gatherings moved outdoors. Second, as organizations manage stakeholder relationships, whether to coordinate their crisis response (e.g., CERC) or for reputational management (e.g., SSCT, image repair) important identities are affirmed through these communication networks. Third, alternative paths forward and foregrounding productive and positive feelings resonate with aspects of DR (e.g., organizational learning for new ways to organize post-crisis) and communicating gratitude for the individual and collective capacity to change behaviors to protect a healthier future. Communication practitioners draw on the empirical frameworks within this chapter and others to prepare, respond, and recover with resilience.

4.15 Operative Power of Pandemic Communication Theoretical Models

Final thoughts bring us to a quote from Bandura (2005) on the power of theory: "The value of a psychological theory is judged not only by its explanatory and predictive power, but also ultimately by its operative power to promote changes in human functioning" (p. 12). While referencing psychology as a discipline, the principle is broadly applied across theory. The empirical frameworks reviewed in the chapter, have operative power "to promote changes in human functioning" in pandemics and, consequentially, affect personal and public health outcomes. Operative power only reaches its potential if scholars and practitioners value these frameworks, explore their application to messaging, implement their related practices, and evaluate their effectiveness.

References

Ajzen I (1985) From intentions to actions: a theory of planned behavior. In: Khul J, Beckmann J (eds) Action control: from cognition to behavior. Springer, New York, pp 11–39
Ajzen I (1991) The theory of planned behavior. Organ Behav Hum Decis Process 50:179–211
Bandura A (1992) Social cognitive theory and exercise of control over HIV infection. In: DiClemente R (ed) Adolescents and AIDS: a generation in jeopardy. Sage, Newbury Park, CA, pp 89–116
Bandura A (1997) Self-efficacy: the exercise of control. Freeman, New York, NY
Bandura A (2002) Social cognitive theory in cultural context. Appl Psychol Int Rev 51(2):269–290

Bandura A (2005) The evolution of social cognitive theory. In: Smith KG, Hitt MA (eds) Great minds in management. Oxford University Press, Oxford, pp 9–35

Benoit WL, Drew S (1997) Appropriateness and effectiveness of image repair strategies. Commun Rep 10(2):153–163. https://doi.org/10.1080/08934219709367671

Benoit WL (1997) Image repair discourse and crisis communication. Publ Relations Rev 23(2):177–186. https://doi.org/10.1016/S0363-8111(97)90023-0

Benoit WL (2018) Crisis and image repair at United Airlines: fly the unfriendly skies. J Int Crisis Risk Commun Res 1(1):11–26. https://doi.org/10.30658/jicrcr.1.1.2

Benoit WL (2020) Image repair theory. In: Frandsen F, Johansen W (eds) Crisis communication, pp 105–119. DeGruyter, Boston, MA

Brinson SL, Benoit WL (1999) The tarnished star: restoring Texaco's damaged public image. Manag Commun Q 12(4):483–510

Booth-Butterfield S, Welbourne J (2002) The elaboration likelihood model: its impact on persuasion theory and research. In: Dillard JP, Shen L (eds) The SAGE handbook of persuasion: developments in theory and practice, 2nd edn. Sage, Thousand Oaks, CA, pp 155–172

Buzzanell PM (2010) Resilience: talking, resisting, ad imagining new normalcies into being. J Commun 60:1–14. https://doi.org/10.1111/j.1460-2466.2009.01469

Buzzanell PM (2018) Communication theory of resilience: enacting adaptive-transformative processes when families experience loss and disruption. In: Braithwaite D, Suter E, Floyd K (eds) Engaging theories in family communication, 2nd edn. Routledge, New York, NY, pp 98–109

Buzzanell PM (2018) Organizing resilience as adaptive-transformational tensions. J Appl Commun Res 46(1):14–18. https://doi.org/10.1080/00909882.2018.1426711

Centers for Disease Control and Prevention (2020) Surgeon general shows how to make your own face covering: PSA [YouTube video]. Retrieved from https://www.youtube.com/watch?v=PI1GxNjAjlw. Accessed 10 Jan 2021

Coombs WT, Holladay SJ (2002) Helping crisis managers protect reputational assets: initial tests of the situational crisis communication theory. Manag Commun Q 16(2):165–186

Coombs WT (2004) Impact of past crises on current crisis communication: Insights from situational crisis communication theory. J Business Commun (1973), 41(3):265–289

Coombs WT (2007) Protecting organization reputations during a crisis: the development and application of situational crisis communication theory. Corp Reput Rev 10(3):163–176

Coombs WT (2010) Conceptualizing crisis communication. In: Heath RL, O'Hair HD (eds) Handbook of risk and crisis communication. Routledge, New York, NY, pp 99–118

Crisis & Emergency Risk Communication (CERC) (2018) Retrieved from https://emergency.cdc.gov/cerc/index.asp. Accessed 10 Jan 2021

Freeman RE (1984) Strategic management: a stakeholder approach. Pitman Publishing, Marshfield, MA

Freeman RE, Gilbert DR (1987) Managing stakeholder interest. In: Sethi SP, Fable CM (eds) Business and society: dimensions of conflict and cooperation. Lexington Books, Lexington, MA, pp 379–422

Jong W (2021) Evaluating crisis communication: a 30-item checklist for assessing performance during COVID-19 and other pandemics. J Health Commun 1–9. https://doi.org/10.1080/108010730.2021.1871791

Kreuter MW, Farrell D, Olevitch L, Brennan L (2000) Tailoring health messages: customizing communication with computer technology. Routledge, New York, NY

Kreuter MW, Wray RJ (2003) Tailored and targeted health communication: strategies for enhancing information relevance. Am J Health Behav 27:S227–S232

Lithopoulos A, Liu S, Zhang C, Rhodes RE (2021) Predicting physical distancing in the context of COVID-19: a test of the extended parallel process model among Canadian adults. Canadian Psychol 1–9. https://doi.org/10.1037/cap0000270

Low YSY, Varughese J, Pang A (2011) Communicating crisis: how culture influences image repair in Western and Asian governments. Corporate Commun Int J 16(3):218–242. https://doi.org/10.1108/13563281111156880

Madden TJ, Ellen PS, Ajzen I (1992) A comparison of the theory of planned behavior and the theory of reasoned action. Pers Soc Psychol Bull 18(1):3–9

National Consortium for the Study of Terrorism and Responses to Terrorism (2012) Understanding risk communication theory: a guide for emergency managers and communicators. Department of Homeland Security Science and Technology Center of Excellence, College Park, MD. Retrieved from https://www.start.umd.edu/sites/default/files/files/publications/UnderstandingRiskCommunicationTheory.pdf#:~:text=Understanding%20Risk%20Communication%20Theory%3A%20A%20Guide%20for%20Emergency,provide%20risk%20messages%2C%20warning%2C%20and%20guidance%20regarding%20preparation. Accessed 10 Jan 2021

Nazione S, Perrault E, Pace K (2021) Impact of information exposure on perceived risk, efficacy, and preventative behaviors at the beginning of the COVID-19 pandemic in the United States. Health Commun 36(1):23–31. https://doi.org/10.1080/10410236.2020.1847446

Noar SM (2006) A 10-year retrospective of research in health mass media campaigns: where do we go from here? J Health Commun 11(21):21–42. https://doi.org/10.1080/10810730500461059

O'Keefe DJ (2008) Elaboration likelihood model. In: Donsbach W (ed) The international Encyclopedia of communication, vol IV, pp 1475–1480. Blackwell Publishing, Malden, MA

O'Keefe DJ (2012) From psychological theory to message design: lessons from the story of gain-framed and loss-framed persuasive messages. In: Cho H (ed) Health communication message design: theory and practice, pp 3–20. Sage, Thousand Oaks, CA

Paek H, Hilyard K, Freimuth V, Barge JK, Mindlin M (2010) Theory-based approaches to understanding emergency preparedness: Implications for effective health and risk communication. J Health Commun 14(4):428–444. https://doi.org/10.1080/10810731003753083

Petty RE, Cacioppo JT (1986) Communication and persuasion: central and peripheral routes to attitude change. Springer, New York

Petty RE, Wegener DT (1999) The elaboration likelihood model: current status and controversies. In: Chaiken S, Trope Y (eds) Dual-process theories in social psychology. Guilford Press, New York, pp 41–72

Prochaska JO, DiClemente CC (1982) Transtheoretical therapy: toward a more integrative model of change. Psychotherapy Theory Res Pract 276–288

Prochaska JO, DiClemente CC (1983) Stages and processes of self-change of smoking: toward an integrative model of change. J Consult Clin Psychol 51:390–395

Prochaska JO, Redding CA, Evers KE (1996) The transtheoretical model and stages of change. In: Glanz K, Lewis FM, Rimer BK (eds) Health behavior and health education: theory, research, and practice, 2nd edn. Jossey-Bass, San Francisco, CA, pp 60–84

Prochaska JO, Velicer WF (1997) The transtheoretical model of health behavior change. Am J Health Promot 12(1):38–48

Prochaska JO (2008) Decision making in the transtheoretical model of behavior change. Med Decis Making 845–849. https://doi.org/10.1177/0272989X08327068

Roberto AJ, Goodall CE, Witte K (2010) Raising the alarm and calming fears: perceived threat and efficacy during risk and crisis. In: Heath RL, O'Hair HD (eds) Handbook of risk and crisis communication. Routledge, New York, pp 285–301

Reynolds B, Seeger MW (2005) Crisis and emergency risk communication as an integrative model. J Health Commun 10:43–55. https://doi.org/10.1080/10810730590904571

Seeger MW (2018) Answering the call for scholarship: the J Int Risk Crisis Commun Res 1(1):7–10. https://doi.org/10.30658/jicrcr.1.1.1

Seeger MW, Reynolds B, Sellnow TL (2010) Crisis and emergency risk communication in health contexts: applying the CDC model to pandemic influenza. In: Heath RL, O'Hair HD (eds) Handbook of risk and crisis communication. Routledge, New York, NY, pp 493–506

Sellnow TL, Seeger MW (2013) Theorizing crisis communication. Wiley

Schiavo R (2007) Health communication: from theory to practice. Jossey-Bass, San Francisco, CA

Spradley RT, Spradley EL (2020) Contradiction and double binds in COVID-19 face covering recommendations. Int Res J Health Educ 3(13). https://escipub.com/Articles/IRJHE/IRJHE-2020-09-1205.pdf. Accessed 10 Jan 2021

Spradley RT (2017) Crisis communication in organizations. In: Scott CR, Lewis L (eds), Barker J, Keyton J, Kuhn T, Turner P (Assoc. eds) The international Encyclopedia of organizational communication. Wiley, Inc
Spradley RT (2020) Image restoration for university leaders' public health COVID-19 response: a case study of Notre Dame. Int Res J Public Health
Springston JK, Avery EJ, Sallot LM (2010) Influence theories: rhetorical, persuasion, and informational. In: Heath RL, O'Hair HD (eds) Handbook of risk and crisis communication. Routledge, New York, pp 268–284
Stephenson M, Palmgreen P (2001) Sensation seeking, perceived message sensation value, personal involvement, and processing of anti-marijuana PSAs. Commun Monogr 68(1):49–71. https://doi.org/10.1080/03637750128051
Ulmer RR, Sellnow TL, Seeger MW (2010) Post-crisis communication and renewal. In: Heath RL, O'Hair HD (eds) Handbook of risk and crisis communication. Routledge, New York, pp 302–322
Ulmer RR, Sellnow TL, Seeger MW (2019) Effective crisis communication: moving from crisis to opportunity, 4th edn. Sage, Los Angeles, CA
Ulmer RR (2001) Effective crisis management through established stakeholder relationships: Malden Mills as a case study. Manag Commun Q 14(4):590–615
von Gottberg C, Krumm S, Porzsolt F, Kilian R (2016) The analysis of factors affecting municipal employees' willingness to report to work during an influenza pandemic by means of the extended parallel process model (EPPM). BMC Public Health 16(26). https://doi.org/10.1186/s12889-015-2663-8
Weick KE (1995) Sensemaking in organizations. Sage Publications, Thousand Oaks, CA
Witte K (1992) Putting the fear back into fear appeals: the extended parallel process model. Commun Monogr 59(4):329–349
Witte K (1994) Fear control and danger control: a test of the extended parallel process model (EPPM). Commun Monogr 61:113–134
Yzer M, Nagler RH (2021) Theories of behavior. In: Thompson TL, Schulz PJ (eds) Health communication theory, pp 178–199. Wiley, Hoboken, NJ

Chapter 5
Pandemics and Resiliency: Psychometrics and Mental Models

Meghnaa Tallapragada

Abstract This chapter examines how people make sense of the uncertainty, the science, and their role in pandemics through the lens of mental models and cognitive biases such as mortality salience, optimism bias, and identity protection. The chapter will present definitions, the similarities and differences between various models, relevant evidence from literature pertaining to past and current pandemics, methods of assessments, and future directions for practitioners, students, and scholars to build resilience through communication driven by an understanding of mental models and cognitive biases.

Keywords Mental models · Cognitive biases · Optimism bias · Mortality salience · Identity protection

5.1 Introduction

Mental models are cognitive structures or maps that help people define, reason, and respond to an experience in the external world (Bostrom 2011; Johnson-Laird 1983, 2010; Weick 1990). They are dynamic, in that people build and rebuild their mental models about specific issues based on their perceptions of daily experiences (Bostrom 2011; Freyd 1987; Johnson-Laird 2010; Jones et al. 2011). As people receive information or encounter situations, their existing mental models are likely to change and affect their decision-making (Johnson-Laird 1983, 2010).

Each pandemic brings with it high levels of uncertainty and unpredictability. People attempt to understand a novel pandemic by relying on their existing mental models (Bostrom 2008; Norman 1983; Visschers et al. 2007). For example, studies have found that several people have used their existing understanding of other mosquito-borne diseases to make sense of the Zika virus, and several others are using their understanding of the influenza vaccine to make sense of the COVID-19

M. Tallapragada (✉)
Department of Advertising & Public Relations, Klein College of Media & Communication, Temple University, Philadelphia, PA, USA

D. M. Berube (ed.), *Pandemic Communication and Resilience*, Risk, Systems and Decisions, https://doi.org/10.1007/978-3-030-77344-1_5

vaccines (Southwell et al. 2018, 2020). The process of relying on previous experiences, existing knowledge, and learning new information coupled with people's tendency to simplify their reality, results in some individuals having inaccurate or harmful mental models about a deadly pandemic (Bostrom 2008; Johnson-Laird 1983; Norman 1983; Visschers et al. 2007). Communication students, scholars, and practitioners have the ability and responsibility to ensure that flawed mental models about a pandemic are corrected to ensure their individual and community safety.

Dual process theories such as the elaboration likelihood model (Petty and Cacioppo, 1986), heuristic systematic model (Chaiken et al. 1989), and the system 1 - system 2 thinking model (Kahneman 2011) would suggest that mental models can be altered through a central/systematic/system 2 route or through a peripheral/heuristic/system 1 route. Although these models have different assumptions of how these two routes affect information processing (see Xu 2017), they all however would assert that mental models formed through the central/systematic/system 2 route i.e., where an individual processes a message intentionally by elaborating on the details presented to them, would often be more accurate and thorough compared to those formed through the peripheral/heuristic/system 1 route (see, Greenwald and Leavitt 1984; Kahneman 2011; Lyons et al. 2019). For example, Fischer et al. (2020) found that prompting people to elaborate on the effectiveness of preventative behaviors of COVID-19 was effective in persuading people to adhere to them. The challenge however is to motivate individuals to always process pandemic communication centrally, while reducing their urge to take mental shortcuts that result in biased information processing. I will discuss some of the common biases that can interfere with shaping or correcting mental models around pandemics later in this chapter. But, first it is important to note that while most risk communication efforts have been focused on designing strategic campaigns to understand and correct mental models of lay individuals, it should not be assumed that experts are somehow immune from having a biased mental model (Jasanoff 1989). The following section outlines the differences and similarities between expert and lay mental models.

5.2 Expert and Lay Mental Models

Experts and lay individuals at times have different mental models pertaining to the same issues or situations involving a risk (Bostrom 1997; Slovic 2000). Sometimes the difference is attributed to their varying levels of access to and comprehension of evidence associated with the risk (Bostrom 1997; Slovic 2000; Visschers et al. 2007). Lay individuals sometimes have the inaccurate mental model that can result in poor decision-making. For example, Raude and Setbon (2009) found that several lay individuals believed that the infectious agent of the H1N1 subtype influenza A virus could not survive or maintain its infectious nature outside of humans, when studies reportedly found that the virus could survive long periods of time outside and on several materials including doorknobs (cited, Abad et al. 1994). In this, and similar instances, knowledge gaps can certainly be harmful if not intervened to help correct any misinformed notions that determine protective behaviors especially during a pandemic. However, gaining more knowledge does not automatically translate to a unified or corrected risk perception, because "individual contexts, motives, and even values" can shape mental models and the risk itself (p. 113, Bostrom 1997). For example, Zingg and Siegrist (2012) found that lay individuals, farmers, and veterinarians all understood and approved that vaccinations were more preferrable over culling as a strategy to address animal infectious diseases, but they all varied in how risky they perceived these strategies to be, because the farmers and lay individuals likely had a different relationship to the animals compared to the veterinarians.

Even experts are not immune from having context and values influence their understanding of a risk. For example, Karasneh et al. (2020) found that although majority of the pharmacists in their study were knowledgeable on the coronavirus disease (COVID-19), those who lived in a city, were male, had children, worked in a hospital, and consumed media regularly had higher risk perceptions of the disease compared to their counterparts. Expert mental models are also susceptible to be biased or incomplete (see Brown 1993; Wynne 1992). For scholars and practitioners of communication, the goal then is to not assume that somehow experts are immune to compromised mental models or assume that lay individuals always carry inaccurate mental models due to insufficient information, but to assess each of their mental models and use strategic messaging to align their perceptions with respect to the risk to ensure safe behaviors are practiced resulting in a healthy and resilient community. One of the obstacles to developing an accurate and consistent mental model involves biases that create a filter that, often unknowingly or unwillingly, contribute to inaccurate information seeking and processing (Chermack 2003; Ford and Sterman 1998).

5.3 Cognitive Biases

The following sections outline biases stemming from unreasonable optimism, mortality salience, and identity protection cognition. These biases typically affect judgments and decision-making during pandemics. Understanding how these are formed and addressed can be crucial in building resiliency among individuals and their communities during pandemics.

5.3.1 Optimism Bias

It's been reported that some people have unrealistic optimism about their susceptibility of contracting COVID-19 (Dolinski et al. 2020) and some even worry that the persistent nature of the threat from COVID-19 has only further stoked that bias (see, Bottemanne et al. 2020). Being unrealistically optimistic when considering the susceptibility or severity of a risk can prevent some individuals from practicing preventive behaviors (Becker 1974; Kim and Niederdeppe 2013; Weinstein 1982). Often people are unrealistically more optimistic about their own invulnerability compared to others, especially when they perceive the risk to be controllable as they tend to overestimate their own protective behaviors (Hoorens and Buunk 1993; Weinstein 1982). For example, Larwood (1978) noticed that participants in her study who felt confident about their own health and in their ability to prevent contracting the swine flu, were less likely to get vaccinated against it.

Once formed, optimism bias can be hard to change (Cho et al. 2013; Weinstein and Klein 1996), especially in conditions where people are engaging in conversations with others who are potentially reaffirming those biased beliefs (Cho et al. 2013). Optimism bias has been seen to decrease with age, but increase with education (Arnett 2000; Chapin 2008; Klacynski and Fauth 1996). Providing information using probabilities of being susceptible to risk has been found to be ineffective in eliminating optimism bias (see French and Hevey 2008). Some studies have shown that unrealistic optimism can be dampened by asking individuals to compare their vulnerability, towards a risk, with their good friends or others in their social network (as opposed to other abstract beings in general) (Perloff and Fetzer 1986), because people seem to perceive individuals within a group more positively and care about them more, compared to an unknown group of individuals (Sears 1983).

Those who have personally experienced a consequence from the risk or are close to those experiencing the risk are also less likely to experience optimism bias (Paek et al. 2008; Weinstein 1982) and seem to display a more calm and collected effort towards handling the risk in a responsible manner (Xie et al. 2011), because they seem to have a more accurate understanding of the risk which is neither overly optimistic nor pessimistic (Halpern-Felsher et al. 2001). Experiencing the risk directly can be a very harmful way of reducing or eliminating optimism bias. Using the theoretical foundation of embodied cognition to simulate personal experiences with the risk, with or without using virtual reality, offers a potential solution to address optimism bias (for a thorough explanation of embodied cognition, see Hardy 2020).

5.3.2 Mortality Salience

Terror management theory (TMT) posits that the terror rising from the realization that death is inevitable is one of the motivations that drives human behavior (Greenberg et al. 1986; Pyszczynski et al. 2015). To cope with the terror of death, people rely on their cultural worldviews to help find validation that their lives though finite can be meaningful and purposeful (Greenberg et al. 1986; Pyszczynski et al. 2015). Reminders of one's mortality i.e., mortality salience can thus trigger TMT-proposed protective responses which can be positive, neutral, or negative (Burke et al. 2049; Pyszczynski et al. 2015). The responses to mortality salience varies based on age, religion, and conditions of one's mental and physical health (Maxfield et al. 2007; Neimeyer et al. 2004).

Pandemics have the potential to trigger mortality salience differently for individuals thereby producing varied health-behavior outcomes (Courtney et al. 2020; Kellaris et al. 2020; Sole 2007). Developed to investigate terror management in the context of health outcomes, the terror management health model (TMHM) posits that when one's mortality is made salient in a health scenario, people experience a conscious or unconscious death thought activation, which will likely lead them to engage in either threat-avoidance or health-behavior responses, which can involve health-defeating or health-facilitating actions (Goldenberg and Arndt 2008). When the death activation is conscious, people are likely to experience their "proximal defenses" becoming activated to confront the threat and minimize their vulnerability to it whereas if the activation is unconscious, then they engage in "distal defenses" that rely on keeping faith or enhancing their self-esteem to cope with the threat (Pyszczynski et al. 2020). With pandemics such as COVID-19 typically receiving widespread media attention, which very likely instigates interpersonal conversations, it is very likely for individuals to experience a conscious death activation leading to the onset of proximal defenses to protect themselves (Gramlich 2020; Mitchell et al. 2020). However, with the push towards reopening the country's businesses and economies and in some instances minimizing the threat from the pandemic, some people could be struggling with a tension between implementing proximal and distal defenses to protect themselves (Pyszczynski et al. 2020) as they strive to balance their preferences put forth by those who support and challenge their worldviews (Harmon-Jones et al. 1997). Should there be a conscious death thought activation prompting proximal defenses, individuals are more likely to engage in behaviors to actively remove that threat by either denying/avoiding it or by practicing healthy behaviors to reduce the threat versus when there is an unconscious death activation people are likely to apply distal defenses prompting them to find a way to live meaningful lives through this all (Courtney et al. 2020; Goldenberg and Arndt 2008; Pyszczynski et al. 2020). This suggests that just because a death activation was set into action, there is no assurance that one will engage in recommended healthy behaviors and that sometimes there can be tension in trying to decide one's course of action to respond to a pandemic.

There is also evidence that some cope with a death activation by believing that they have access to literal immortality, where they will continue to live by transcending the physical body, and others focus on establishing a symbolic immortality where although they will be gone, a part of them will symbolically live on (Pyszczynski et al. 2015). Studies have found that some people achieve symbolic immortality by believing that the group they belong to, is likely to exist long after they are gone and thus feel like their values and visions will continue to survive (Greenberg and Kosloff 2008; Sani et al. 2009). Others achieve symbolic immortality by engaging in self-sacrifice as a means of "defending against the awareness of death" (p. 538, Routledge and Arndt 2007). Comments about opening up businesses for the sake of the economic health of their country even if it means risking their own lives illustrates this self-sacrifice for symbolic immortality in the context of the COVID-19 pandemic (Courtney et al. 2020).

The goals of scholars and practitioners addressing pandemics is to motivate individuals experiencing mortality salience to adopt healthy behaviors by providing them with concrete steps to help manage their terror arising from the threat of contracting the disease. Kellaris et al. (2020) found that raising mortality salience alone did not guarantee compliance with the recommended behaviors for protection against COVID-19, but instead having rhyming messages, that can be positive and memorable, along with firmer directions that convey authority to be effective in motivating compliant behaviors (i.e., proximal defenses) when experiencing mortality salience with a pandemic. Along with providing concrete steps of management to aid implementing proximal defenses, researchers also recommend addressing distal defenses by using compassion-based communication and efforts to develop a better sense of self-worth as means to encouraging living a meaningful healthy life by protecting themselves against a pandemic (Greenberg and Arndt 2012; Harmon-Jones et al. 1997; Ju et al. 2020).

5.3.3 Identity Protection

The media coverage of COVID-19 pandemic, especially in the US, has been politically polarized with newspaper coverage having more politicians than scientists speaking about the issue (Hart et al. 2020). People consuming these narratives are likely adjusting their mental models to take into account the politics of pandemics (McLaughlin et al. 2019). Americans vary in how they perceive the risks from COVID-19, with Democrats more worried about public health and workers while Republicans about the threat to businesses (Green et al. 2020). More Republicans report being skeptical about the effectiveness of masks while Democrats are reporting a concern about how several are not taking the pandemic seriously and not wearing masks regularly (Kessel and Quinn 2020). Among people who harbor intense dislike for their competing party supporters, they begin associating the pandemic (which should be apolitical) with highly politicized positions (Druckman et al. 2020). The politicization and polarization of COVID-19 pandemic has become a global concern (Mordecau and Connaughton 2020), and illustrates the power that bias stemming from identity protection can have on mental models about the susceptibility, severity, and solutions to address a pandemic.

When an individual's specific identity is primed, political or otherwise, it becomes hard for an individual to not analyze the information they are presented with through their primed identity filter. For example, Adida et al. (2020) found that when simply presented with messages of Ebola, there was no effect on immigration attitudes, but when primed with their political identity their immigration attitudes began to align more closely with their political sentiments. Identity salience thus can be used to amplify (see Diamond 2020) or attenuate polarized information processing (see Levendusky 2017). Communication professionals have to consider strategically framing their messages by deciding which identity to make salient (or not) at the moment of information processing. For example, collectivist cultures compared to individualistic ones have been found to be more likely to protect against and lessen the transmission of infectious diseases (Fincher et al. 2008; Morand and Walther 2018). Kim et al. (2016) found that in the case of xenophobia with Ebola, the more people felt vulnerable the more likely they were to feel xenophobic, but this relationship was more pronounced among individualists rather the collectivists. Typically, an entire country is deemed as being more or less collectivist or individualistic, but evidence also indicates that there could be individualistic/collectivist tendencies within the same country (Vandello and Cohen 1999) and that priming can help activate the tendency to adopt a more individual or collective mindset (Hoersting et al. 2020; Knyazev et al. 2018). Deciding which identity to make more or less salient can become pivotal during messaging around pandemics.

People engage in identity protective behaviors when they seek information and process the findings presented to them. It might seem logical to assume that people consume information, reason with it, and then form an opinion, but it's been found that people are more likely to have opinions and then motivate themselves to seek information that supports their existing stances i.e., they are more likely to engage in motivated reasoning (Kunda 1990). Motivated reasoning goes hand-in-hand with selective exposure i.e., selectively choosing to consume information that supports your existing beliefs (Lazarsfeld et al. 1948). Engaging in motivated reasoning and selective exposure is likely to help minimize any attitude inconsistencies (Heider 1946) or cognitive dissonances (Festinger 1964) one may feel from being exposed to contradictory messages.

Motivated reasoning and selective exposure are often associated with misperceptions leading to flawed mental models that deviate from expert models (Bode and Vraga 2015; Vicario et al. 2016). The COVID-19 pandemic has also been called an infodemic where there is extreme amounts of information and misinformation (Zarocostas 2020). Roozenbeek et al. (2020) found that internationally, people accessing information through social media, and those who are politically conservative are more susceptible to misinformation through social media whereas those who trust scientists and scientific organizations, and who have the ability to comprehend quantitative information were less likely to be susceptible to misinformation. Both factors – political ideology and social media channels which are notorious for creating echo chambers – are associated with identity protection leading to misinformed mental models. But, Chan et al. (2020) found that conversations with friends and family can also help counter misinformation that could have been found through social media. Additionally, Pennycook et al. (2020) found that a nudge to consider the accuracy of the message being presented helps people to better discern misinformation.

5.4 Mental Models Measurements

Several methods have been used to understand, assess, and affect mental models. Scholars and practitioners have used interviews or case studies to understand mental models of experts and/or lay individuals (Bostrom 2011; Cousin and Siegrist 2010;

de Bruin and Bostrom 2013; Kessel and Quinn 2020). Having an interview guide to help design semi-structured interviews have been helpful in understanding the constitution of one's mental models (de Bruin and Bostrom 2013). Think-aloud procedures have also been used to help understand the conceptual making of one's mental models (French and Hevey 2008; Genie et al. 2020). Content analysis of social media has also been used to assess the makeup of mental models (Jiang et al. 2020). For those interested in assessing how a specific aspect within one's mental model may be affecting their decision-making processes, while controlling for other related factors, surveys have been implemented (Adida et al. 2020, Chan et al. 2020, Cho et al. 2013; Dolinski et al. 2020; Zingg and Siegrist 2012). Online experiments embedded within surveys have been used to assess the effects of manipulations on influencing mental models (Bode and Vraga 2015; Druckman et al. 2020; Hoersting et al. 2020; Pennycook et al. 2020).

The method must align with the goals of the researchers or practitioners, because some are able to help make causal arguments (such as experiments) while others might be more suitable to provide descriptive information (such as content analysis or think-aloud procedures). Triangulation by using multiple methods can be beneficial to assessing and manipulating mental models (Kirk and Miller 1986). It is advised to be as concrete as possible in the question wordings of interview guides, think-aloud guides, content analysis codebooks, and survey and experiment questions to ensure that the measurement is valid and can be reliable (Tallapragada et al. 2020).

Typical quantitative statistical analyses can be used to assess mental model data when gathered through online experiments or surveys. Theoretical sampling and saturation can guide the data collection process that involve qualitative methods of interviews or think aloud procedures, and can be analyzed using grounded theory (Charmaz 2006). Data gathered through other qualitative methods can also be assessed using the process outlined by Carley and Palmquist (1992) which recommends identifying sets of concepts to code in texts (which can include transcriptions or other textual content), hypothesize relationships between concepts, and use software programs such as R (or R studio) to run statistical analyses to test those relationships.

5.5 Conclusion

Pandemics unexpectedly and with little warning affect individuals and their communities. An individual's decision during a pandemic translates into consequences for their communities. To understand why and how people within the same community can make different decisions about the same pandemic, it is important to assess existing mental models and the cognitive biases that could be shaping their perceptions about the causes, susceptibility, severity, prevention, and cures of the pandemic. As discussed in this chapter, experts and lay individuals are both vulnerable to biased information processing. The chapter outlined how optimism bias, mortality salience, and identity protective cognition could interfere with developing accurate mental models of pandemics, and presented potential strategic solutions to assess and correct for those psychological biases. Being proactive about measuring and remedying mental models of individuals can have consequences for how the individual and their communities prepare, respond, and recover from pandemics.

References

Abad FX, Pintó RM, Bosch A (1994) Survival of enteric viruses on environmental fomites. Appl Environ Microbiol 60(10):3704–3710

Adida CL, Dionne KY, Platas MR (2020) Ebola, elections, and immigration: How politicizing an epidemic can shape public attitudes. Politics Groups Identities 8(3):488–514. https://doi.org/10.1080/21565503.2018.1484376

Arnett J (2000) Optimistic bias in adolescent and adult smokers. Addict Behav 25:625–632

Becker MH (1974) The health belief model and personal health behavior. Health Educ Monogr 2:326–473

Bode L, Vraga EK (2015) In related news, that was wrong: The correction of misinformation through related stories functionality in social media. J Commun 65:619–638. https://doi.org/10.1111/jcom.12166

Bostrom A (1997) Risk perceptions: "Experts" vs. "lay people." Duke Environ Law Policy Forum 8:101–113

Bostrom A (2008) Lead is like mercury: Risk comparisons, analogies and mental models. J Risk Res 11(1–2):99–117

Bostrom A (2017) Mental models of risk. In: Oxford research encyclopedia, communication. https://doi.org/10.1093/acrefore/9780190228613.013.533

Bottemanne H, Morlaàs O, Fossati P, Schmidt L (2020) Does the coronavirus epidemic take advantage of human optimism bias? Front Psychol 11:1–5. https://doi.org/10.3389/fpsyg.2020.02001

Brown P (1993) When the public knows better: popular epidemiology challenges the system. Environment, pp 16–41

Burke BL, Martens A, Faucher EH (2010) Two decades of terror management theory: a meta-analysis of mortality salience research. Pers Soc Psychol Rev 14:155–195. https://doi.org/10.1177/1088868309352321. Accessed 20 Feb 2021

Carley K, Palmquist M (1992) Extracting, representing, and analyzing mental models. Soc Forces 70(3):601–636

Chaiken S, Liberman A, Eagly A (1989) Heuristic and systematic processing within and beyond the persuasion context. In: Veleman JS, Bargh JA (eds) Unintended thought. Guilford, New York, pp 212–252

Chan M-PS, Jamieson KH, Albarracin D (2020) Perspective associations of regional social media messages with attitudes and actual vaccination: a big data and survey study of the influenza vaccine in the United States. Vaccine 38(40):6236–6247. https://doi.org/10.1016/j.vaccine.2020.07.054. Accessed 20 Feb 2021

Chapin J (2008) Youth perceptions of their school violence risks. Adolescence 171:461–471

Charmaz K (2006) Theoretical sampling, saturation, and sorting. In: Constructing grounded theory: a practical guide through qualitative analysis, pp 96–112. Sage, London

Chermack TJ (2003) Mental models in decision making and implications for human resource development. Adv Dev Hum Resour 5(4):408–422. https://doi.org/10.1177/1523422303257373

Cho H, Lee J-S, Lee S (2013) Optimistic bias about H1N1 flu: Testing the links between risk communication, optimistic bias, and self-protection behavior. Health Commun 28(2):146–158. https://doi.org/10.1080/10410236.2012.664805

Courtney EP, Goldenberg JL, Boyd P (2020) The contagion of mortality: A terror management health model for pandemics. Br J Soc Psychol 59(3):607–617. https://doi.org/10.1111/bjso.12392. Accessed 20 Feb 2021
Cousin M-E, Siegrist M (2010) Risk perception of mobile communication: a mental models approach. J Risk Res 13(5):599–620
de Bruin WB, Bostrom A (2013) Assessing what to address in science communication. Proc Natl Acad Sci 110:14062–14068. https://doi.org/10.1073/pnas.1212729110. Accessed 20 Feb 2021
Diamond EP (2020) The influence of identity salience on framing effectiveness: an experiment. Polit Psychol 41(6):1133–1150. https://doi.org/10.1111/pops.12669. Accessed 20 Feb 2021
Dolinski D, Dolinski B, Zmaczynska-Witek B, Banach M, Kulesza W (2020) Unrealistic optimism in the time of coronavirus pandemic: may it help to kill, if so—whom: Disease or the person? J Clin Med 9:1464
Druckman JN, Klar S, Krupnikov Y, Levendusky M, Ryan JB (2020) How affective polarization shapes Americans' political beliefs: a study of response to the COVID-19 pandemic. J Exp Polit Sci, pp 1–12. https://doi.org/10.1017/XPS.2020.28
Festinger L (1964) Conflict, decision, and dissonance. Stanford University Press, Stanford, CA
Fincher CL, Thornhill R, Murray DR, Schaller M (2008) Pathogen prevalence predicts human cross-cultural variability in individualism/collectivism. Proc R Soc B Biol Sci 275:1279–1285. https://doi.org/10.1098/rspb.2008.0094. Accessed 20 Feb 2021
Fischer I, Avrashi S, Oz T, Fadul R, Gutman K, Rubenstein D, Kroliczak G, Goerg S, Glöckner A (2020) The behavioural challenge of the COVID-19 pandemic: Indirect measurements and personalized attitude changing treatments (IMPACT). R Soc Open Sci 7:201131. https://doi.org/10.1098/rsos.201131. Accessed 20 Feb 2021
Ford DN, Sterman JD (1998) Expert knowledge elicitation to improve formal and mental models. Syst Dyn Rev 14(4):309–340
French DP, Hevey D (2008) What do people think about when answering questionnaires to assess unrealistic optimism about skin cancer? A think aloud study. Psychol Health Med 13(1):63–74. https://doi.org/10.1080/13548500701243959. Accessed 20 Feb 2021
Freyd JJ (1987) Dynamic mental representations. Psychol Rev 94(4):427–438
Genie MG, Loría- Rebolledo LE, Paranjothy S, Powell D, Ryan M, Sakowsky RA, Watson V (2020) Understanding public preferences and trade-offs for government responses during a pandemic: a protocol for a discrete choice experiment in the UK. BMJ Open 10:e043477. https://doi.org/10.1136/bmjopen-2020-043477. Accessed 20 Feb 2021.
Goldenberg JL, Arndt J (2008) The implications of death for health: a terror management model of behavioral health promotion. Psychol Rev 115: 1032–153. https://doi.org/10.1037/a0013326. Accessed 20 Feb 2021
Gramlich J (2020). 44% of Americans say they talk about the coronavirus outbreak most or almost all of the time. Pew Research Center. https://www.pewresearch.org/fact-tank/2020/05/13/44-of-americans-say-they-talk-about-the-coronavirus-outbreak-most-or-almost-all-of-the-time/. Accessed 20 Feb 2021
Greenberg J, Arndt J (2012) Terror management theory. In Lange PAV, Kruglanski AW, Higgins ET (eds) Handbook of theories of social psychology, vol 1. Sage, London, UK, pp 398–415. https://doi.org/10.4135/9781446249215.n20. Accessed 20 Feb 2021
Greenberg J, Kosloff S (2008) Terror management theory: implications for understanding prejudice, stereotyping, intergroup conflict, and political attitudes. Soc Pers Psychol Compass 2:1881–1894
Greenberg J, Pyszczynski T, Solomon S (1986) The causes and consequences of the need for self-esteem: a terror management theory. In: Self P, Self P (eds) Baumeister RF. Springer, New York, pp 189–212
Greenwald AG, Leavitt C (1984) Audience involvement in advertising: four levels. J Cons Res 11:581–592
Green J, Edgerton J, Naftel D, Shoub K, Cranmer SJ (2020) Elusive consensus: polarization in elite communication on the COVID-19 pandemic. Sci Adv 6(eabc2717):1–5

Halpern-Felsher BL, Millstein SG, Ellen JM, Adler NE, Tschann JM, Biehl M (2001) The role of behavioral experience in judging risks. Health Psychol 20:120–126

Hardy BW (2020) Embodied cognition in communication science. Commun Theor. https://doi.org/10.1093/ct/qtaa003. Accessed 20 Feb 2021

Harmon-Jones E, Simon L, Greenberg J, Pyszczynski T, Solomon S, McGregor H (1997) Terror management theory and self-esteem: Evidence that increased self-esteem reduces mortality salience effects. J Pers Soc Psychol 72(1):24–36

Hart PS, Chinn S, Soroka S (2020) Politicization and polarization in COVID-19 news coverage. Sci Commun 42(5):679–697. https://doi.org/10.1177/1075547020950735

Heider F (1946) Attitudes and cognitive organization. J Psychol 21(1):107–112

Hoersting RC, Phillips NL, Murrell AR (2020) Culture and priming in the perception of facial emotion, self-representation and thought: Brazil and United States. Culture and Brian. https://doi.org/10.1007/s40167-020-00090-6. Accessed 20 Feb 2021

Hoorens V, Buunk BP (1993) Social comparison of health risks: Locus of control, the person-positivity bias, and unrealistic optimism. J Appl Soc Psychol 23(4):291–302

Hu J, He W, Zhou K (2020) The mind, the heart, and the leader in times of crisis: how and when COVID-19-triggered mortality salience relates to state anxiety, job engagement, and prosocial behavior. J Appl Psychol 105(11):1218–1233. https://doi.org/10.1037/apl0000620

Jasanoff S (1989) The political science of risk perception. Reliab Eng Syst Saf 59(1):91–99. https://doi.org/10.1016/S0951-8320(97)00129-4. Accessed 20 Feb 2021

Jiang J, Chen E, Yan S, Lerman K, Ferrara E (2020) Political polarization drives online conversations about COVID-19 in the United States. Hum Behav Emerg Technol 2(3):200–211. https://doi.org/10.1002/hbe2.202. Accessed 20 Feb 2021

Johnson-Laird PN (1983) Mental models: towards a cognitive science of language, inference, and consciousness. Harvard Univ Press, Cambridge, MA

Johnson-Laird PN (2010) Mental models and human reasoning. Proc Natl Acad Sci U S A 107(43):18243–18250. https://doi.org/10.1073/pnas.1012933107. Accessed 20 Feb 2021

Jones NA, Ross H, Lyman T, Perez P, Leitch A (2011) Mental models: An interdisciplinary synthesis of theory and methods. Ecol Soc 16(1):46–58

Kahneman D (2011) Thinking, fast and slow. Farrar, Straus and Giroux

Karasneh R, Al-Azzan S, Muflih S, Soudah O, Hawamdeh S, Khader Y (2020) Media's effect on shaping knowledge, awareness risk perceptions and communication practices of pandemic COVID-19 among pharmacists. Res Soc Admin Pharm. https://doi.org/10.1016/j.sapharm.2020.04.027. Accessed 20 Feb 2021

Kellaris JJ, Machleit KA, Gaffney DR (2020) Sign evaluation and compliance under mortality salience: lessons from a pandemic. Interdisc J Signage Wayfinding 4(2):51–66

Kessel PV, Quinn D (2020) Both Republicans and Democrats cite masks as a negative effect of COVID-19 but for very different reasons. Pew Research Center. https://www.pewresearch.org/fact-tank/2020/10/29/both-republicans-and-democrats-cite-masks-as-a-negative-effect-of-COVID-19-but-for-very-different-reasons/. Accessed 20 Feb 2021

Kim HK, Niederdeppe J (2013) Exploring optimistic bias and the integrative model of behavioral prediction in the context of a campus influenza outbreak. J Health Commun 18(2): 206–222. https://doi.org/10.1080/10810730.2012.688247. Accessed 20 Feb 2021

Kim HS, Sherman DK, Updegraff JA (2016) Fear of Ebola: the influence of collectivism on xenophobic threat responses. Psychol Sci 27(7):935–944. https://doi.org/10.1177/0956797616642596

Kirk J, Miller ML (1986) Reliability and validity in qualitative research. Sage, Newbury Park, CA

Klacynski P, Fauth J (1996) Intellectual ability, rationality, and intuitiveness as predictors of warranted and unwarranted optimism for future life events. J Youth Adolesc 25:755–819

Knyazev GG, Merkulova EA, Savostyanov AN, Bocharov AV, Saprigyn AE (2018) Effect of cultural priming on social behavior and EEG correlates of self-processing. Front Behav Neurosci. https://doi.org/10.3389/fnbeh.2018.00236. Accessed 20 Feb 2021

Kunda Z (1990) The case for motivated reasoning. Psychol Bull 108:480–498. https://doi.org/10.1037/0033-2909.108.3.480
Larwood L (1978) Swine flu: a field study of self-serving biases. J Appl Soc Psychol 8(3):283–289
Lazarsfeld PF, Berelson B, Gaudet H (1948) The people's choice: how the voter makes up his mind in a Presidential Election. Columbia University Press, New York
Levendusky MS (2017) Americans, not partisans: can priming American national identity reduce affective polarization? J Polit 80(1):59–70. https://doi.org/10.1086/693987. Accessed 20 Feb 2021
Lyons BA, Hasell A, Tallapragada M, Jamieson KH (2019) Conversion messages and attitude change: Strong arguments, not costly signals. Pub Underst Sci 28(3):320–338. https://doi.org/10.1177/0963662518821017. Accessed 20 Feb 2021
Maxfield M, Kluck B, Greenberg J, Pyszczynski T, Coz CR, Solomon S, Weise D (2007) Age-related differences in responses to thoughts of one's own death: Mortality salience and judgments of moral transgressions. Psychol Aging 22(2):3410353. https://doi.org/10.1037/0882-7974.22.2.341. Accessed 20 Feb 2021
McLaughlin B, Velez JA, Dunn JA (2019) The political world within: how citizens process and experience political narratives. Ann Int Commun Assoc 43(2):156–172. https://doi.org/10.1080/23808985.2019.1597635. Accessed 20 Feb 2021
Mitchell A, Oliphant JB, Shearer E (2020) Americans are turning to media, government and others for COVID-19 news. Pew Research Center. https://www.journalism.org/2020/04/29/1-americans-are-turning-to-media-government-and-others-for-COVID-19-news/. Accessed 20 Feb 2021
Morand S, Walther BA (2018) Individualistic values are related to an increase in the outbreaks of infectious diseases and zoonotic diseases. Nat Sci Rep **8**:3866. https://doi.org/10.1038/s41598-018-22014-4. Accessed 20 Feb 2021
Mordecau M, Connaughton A (2020) Public opinion about coronavirus is more politically divided in U.S. than in other advanced economies. Pew Research Center. https://www.pewresearch.org/fact-tank/2020/10/28/public-opinion-about-coronavirus-is-more-politically-divided-in-u-s-than-in-other-advanced-economies/. Accessed 20 Feb 2021
Neimeyer RA, Wittkowski J, Moser RP (2004) Psychological research on death attitudes: an overview and evaluation. Death Stud 28(4):309–340
Norman D (1983) Some observations on mental models. In: Gentner D, Stevens A (eds) Mental models. Lawrence Erlbaum Associates Inc., Mahwah, NJ, pp 299–324
Paek H-J, Hilyard K, Freimuth VS, Barge JK, Mindlin M (2008) Public support for government actions during a flu pandemic: lessons learned from a statewide survey. Health Promotion Practces 9(4):60S-72S. https://doi.org/10.1177/1524839908322114
Pennycook G, McPhetres J, Zhang Y, Lu JG, Rand DG (2020) Fighting COVID-19 misinformation on social media: experimental evidence for a scalable accuracy-nudge intervention. Psychol Sci 31(7):770–780. https://doi.org/10.1177/0956797620939054
Perloff LS, Fetzer BK (1986) Self-other judgments and perceived vulnerability to victimization. J Pers Soc Psychol 50:502–510
Petty RE, Cacioppo JT (1986) The elaboration likelihood model of persuasion. Adv Exp Soc Psychol 19:123–205
Pyszczynski T, Solomon S, Greenberg J (2015) Thirty years of terror management theory: From genesis to revelation. Adv Exp Soc Psychol 52(1):1–70. https://doi.org/10.1016/bs.aesp.2015.03.001. Accessed 20 Feb 2021
Pyszczynski T, Lockett M, Greenberg J, Solomon S (2020) Terror management theory and the COVID-19 pandemic. J Humanist Psychol. https://doi.org/10.1177/0022167820959488
Roozenbeek J, Schneider CR, Dryhurst S, Kerr J, Freeman ALJ, Recchia G, van der Bles AM, van der Linden S (2020). Susceptibility to misinformation about COVID-19 around the world. R Soc Open Sci 7:201199. https://doi.org/10.1098/rsos.201199. Accessed 20 Feb 2021
Routledge C, Arndt J (2007) Self-sacrifice as self-defence: Mortality salience increases efforts to affirm a symbolic immortal self at the expense of the physical self. Eur J Soc Psychol 38:531–41. https://doi.org/10.1002/ejsp.442. Accessed 20 Feb 2021

Sani F, Herrera M, Bowe M (2009) Perceived collective continuity and ingroup identification as defense against death awareness. J Exp Soc Psychol 45:242–245

Sears DO (1983) The person-positivity bias. J Pers Soc Psychol 44:233–250

Slovic P (2000) The perception of risk. Earthscan Publications

Sole JW (2007) Terror management and pandemic influenza: social perception and response [Unpublished Master's thesis]. Lakehead University

Southwell BG, Ray SE, Vazquez NN, Ligorria T, Kelly BJ (2018) A mental models approach to assessing public understanding of Zika virus, Guatemala. Emerg Infect Dis 24(5):938–939. https://doi.org/10.3201/eid2405.171570. Accessed 20 Feb 2021

Southwell BG, Kelly BJ, Bann CM, Squiers LB, Ray SE, McCormack LA (2020) Mental models of infectious diseases and public understanding of COVID-19 prevention. Health Commun 35(14):1707–1710. https://doi.org/10.1080/10410236.2020.1837462. Accessed 20 Feb 2021

Tallapragada M, Hardy BW, Lybrand E, Hallman WK (2020) Impact of abstract versus concrete conceptualization of genetic modification (GM) technology on public perceptions. Risk Anal. https://doi.org/10.1111/risa.13591. Accessed 20 Feb 2021

Vandello JA, Cohen D (1999) Patterns of individualism and collectivism across the United States. J Pers Soc Psychol 77:279–292

Vicario MD, Bessi A, Zollo F, Petroni F, Scala A, Calderlli G, Stanley E, Quattrociocchi (2016) The spreading of misinformation online. Proc Natl Acad Sci 113(3):554–559. https://doi.org/10.1073/pnas.1517441113. Accessed 20 Feb 2021

Visschers VHM, Meertens RM, Passchier WF, deVries NK (2007) How does the general public evaluate risk information? the impact of associations with other risks. Risk Anal 27(3):715–727

Weick KE (1990) Introduction: cartographic myths in organizations. In Sigismund Huff A (Ed), Mapping strategic thought. Wiley , New York, pp 1–10

Weinstein ND (1982) Unrealistic optimism about susceptibility to health problems. J Behav Med 5(4):441–460

Weinstein ND, Klein WM (1996) Unrealistic optimism: present and future. J Soc Clin Psychol 15:1–8

Wynne B (1992) Misunderstood misunderstanding: social identities and public uptake of science. Public Underst Sci 1:281–304

Xie X-F, Stone E, Zheng R, Zhang R-G (2011) The 'Typhoon Eye Effect': determinants of distress during the SARS epidemic. J Risk Res 14(9):1091–1107. https://doi.org/10.1080/13669877.2011.571790

Xu Q (2017) Dual process models of persuasion. Int Encyclopedia Media Effects. https://doi.org/10.1002/9781118783764.wbieme0074. Accessed 20 Feb 2021

Zarocostas J (2020) How to fight an infodemic. Lancet 395:676. https://doi.org/10.1016/S0140-6736(20)30461-X

Zingg A, Siegrist M (2012) Lay people's and experts' risk perception and acceptance of vaccination and culling strategies to fight animal epidemics. J Risk Res 15(1):53–66. https://doi.org/10.1080/13669877.2011.601320

Chapter 6
Vaccine Hesitancy and Secondary Risks

Christopher L. Cummings, Shreya Gopi, and Sonny Rosenthal

Abstract One facet of pandemic preparedness and resiliency planning is to anticipate that a significant portion of the population will not understand, or be willing, to adopt advocated risk mitigation responses. Communication plays a central and vital role in creating salience and stoking motivations to respond effectively to pandemics and other public health crises. Widespread adoption of variolation, inoculation, and vaccination has historically improved community resilience to disease, but with limited effectiveness due to a growing community who are unwilling to vaccinate. Vaccine hesitancy is a decision-making outcome stemming from diverse motives and is often related to a lack of vaccine confidence and perceived risks. In the race to develop vaccines to mitigate pandemic risks, there is a need to understand factors influencing vaccine hesitancy. Secondary risk theory (SRT) is a useful framework to explain this, accounting for concerns about vaccine efficacy and safety. This chapter unpacks how vaccine hesitancy should be of critical concern to health and risk communicators and introduces SRT as a foundational theoretical framework to explain and predict vaccination decisions.

Keywords Vaccines · Vaccination · Vaccine hesitancy · Secondary risks · Secondary risk theory

6.1 Introduction

The greatest death tolls in human history have been caused by infectious diseases. For instance, the bubonic plague decimated more than 25% of the European population (Scott & Duncan 2001). At the time of the writing for this chapter, Covid-19 has caused more than 2.16 million deaths. Infectious diseases can be caused by a host of agents including viruses, bacteria, fungi, and parasites, and depending on the type,

C. L. Cummings (✉)
North Carolina State University, Raleigh, NC, USA

Iowa State University, Ames, IA, USA

S. Gopi · S. Rosenthal
Nanyang Technological University, Singapore, Singapore

D. M. Berube (ed.), *Pandemic Communication and Resilience*, Risk, Systems and Decisions, https://doi.org/10.1007/978-3-030-77344-1_6

can be passed from person to person or may be transmitted by insects or other animals or via ingestion of contaminated food or water or being exposed to organisms in the environment.

For millennia, societies around the world have sought to prevent infectious diseases before they can virulently spread. Preventative measures have taken many forms during human history and some persist today. Such measures have included theology-based prayer and sacrifice hoping to garner divine beneficence; pseudo-scientific treatments like nosegays (a small bundle of fragrant flowers thought to overpower the bad smells that "cause" disease); observation-based strategies like patient isolation and cremation of the dead; and widespread adoption of variolation and vaccination which contributed to improved community resilience. One facet of pandemic preparedness and resiliency planning is to anticipate that a significant portion of the population will not understand, or be willing, to adopt advocated risk mitigation responses and may believe that alternate forms of preventative behaviour, or no behaviour at all, are superior. Of all preventative treatments, public health experts recommend vaccination as the most effective mitigation strategy, noting that it has contributed a three percent decline annually in global deaths from preventable infectious disease (WHO 2018).

Vaccines train the human immune system to produce the antibodies specific to the target disease and provide first-line disease prevention and are celebrated as both highly efficacious and cost-effective among public health experts (ECDC n.d.). Unfortunately, the expert community's high valuation of vaccination is not met equally by the public. Thanks to vaccination efforts, mortality rates of several infectious diseases have plummeted (Facciolà et al. 2019), but global vaccine coverage is incomplete, and progress in some countries has slowed or even reversed. There is ample evidence that mass vaccination uptake reduces incidence and can reduce the virulent spread of disease due to increased herd immunity and, as a corollary, non-vaccination behavior may exacerbate the probability of disease outbreaks (Canning et al. 2005).

Sadly, immunization rates of preventable diseases remain low in many parts of the world. This signals a high-priority need to improve vaccine decision-making processes among the public to improve health literacies and promote optimal vaccination uptake around the world (Larson et al. 2014). While there are many logistical reasons to not vaccinate (e.g., access, cost, etc.) as well as some rare physical risks to those already significantly immunocompromised, there are also many individuals who are hesitant to vaccinate against infectious diseases.

Vaccine hesitancy is a decision-making outcome stemming from diverse motives and is often related to a lack of vaccine confidence and perceived risks. In the race to develop vaccines to mitigate pandemic risks, there is a need to understand factors influencing vaccine hesitancy. Secondary risk theory (SRT) is a useful framework to explain this, accounting for concerns about vaccine efficacy and safety (Cummings et al. 2020a). This chapter first unpacks the history of vaccine hesitancy and then reviews modern theoretical understanding of the concept. We then introduce SRTas a foundational theoretical framework to explain and predict vaccination decisions for use by public health officials, risk communicators and policymakers. The application of that theory can improve coordinated communication campaigns that play a central and vital role in creating salience and stoking motivations to adopt advocated risk mitigation responses to pandemics and other public health crises.

6.2 Vaccine Hesitancy: A History Lesson

Vaccine hesitancy is as old as vaccination itself and perhaps even older as concerns arose about vaccines during their conceptualization and development. The term 'vaccine hesitancy' refers to a spectrum of behavioral and psychological responses to vaccination ranging from outright refusal to delayed acceptance, and even post-vaccination concerns about the vaccine already taken (Salmon et al. 2015). The motivations and rationales for vaccination hesitancy attitudes and behaviors are diverse and stem from physical risk perceptions as well as values-based objections. It was seen after the discovery of the very first vaccination and appeared even earlier when European and American advocates attempted to introduce inoculation from Asia, the Middle East, and Africa, to Western societies (Gearon n.d.; McHugh 2020). In 1796, Edward Jenner achieved an experimental breakthrough by inserting pus from a cowpox sore (instead of smallpox matter) into a boy's arm, thus discovering a safer method of achieving immunisation than Eastern or African inoculation. Jenner published his results in 1798 after gathering more evidence and coining the term 'vaccine', but was widely ridiculed (BBC n.d.). He received religious opposition to the ungodly idea of injecting oneself with matter from a dead animal, and many perceived the idea of being told what was good for them as an imposition on their liberty (Watson 2019). The discovery was even mocked by the *Publications of the Anti-Vaccine Society* in Gillray's (1802) satirical comic featuring sprouting bovines from the appendages of the recently vaccinated as viewed in Fig. 6.1.

Nevertheless, with mounting evidence for the advantages of smallpox vaccination in England, the Vaccination Act of 1853 made it compulsory for babies to be vaccinated within three months of birth. Fines for noncompliance were imposed in subsequent legislation. The public, especially working-class citizens, reacted strongly to what was perceived as a draconian measure, leading to the formation of the Anti-Vaccination League and the Anti-Compulsory Vaccination League, as well as numerous anti-vaccination publications (History of Vaccines n.d., a). Issues of oppression arose as poorer citizens could not afford to pay fines for noncompliance. At the same time, with regards to the arm-to-arm vaccination technique then employed, poorer citizens did not have the means to verify the purity of the component materials used in the vaccines ("Conference on Vaccination" 1869).

Vaccination was a painful process and citizens were anxious about the technology and the possible spread of other conditions like syphilis, mental illness, or animal diseases (Durbach 2000). Further, "working-class anti-vaccinationists distinguished themselves from a lower stratum of uninformed, negligent, or lazy parents by identifying themselves instead as 'respectable' and 'conscientious' citizens" (Durbach

Fig. 6.1 "The COW-POCK_ or _ The Wonderful Effects of the New Inoculation!" (Gillray 1802) Smallpox and Inoculation Hospital (St. Pancras, London, England). Wellcome library no. 11755i

2000, p. 46), highlighting how a fear of erosion of class consciousness could induce opposition to vaccination.

The American counterpart, the Anti Vaccination Society of America, was founded in 1879 and two other leagues were established in the next six years. The city of Cambridge, Massachusetts instituted mandatory smallpox vaccination in 1902 and forcibly vaccinated workmen at the railroad yards, injuring public trust in the authorities (Parmet et al. 2005). Reverend Henning Jacobson refused vaccination; his lawyers argued that the state had overreached and that the mandate infringed on natural rights (Parmet et al. 2005). The case was taken to the Supreme Court, which ruled in the state's favour.

Meanwhile in Asia, the bubonic plague spread from China and broke out in Bombay, India in 1896, causing immense upheaval. The British government's initial response measures failed to comprehend the role of rats and rodent fleas in the transmission of the disease and were so heavy-handed that they were "resisted intensely and stubbornly by many communities and were mostly abandoned or modified before the scourge was controlled" (Klein 1988), but not before more than 12 million died in India. Identification of infected persons was highly delayed—most people died the day they were hospitalised, spreading fears that plague officers were killing people at hospitals (Klein 1988). A vaccine was developed but it was eyed with suspicion in this context of disease mismanagement, and as its use came with the occasional death, adverse effects, and mistaken doses.

Smallpox was a known disease across South Asia—Sanskrit texts dating from around 1500 BCE described a smallpox-like condition (History of Vaccines n.d., b). The indigenous practice of inoculation was established from at least the sixteenth century CE and involved "dipping a sharp iron needle into a smallpox pustule and then puncturing the skin repeatedly in a small circle, usually on the upper arm" (Boylston 2012, p. 312). Inoculation was ritualised, "practically a religious ceremony" (Arnold 2008, p. 33). Despite this history of inoculation acceptance, however, hesitancy emerged when vaccines reached India in the early nineteenth century and colonial officers were eager to replace inoculation with the new technique. A British civil servant went to the extent of falsely claiming that vaccination was mentioned in Sanskrit texts to boost its acceptance (Brimnes 2017). Because of its connection to the revered cow, the British expected quick uptake of the smallpox vaccine in India but, like elsewhere, were met with reluctance as vaccination methods in the nineteenth century were harmful, painful, and not yet proven to be better than indigenous variolation (Brimnes 2017). From the 1920s, vaccine rejection in India stemmed from several factors, including the pain caused to calves in the process of making the smallpox vaccine; the apparent ineffectiveness of vaccines; the interference of dubious medical techniques imported wholesale from the West; a fear of devaluation and potential extinction of centuries-old indigenous religious practices; displeasure that the new, contested Bacille Calmette-Guerin (BCG) vaccine was deployed on a mass scale only in India, like on guinea pigs, undermining India's newfound independence; and from the coercive methods employed by officials (Brimnes 2017).

In mid-1970s Britain, some doctors raised doubts about the safety of the pertussis (whooping cough) vaccine, stating that it could cause brain damage in young children. Members of the medical community spoke out against this claim, but despite this, public trust in this vaccine plummeted and vaccination rates fell from 78.5% in England and Wales in 1971, to just 37% in 1974 (Millward 2019). Vaccination rates would only rise again in the mid-1980s, after a sserious pertussis outbreak in the winter of 1978–1979 (Millward 2019).

Talwar et al. (1994) discussed an antipregnancy vaccine, in which the mention of tetanus toxoid used as a carrier protein was misinterpreted. A pro-life Catholic group consequently suggested that tetanus vaccines could cause sterilisation, resulting in rumours in Mexico, the Philippines, Tanzania, and Nicaragua that women were being used as guinea pigs to test a contraceptive, administered under the guise of a tetanus vaccine (Milstein et al. 1995). These false claims had a strong negative impact on vaccination programmes in these countries.

In 1998, former British doctor Andrew Wakefield and his co-authors published a study linking the measles, mumps, and rubella (MMR) vaccine with a predisposition to autism. The paper was widely publicised despite flawed methods, subsequent refutations, a complete retraction of the published article, and the discovery that Wakefield had used unethical methods and selectively published the data (Rao and Andrade 2011). Regardless of pro-vaccine advocacy, Wakefield's disinformation and fearmongering persists in the 21st Century, with added publicity from celebrity endorsements (such as from Robert De Niro) and films, setting off vaccine denials and a resurfacing of measles in the UK and the US (Quick and Larson 2018). Wakefield is now “frighteningly influential” (Boseley)—to anti-vaccination supporters, he has been victimised by medical establishments and drug companies who prioritise profits over people. “He and those around him now believe there is a massive conspiracy to force vaccines upon our children, driven and funded by the wealthy pharmaceutical companies” (Boseley 2018).

In France, the government initiated a hepatitis B vaccination campaign in 1994, targeted mainly at infants but including a 10-year, school catch-up vaccination programme, as well as outreach to high-risk groups. However, hepatitis B vaccines in schools were suspended in 1998 after reports surfaced of individuals suffering from multiple sclerosis after vaccination. This caused a significant loss of confidence in the vaccine among both parents and health professionals (Balinska 2009), affecting not just students but all target groups. Although most scientific studies found no relation between the vaccine and multiple sclerosis, hepatitis B-vaccinated individuals suffering from multiple sclerosis have been compensated since the late-1990s, and even 10 years after the school suspension, it was estimated that less than a third of French infants were vaccinated for hepatitis B (Balinska 2009).

In 1999, in a cautionary move to assure the US public about the safety of vaccines, manufacturers were asked to remove the ingredient thimerosal (a preservative containing ethyl mercury that has been used to prevent bacterial contamination) as soon as practical (Larson et al. 2011). However, poor communication about the rationale for removal led to decreased confidence and an increase in vaccine hesitancy (Goldstein et al. 2015). Some hospitals suspended use of the hepatitis B vaccine for all new-borns, regardless of their level of risk, resulting in the death of a three-month-old from infection (Offit, 2007). Parents of some autistic children were convinced that the autism had been caused by the mercury in vaccines and prepared to apply to the US National Vaccine Injury Compensation Program (Larson et al. 2011). Although several epidemiologic studies have disproved the link between thimerosal and autism, the “cottage industry of charlatans offering false hope” continues to thrive, providing mercury-chelating agents (used to treat mercury poisoning) to about 10,000 autistic children annually (Offit 2007, p. 1279).

A phase of the World Health Organization’s Global Polio Eradication Initiative was launched in 2003 in Africa. In three northern Nigerian states, leaders and citizens began a boycott of this polio vaccine. In a reversal of England’s early distrust of Turkish inoculation, the northern Nigerian opposition was based on fears that these vaccines from the “modern-day Hitlers” contained anti-fertility agents, HIV, and cancerous agents (Jegede 2007, p. 418), a perception exacerbated by America’s then-ongoing war with Iraq. Further, the aggressive, mass, door-to-door nature of the immunisation program created distrust in a region with limited access to healthcare or free medicine, and made some mothers hide their children under the beds (Murphy 2004). Prior tragedy from pharmaceutical company Pfizer’s 1996 trials in Nigeria also left a lingering atmosphere of distrust of western medicine. The boycott lasted 15 months, but the suspicions never left—in 2008, Nigeria accounted for 86% of all African polio cases (Tomori 2018).

During the H1N1 pandemic of 2009, there was considerable opposition to the use of the vaccine, based on claims that the vaccine approval was rushed, that it had higher levels of preservatives, that it would result in complications, and that the pharmaceutical industry was not to be trusted (Greenberg 2009). On the other hand, due to the delayed rolling out of the vaccine in the US, concern waned, and the disease was no longer seen as a serious risk that required vaccination (Jacobsen et al. 2015).

The Indian government suspended demonstration projects for the human papilloma virus in two Indian states in 2010, after mounting pressure from advocacy groups that raised concerns about the vulnerability of researched populations and the ethics of the demonstrations, the safety and efficacy of the vaccine in preventing cervical cancer, the interference of multinational companies, and the commercial interests of manufacturers (Larson et al. 2014; Ramanathan and Varghese 2010).

Ukraine has faced a measles epidemic since 2017, rising from a mistrust of vaccines. Parents have little confidence in the healthcare system and only 50% the population agrees that vaccines are effective (Gallup 2019). Their hesitancy towards vaccines also stemmed from concerns about the quality of ingredients used, fears that vaccination is merely a profit-making scheme involving pharmaceutical companies and governments, and experiences of ineffective vaccinations due to unreliable power supplies at hospitals that result in vaccines being unrefrigerated (Kelland and Polityuk 2019).

Influenza vaccine hesitancy in Singapore, a developed country with easy access to the vaccine, is fostered in part by misperceptions about the vaccine (Cummings et al. 2020b). These include concerns that the influenza vaccine is unsafe because alterations are made to it every year, displaying a lack of public knowledge of the mutability of the influenza virus. Other reasons for influenza vaccine hesitancy included influenza-related reasons such as a lack of familiarity with influenza, incorrect knowledge about the disease and its severity, and the inaccurate notion that having a strong immune system would protect one from contracting the disease. Apart from concerns about its safety, other vaccine-specific misperceptions included the impression that the influenza vaccine was only for people travelling out of Singapore, that it causes the disease itself, and that it is expensive.

Cummings and Kong (2019) also found that Singaporeans display greater hesitance to get vaccinated against influenza when this option is called a "flu shot," as compared to when the formal term "influenza vaccine" is used. This is in part because the use of the conversational term "flu" may cause the disease to be perceived as a low risk one, thus reducing motivations to engage in precautionary behavior. This hesitance stemming from complacency is mirrored is other countries and populations. A survey of undergraduate students majoring in public health in an American university found that only 43% had received the seasonal influenza vaccine (Rogers et al. 2018). Among the non-vaccinated students, close to 30% believed they were not at risk from influenza. A focus group study of Aboriginal healthcare workers who facilitate immunization of Aboriginal people also revealed that "substantial proportions (of participants) believed that influenza is not a serious disease and that natural immunity or other non-vaccine methods of protection were better" (Menzies et al. 2020, p. 282). A study of European Union members' influenza vaccination rates for the 2011–2012 and 2012–2013 seasons found that only three nations hit the 75%

target for older age groups (de Lusignan et al. 2016). In the UK, it is believed that people underestimate the severity of this disease (Lynch and Manning 2020). This influenza example and the recurring global complacency that emerges as a strong reason for not getting vaccinated stands in contrast to examples of other diseases that are perceived as more severe, where secondary risk perception presents itself as a strong factor for people displaying hesitancy. In the case of influenza, thanks to the perceived low severity of the primary risk it poses, it may be that people are less motivated to consider disease prevention measures and are thus less concerned about potential risks arising from the vaccine.

A survey of adults in the US found that around 20% would decline a hypothetical, FDA-approved Covid-19 vaccine, citing reasons that include its novelty compared to more familiar vaccines, which casts doubts on its safety and effectiveness, and distrust of the government (Thunstrom et al. 2020). Neumann-Böhme et al. (2020) found similar results in a survey of Europeans, where over 26% stated that they would not want to get vaccinated or were not sure, due to concerns over potential side effects, questionable safety, lack of studies on applicability to specific groups such as pregnant women, and fears stemming from conspiracy theories.

Rosenthal and Cummings (under review) found that among Americans who already plan to get vaccinated against Covid-19, rapid development and deployment of the vaccine would not be a dissuading factor. However, conspiracy beliefs about vaccines—for example, that the harm faced by immunised children has been covered up—were found to affect the perceived efficacy of vaccines, perceived risks associated with vaccines, as well as willingness to be vaccinated. A survey of residents in Turkey and the UK found that Covid-19 vaccine acceptance rates were lower among participants who, in line with conspiracy theories, believed that the origin of the disease was not natural or who were not sure of the origin of the disease, whereas a belief that the virus has a natural origin significantly increased the odds of accepting the vaccine (Salali and Uysal 2020).

Given such ample history of vaccine hesitant perceptions and behaviors, many scholars have sought to understand the behavior—theorizing how it manifests in order to proactively diffuse misguided perceptions and encourage vaccination uptake. The next section reviews this literature.

6.3 Vaccine Hesitancy

In 2015, the WHO guest-edited a special issue of the journal *Vaccine* in which the SAGE Working Group on Vaccine Hesitancy concluded that "vaccine hesitancy" refers to a "delay in acceptance or refusal of safe vaccines despite availability of vaccination services" ("Vaccine Hesitancy" 2015). This has been an ongoing issue: Andre et al. (2008) demonstrated that in the early twenty-first century, a search of the MEDLINE database produced five times as many hits for "vaccine risks" than for "vaccine benefits," i.e., negative features of vaccines were getting much more publicity. More recently, Facciolà et al. (2019, p. 13) reported that "approximately

1 in 8 children under 2 years old in the United States are under-vaccinated, due to parental choice, and a majority of physicians reports at least one vaccine refusal per month." Given that "effective control of vaccine-preventable diseases requires extremely high rates of timely vaccination" (Salmon et al. 2015, p. 69), vaccine hesitancy is a growing cause for concern. In 2019, the WHO highlighted vaccine hesitancy as an issue demanding attention as it can hinder or reverse the progress made in disease management via vaccination, one of the most cost-effective methods ("Ten Threats" n.d.).

While the definition above refers to people who either refuse vaccines outright, or are hesitant about vaccination, Salmon et al. (2015) noted that the spectrum of vaccine hesitancy extends to include people who proceed with all vaccines despite lingering concerns. Vaccine hesitancy can thus range from an individual who refuses all vaccines, to an individual who may delay or avoid specific vaccines, to an individual who accepts all recommended vaccines but remains worried about "adverse events associated with the vaccine" (Salmon et al. 2015, p. 66). The authors cautioned that people in this third group—the seemingly least hesitant—could be "particularly vulnerable to misinformation, with the potential of being swayed to delay or refuse future vaccines" (p. 67). This suggests that vaccine hesitancy must be examined in terms of a person's attitudes and beliefs, rather than merely based on the action of vaccinating.

Data from a WHO/UNICEF Joint Report Form showed that vaccine hesitancy was reported by over 90% of surveyed countries from 2015 to 2017 (Lane et al. 2018). Complicating the seriousness of this prevalence is the concern that discussing vaccine hesitancy might legitimise it and create a self-fulfilling prophecy, but Goldstein et al. (2015) argued otherwise. They presented evidence from the Global Polio Eradication Initiative showing that good communication that engages community leaders does indeed lead to greater vaccine acceptance. Understanding and discussing the causes and reasons for vaccine hesitancy could thus be a productive and crucial public health measure.

Ironically, the success of vaccination is itself a reason for vaccine hesitancy—by making once-prevalent diseases unfamiliar to current generations of young parents, the focus has shifted from diseases to vaccination risks. Larson et al. (2011) presented other factors that can cause vaccine hesitancy, including fears about safety and genetic predispositions to side effects, confusion over the choice of vaccines and changing vaccine schedules, research suggesting that vaccines are dangerous, as well as the proliferation of misinformation and outlier views that go viral on social media. Focusing on the ease with which information can spread on social media, Wilson and Wiysonge (2020, p. 2) cautioned that vaccine-hesitant groups have an "alarming" online footprint. They explained that misinformation can become viral despite its lack of credibility because, although unlikely, the results it claims sound horrifying. These negative results are incentivised, leading to "a spiral of threat matched by public fear" (p. 2). Adding to this is the fact that disinformation, or the intentional spread of misinformation, is rampant and "Russian bots and troll farms, in conjunction with Russia's foreign broadcast network RT formerly Russia

Today), have pushed anti-vaccination messages on a large scale on Western social media" (Wilson and Wiysonge 2020, p. 2) to undermine American influence in the world, regardless of the cost on public health. Their cross-national analysis of tweets from 2018 to 2019 showed that the use of social media to organise offline action is highly associated with an increase in public belief that vaccines are unsafe. Further, it showed that foreign-sourced disinformation on domestic social media in each country is associated with declines in mean vaccination rates. These findings suggest that to dispel vaccine hesitancy, rather than simply blame foreign interference, beliefs that vaccines are unsafe need to be addressed as well.

Salmon et al. (2015, p. 67) added that vaccine concerns may be amplified by "heuristics that impact perceptions of risk," for example, the compulsory nature of certain vaccines may cause parents to perceive vaccinations as having greater risk than if they were voluntary. Other factors that may heuristically lead to high-risk perceptions include the manmade nature of vaccines, in contrast to diseases that are perceived as natural and thus less risky, and the notion that falling ill after being vaccinated is riskier than getting infected from lack of vaccination. Piltch-Loeb and DiClemente (2020, p. 1) defined vaccine hesitancy as "a socio-behavioral and cultural phenomenon" and considered the logistic and perceptual factors to vaccination uptake including awareness of the disease and availability, accessibility, affordability, and acceptability of the vaccine. They emphasised that understanding the reasons behind anti-vaccine sentiments would be instrumental to altering risk perceptions. Others have proposed that knowledge about vaccine effectiveness and safety may not be able to easily override vaccine hesitancy once it has been generated (Okuhara et al. 2020). Risk perceptions are thus an undoubtedly crucial factor in bolstering a whole spectrum of hesitant attitudes and must be better understood.

Lane et al. (2018) demonstrated the prevalence of negative risk perception in their analysis of three years of data (2014–2016) from the WHO/UNICEF Joint Report Form—a standardised questionnaire sent to member states since 1998 to collate annual data on immunization. The authors examined the data for reasons cited for hesitancy and classified them according to the matrix of hesitancy determinants developed by the Strategic Advisory Group of Experts (SAGE) on Immunization. Over a third of the countries reported that the reasons they provided were assessment-based while the rest were opinion-based. The categorization matrix includes 23 determinants of hesitancy in three categories: contextual influences (such as influential leaders), individual and group influences (such as personal experience with vaccination), and vaccine /vaccination-specific influences (such as the mode of vaccine administration) (WHO 2014a). Lane et al. (2018) discovered that consistently from 2014 to 2016, the overall top reason cited for vaccine hesitancy globally was "risk–benefit (scientific evidence)", a determinant falling under the category of vaccine/vaccination-specific influences. The SAGE Working Group explained that "risk–benefit (scientific evidence)" refers to how "scientific evidence of risk/benefit and history of safety issues can prompt individuals to hesitate, even when safety issues have been clarified and/or addressed" (WHO 2014b). This was also the top cited reason across all three years in upper middle-income and high-income countries. The authors added a caveat that this top-cited reason only accounts for less than a quarter of the given responses, highlighting the multitudinous reasons for vaccine hesitancy.

6.4 Overcoming Vaccine Hesitancy: The Need for Secondary Risk Theory (SRT)

In examining people's motivation to engage in recommended protective behaviour, such as getting vaccinated against a disease, Cummings et al. (2020a) demonstrated the importance of factoring in secondary risk—the inherent or perceived hazard that may arise from adopting the recommended protective behavior. Their model extended protection motivation theory (Rogers 1975) to offer a more complete prediction of intentions to avoid risk. Whereas protection motivation theory explains behaviours in terms of perceived threat and perceived ability to cope with the threat, SRT uniquely accounts for perceived threats associated with a recommended coping mechanism. This model is useful in behavioural contexts where people weigh the benefits of avoiding a primary threat against the risks or perceived risks related to engaging in protective behaviour. The conceptual model is provided in Fig. 6.2.

Vaccination provides an ideal context for studying secondary risk because vaccines prevent disease and have concomitant real or perceived secondary risks. In some instances, the public may have legitimate concerns about secondary risks, such as when there is scant evidence surrounding a new and controversial medical technique mandated for public uptake ("Conference on Vaccination" 1869), when medical professionals make claims about the dangers of vaccines—despite those claims being subsequently refuted (Millward 2019; Rao and Andrade 2011), or when improperly stored vaccines are implicated in the deaths of dozens of children (Yang et al. 2020). In contrast, other perceptions of secondary risk may have no evidentiary basis, such as when members of the public are worried about additives like mercury and aluminium (McCarthy 2019) or when they believe vaccination efforts are a cover story for systematic sterilisation of local populations by foreign governments (Milstein et al. 1995; Jegede 2007). Whether individuals hold reasonable or unreasonable perceptions of secondary risks, vaccine hesitancy often comes about through a personal deliberative process. Individuals have reasons for avoiding vaccines and influencing them away from hesitancy requires respect as a starting point for dialogue (Velan 2016). By the same token, it is critical to understand public beliefs about secondary risks, which can inform strategic communication efforts to address public vaccine hesitancy (Cummings et al. 2020a).

SRT conceptualizes secondary risk perceptions in two parts. The first part regards the perceived severity of the secondary risk. Perceived severity refers to the noxiousness of an event. In the case of vaccines, the perceived harm of a experiencing a side-effect would correspond with the perceived secondary risk severity. The

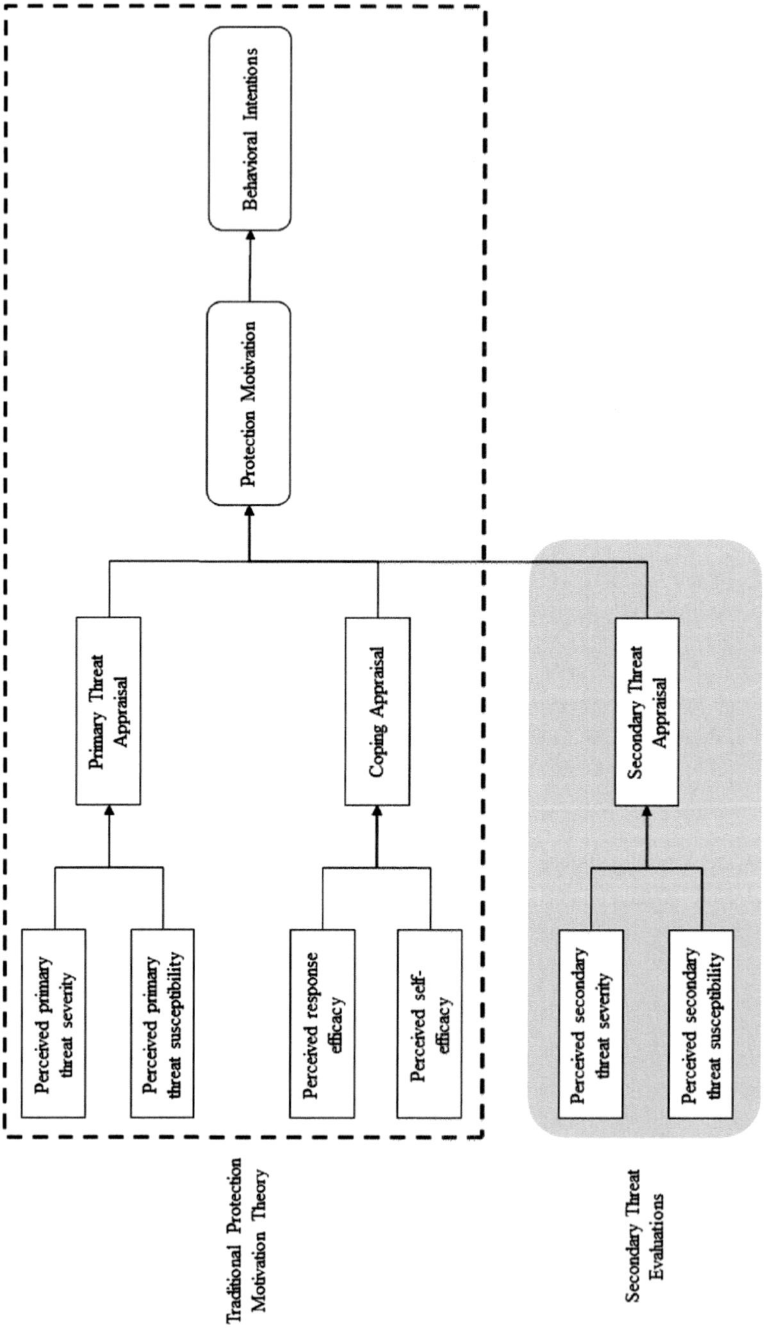

Fig. 6.2 Secondary risk theory

second part regards the perceived susceptibility to experiencing the secondary risk. In parallel with the modelling of primary threat appraisal in protection motivation theory (Rogers 1975), these two parts are multiplicative and create an assessment of secondary threat. Secondary threat will be high only if there is high perceived severity and high perceived susceptibility. If individuals believe either the potential side effects are mild or they are unlikely to experience any side effects, then the secondary threat will be low.

Rosenthal and Cummings (under review) used SRT to explain vaccine hesitancy. They conducted an online between-subjects experiment in which 216 American adults read one of three statements about a hypothetical FDA-approved Covid-19 vaccine. The statements were identical except how soon the vaccines would be released, which included "next week," "in one year," and "in two years." Among several dependent measures were perceived secondary risk and vaccination willingness. The measure of perceived secondary risk included items measuring both the perceived severity of and susceptibility to vaccine side-effects. This conflation was the result of a factor analysis failing to differentiate the two dimensions of secondary threat appraisal. The results showed perceived secondary risk of the "next week" option was higher than of the "one-year" option ($p = 0.01$) and debatably higher than that of the "two-year" option ($p = 0.09$). Despite the treatment effect on perceived secondary risk, vaccination willingness was not different among the conditions. This finding is in line with prior research differentiating attitudinal and behavioural vaccine hesitancy (Salmon et al. 2015). It suggests that, although individuals may have safety concerns about a hastily developed vaccine, they may opt to receive it, nonetheless. This does not mean individuals feel comfortable about receiving it, and they may ultimately begrudge the act. If the only goal is to get individuals to take the vaccine, then a bit of begrudging is tolerable. But if there is also a goal to address and reduce vaccine hesitancy, then there must be strong efforts to demonstrate both the efficacy and safety of vaccines. This is especially true in instances like the Covid-19 vaccine, where the rapid development creates special safety concerns (Jiang 2020).

Overcoming vaccine hesitancy will remain a primary public health challenge in the twenty-first century as societies around the world continue to seek means to improve their public health infrastructure and resiliency to disease. Research applications of SRT can improve community preparedness through a more robust understanding of the extent and cause for vaccine hesitancy. From this foundation, policy makers and government agencies, practitioners, and health campaigners can better prioritize communication and interventions regarding the expected utility of vaccines within target audiences to prevent disease. This robust and granular accounting of disease threat perceptions as well as perceived secondary risks posed by taking a vaccine can help predict and promote protection motivated behavior. Public health campaigns that do not account for secondary risks of vaccines are likely to put substantial time and resources into promoting vaccination only to see sustained rejection of that behavior due to secondary risk perceptions. Use of this theory at a formative level can provide a stronger basis for strategic planning and can facilitate better informed interventions and communication campaigns to promote vaccine uptake.

References

Andre FE, Booy R, Bock HL, Clemens J, Datta SK, John TJ, Lee BW, Lolekha S, Peltola H, Ruff TA, Santosham M, Schmitt HJ (2008) Vaccination greatly reduces disease, disability, death and inequity worldwide. Bull World Health Organ 86(2). https://www.who.int/bulletin/volumes/86/2/07-040089/en/

Arnold D (2008) Epidemic smallpox in India. Hist Speak 9(7):31–33. https://doi.org/10.1353/hsp.2008.0019

Balinska MA (2009) Hepatitis B vaccination and French society ten years after the suspension of the vaccination campaign: how should we raise infant immunization coverage rates? J Clin Virol 46(3):202–205. https://doi.org/10.1016/j.jcv.2009.07.024

BBC (n.d.) Edward Jenner (1749–1823). http://www.bbc.co.uk/history/historic_figures/jenner_edward.shtml

Boseley S (2018, July 18) How disgraced anti-vaxxer Andrew Wakefield was embraced by Trump's America. Guardian. https://www.theguardian.com/society/2018/jul/18/how-disgraced-anti-vaxxer-andrew-wakefield-was-embraced-by-trumps-america

Boylston A (2012) The origins of inoculation. J R Soc Med 105(7):309–313. https://doi.org/10.1258/jrsm.2012.12k044

Brimnes N (2017) Fallacy, sacrilege, betrayal and conspiracy: the cultural construction of opposition to immunisation in India. In Holmberg C, Blume S, Greenough P (eds) The politics of vaccination: a global history. Manchester University Press, Manchester, pp 51–76. Retrieved 23 Oct 2020, from http://www.jstor.org/stable/j.ctt1wn0s1m.8

Canning HS, Phillips J, Allsup S (2005) Health care worker beliefs about influenza vaccine and reasons for non-vaccination–a cross-sectional survey. J Clin Nurs 14(8):922–925. https://doi.org/10.1111/j.1365-2702.2005.01190.x. PMID: 16102143

Cummings CL, Kong WY (2019) "Influenza" versus "flu": do different medical terms affect vaccination intention? J Health Commun 24(4):456–460. https://doi.org/10.1080/10810730.2019.1630527

Cummings CL, Rosenthal S, Kong WY (2020) Secondary risk theory: validation of a novel model of protection motivation. Risk Anal. https://doi.org/10.1111/risa.13573

Cummings CL, Kong WY, Orminski J (2020) A typology of beliefs and misperceptions about the influenza disease and vaccine among older adults in Singapore. PLoS ONE 15(5):e0232472. https://doi.org/10.1371/journal.pone.0232472

de Lusignan S, Correa A, Ellis J, Pebody R (2016) Influenza vaccination: in the UK and across Europe. Br J Gen Pract J R College Gen Pract 66(650):452–453. https://doi.org/10.3399/bjgp16X686677

Durbach N (2000) 'They might as well brand us': Working-class resistance to compulsory vaccination in Victorian England. Soc Hist Med J Soc Soc Hist Med 13(1):45–63. https://doi.org/10.1093/shm/13.1.45

European Centre for Disease Prevention and Control (ECDC) (n.d.) Immunisation and vaccines. http://ecdc.europa.eu/en/healthtopics/immunisation/Pages/index.aspx.

Facciolà A, Visalli G, Orlando A, Bertuccio MP, Spataro P, Squeri R, Picerno I, Di Pietro A (2019) Vaccine hesitancy: an overview on parents' opinions about vaccination and possible reasons of vaccine refusal. J Public Health Res 8(1):1436. https://doi.org/10.4081/jphr.2019.1436

Gallup (2019) Wellcome global monitor – first wave findings. https://wellcome.org/sites/default/files/wellcome-global-monitor-2018.pdf

Gearon E (n.d.) Who was Lady Mary Wortley Montagu? National Trust, UK. https://www.nationaltrust.org.uk/features/who-was-lady-mary-wortley-montagu

Gillray J (1802) Smallpox and Inoculation Hospital (St. Pancras, London, England). Wellcome Library no. 11755i

Goldstein S, MacDonald NE, Guirguis S, The SAGE Working Group on Vaccine Hesitancy (2015) Health communication and vaccine hesitancy. Vaccine 33(34):4212–4214. https://doi.org/10.1016/j.vaccine.2015.04.042

Greenberg TM (2009, November 1) What's scary about the H1N1 vaccine? Psychol Today. https://www.psychologytoday.com/sg/blog/21st-century-aging/200911/whats-scary-about-the-h1n1-vaccine

"Immunization coverage" (2020, July 15) World Health Organization. https://www.who.int/news-room/fact-sheets/detail/immunization-coverage

Jacobson RM, St. Sauver JL, Finney Rutten LJ (2015) Vaccine hesitancy. Mayo Clinic Proce 90(11):1562–1568. https://doi.org/10.1016/j.mayocp.2015.09.006

Jegede AS (2007) What led to the Nigerian boycott of the polio vaccination campaign? PLoS Med 4(3):e73. https://doi.org/10.1371/journal.pmed.0040073

Jiang S (2020, 16 March) Don't rush to deploy COVID-19 vaccines and drugs without sufficient safety guarantees. Nature. https://www.nature.com/articles/d41586-020-00751-9

Kedl RM, Seder R (2017) Editorial overview: vaccines. Curr Opin Immunol 47:A1–A2. https://doi.org/10.1016/j.coi.2017.09.001

Kelland K, Polityuk P (2019, November 5) Measles and mistrust in Ukraine weaken world's defenses. Reuters. https://www.reuters.com/article/us-health-measles-ukraine-insight-idUSKBN1XE15T

Klein I (1988) Plague, policy and popular unrest in British India. Modern Asian Studi 22(4):723–755. http://www.jstor.org/stable/312523

Lane S, MacDonald NE, Marti M, Dumolard L (2018) Vaccine hesitancy around the globe: analysis of three years of WHO/UNICEF joint reporting form data-2015–2017. Vaccine 36(26):3861–3867. https://doi.org/10.1016/j.vaccine.2018.03.063

Larson HJ, Jarrett C, Eckersberger E, Smith DM, Paterson P (2014) Understanding vaccine hesitancy around vaccines and vaccination from a global perspective: a systematic review of published literature, 2007–2012. Vaccine 32(19):2150–2159. https://doi.org/10.1016/j.vaccine.2014.01.081

Larson HJ, Cooper LZ, Eskola J, Katz SL, Ratzan S (2011) Addressing the vaccine confidence gap. The Lancet (br Ed) 378(9790):526–535. https://doi.org/10.1016/S0140-6736(11)60678-8

Lynch P, Manning J (2020, August 27) UK flu jab rates prompt complacency warning. BBC. https://www.bbc.com/news/uk-53889184

McCarthy C (2019, August 2) Why do parents worry about vaccines? Harvard Health Publishing. https://www.health.harvard.edu/blog/why-do-parents-worry-about-vaccines-2019080217406

McHugh J (2020, March 8) A Puritan minister incited fury by pushing inoculation against a smallpox epidemic. Washington Post. https://www.washingtonpost.com/history/2020/03/07/smallpox-coronavirus-antivaxxers-cotton-mather/

Menzies R, Aqel J, Abdi I, Joseph T, Seale H, Nathan S (2020) Why is influenza vaccine uptake so low among aboriginal adults? Aust N Z J Public Health 44(4):279–283. https://doi.org/10.1111/1753-6405.13004

Millward G (2019) Pertussis. In: Vaccinating Britain: Mass vaccination and the public since the Second World War. [Internet]. Manchester University Press, Manchester (UK). Available from: https://www.ncbi.nlm.nih.gov/books/NBK545994/

Milstein J, Griffin PD, Lee J-W (1995) Damage to immunisation programmes from misinformation on contraceptive vaccines. Reprod Health Matters 3(6):24–28. https://doi.org/10.1016/0968-8080(95)90155-8

Murphy J (2004, January 4) Distrust of U.S. foils effort to stop crippling disease. Baltimore Sun. https://www.baltimoresun.com/news/bal-polio0104-story.html?ctrack=1&cset=true.

Neumann-Böhme S, Varghese NE, Sabat I, Barros PP, Brouwer W, van Exel J, Schreyögg J, Stargardt T (2020) Once we have it, will we use it? A European survey on willingness to be vaccinated against COVID-19. Eur J Health Econ. https://doi.org/10.1007/s10198-020-01208-6

Offit PA (2007) Thimerosal and vaccines—a cautionary tale. N Engl J Med 357:1278–1279. https://doi.org/10.1056/NEJMp078187

Okuhara T, Ishikawa H, Okada H, Ueno H, Kiuchi T (2020) Dual-process theories to counter the anti-vaccination movement. Prev Med Rep 20:101205–101205. https://doi.org/10.1016/j.pmedr.2020.101205

Parmet WE, Goodman RA, Farber A (2005) Individual rights versus the public's health—100 years after Jacobson v. Massachusetts. New England J Med 352(7):652–654. https://doi.org/10.1056/NEJMp048209
Piltch-Loeb R, DiClemente R (2020) The vaccine uptake continuum: applying social science theory to shift vaccine hesitancy. Vaccines (basel) 8(1):76. https://doi.org/10.3390/vaccines8010076
Quick JD, Larson H (2018) The vaccine-autism myth started 20 years ago. Here's why it still endures today. Time. https://time.com/5175704/andrew-wakefield-vaccine-autism/
Ramanathan M, Varghese J (2010) The HPV vaccine demonstration projects: we should wait, watch and learn. Indian J Med Ethics 7(1):43. https://doi.org/10.20529/IJME.2010.017
Rao TS, Andrade C (2011) The MMR vaccine and autism: sensation, refutation, retraction, and fraud. Indian J Psychiatry 53(2):95–96. https://doi.org/10.4103/0019-5545.82529
Rogers R (1975) Protection motivation theory of fear appeals and attitude-change. J Psychol 91(1):93–114. https://doi.org/10.1080/00223980.1975.9915803
Rogers CJ, Bahr KO, Benjamin SM (2018) Attitudes and barriers associated with seasonal influenza vaccination uptake among public health students: a cross-sectional study. BMC Public Health 18:1131. https://doi.org/10.1186/s12889-018-6041-1
Rosenthal S, Cummings C (under review) Influence of rapid COVID-19 vaccine development on vaccine hesitancy
Salali G, Uysal M (2020) COVID-19 vaccine hesitancy is associated with beliefs on the origin of the novel coronavirus in the UK and Turkey. Psychol Med, pp 1–3. doi:https://doi.org/10.1017/S0033291720004067
Salmon DA, Dudley MZ, Glanz JM, Omer SB (2015) Vaccine hesitancy: causes, consequences, and a call to action. Am J Prev Med 49(6 Suppl 4):S391–S398. https://doi.org/10.1016/j.amepre.2015.06.009
Scott S, Duncan CJ (2001) Biology of plagues: Evidence from historical populations. Cambridge University Press
Talwar GP, Singh O, Pal R, Chatterjee N, Sahai P, Dhall K, Kaur J, Das SK, Suri S, Buckshee K (1994) A vaccine that prevents pregnancy in women. Proc Natl Acad Sci USA 91(18):8532–8536. https://doi.org/10.1073/pnas.91.18.8532
"Ten threats to global health in 2019" (n.d.) World Health Organization. https://www.who.int/news-room/spotlight/ten-threats-to-global-health-in-2019
"The conference on vaccination" (1869, July 31) Western mail. Cardiff, Wales. Issue 79
The History of Vaccines (n.d., a) History of anti-vaccination movements. https://www.historyofvaccines.org/content/articles/history-anti-vaccination-movements#Source%203
The History of Vaccines (n.d., b) History of smallpox. https://www.historyofvaccines.org/timeline/smallpox
The World Health Organization (WHO) (2014a, November 12) Report of the SAGE working group on vaccine hesitancy. https://www-who.int/immunization/sage/meetings/2014/october/SAGE_working_group_revised_report_vaccine_hesitancy.pdf
The World Health Organization (WHO) (2014b, October 1) Appendices to the report of the SAGE working group on vaccine hesitancy. https://www.who.int/immunization/sage/meetings/2014/october/2_SAGE_Appendicies_Background_final.pdf
The World Health Organization (WHO) (2018) The power of vaccines: still not fully utilized. http://www.who.int/publications/10-year-review/vaccines/en/
Thunstrom L, Ashworth M, Finnoff D, Newbold S (2020) Hesitancy towards a COVID-19 vaccine and prospects for herd immunity. SSRN (pre-publication). https://doi.org/10.2139/ssrn.3593098
Tomori O (2018, January 26) Nigerian vaccination campaigns may be threatened by new round of rumours. The Conversation. https://theconversation.com/nigerian-vaccination-campaigns-may-be-threatened-by-new-round-of-rumours-90443
"Vaccines and diseases" (n.d.) World Health Organization. https://www.who.int/teams/immunization-vaccines-and-biologicals/diseases

"Vaccine hesitancy: A growing challenge for immunization programmes" (2015, August 18) World Health Organization. https://www.who.int/news/item/18-08-2015-vaccine-hesitancy-a-growing-challenge-for-immunization-programmes

"Vaccines and immunization" (n.d.) World Health Organization. https://www.who.int/health-topics/vaccines-and-immunization

Velan B (2016) Vaccine hesitancy as self-determination: an Israeli perspective. Israel J Health Policy Res 5(1):13–13. https://doi.org/10.1186/s13584-016-0071-x

Vukovic V, Lillini R, Lupi S, Fortunato F, Cicconi M, Matteo G, Arata L, Amicizia D, Boccalini S, Bechini A, Prato R, Stefanati A, Panatto D, de Waure C (2020) Identifying people at risk for influenza with low vaccine uptake based on deprivation status: a systematic review. Eur J Pub Health 30(1):132–141. https://doi.org/10.1093/eurpub/cky264

Watson G (2019, December 28) The anti-vaccination movement that gripped Victorian England. BBC. https://www.bbc.com/news/uk-england-leicestershire-50713991

Wilson SL, Wiysonge C (2020) Social media and vaccine hesitancy. BMJ Glob Health 5(10):e004206. https://doi.org/10.1136/bmjgh-2020-004206

Xiao X, Wong RM (2020) Vaccine hesitancy and perceived behavioral control: a meta-analysis. Vaccine 38(33):5131–5138. https://doi.org/10.1016/j.vaccine.2020.04.076

Yang R, Penders B, Horstman K (2020) Addressing vaccine hesitancy in China: a scoping review of Chinese scholarship. Vaccines (basel) 8(1):2. https://doi.org/10.3390/vaccines8010002

Chapter 7
COVID and Cuomo: Using the CERC Model to Evaluate Strategic Uses of Twitter on Pandemic Communications

Aisha Powell

Abstract Social media has presented itself as an essential tool for messaging and communications during health crises, with multiple sources circulating information to the public to steer the masses into a particular direction. The Crisis and Emergency Risk Communication Model (CERC) (Reynolds and Seeger, J Health Commun 10:43–55, 2005) has been used to evaluate messaging on social media platforms for effectiveness, as well as providing guidelines on how to arrange pertinent information online. This study will look at Twitter messaging from NYS Governor, Andrew Cuomo, during the early stages of the COVID-19 pandemic, particularly looking at the different types of information he gave during five different stages of the pandemic. A total of 406 COVID-related tweets from his official Twitter were collected from January 24, 2020, to April 12, 2020, and thematically analyzed. Tweets were evaluated using the CERC guidelines for relevance, pertinence and overall effectiveness, to "prevent further illness, injury, or death; restores or maintain calm; and engender confidence in the operational response" (Reynolds, J Appl Commun Res 34:249–252, 2006, p. 249). Governor Cuomo followed the majority of the CERC guidelines during different stages of the pandemic but fell the shortest during the initial event phase. Other information like content reactions, shares and likes showed that he received the most engagement during the maintenance phase, but especially when the tweet signified "hope" or reassurance to the public. The implications of this study indicate that public officials can use social media sites, like Twitter, to successfully inform the public in the midst of a modern health crisis. Hopeful messaging is especially well received and utilizing celebrities or well-known figures likeliness in health communications can help reach a broader audience. The use of Twitter for pandemic communications must be coupled and guided with strategic objectives throughout different times that help direct and comfort the public to minimize harm.

Keywords COVID-19 · Twitter · Social media · Health communication · Crisis communication · CERC · Pandemic messaging

---

A. Powell (✉)
Howard University, Washington, USA
e-mail: aisha.powell@bison.howard.edu

D. M. Berube (ed.), *Pandemic Communication and Resilience*, Risk, Systems and Decisions, https://doi.org/10.1007/978-3-030-77344-1_7

7.1 Introduction

And the International Emmy for masterful COVID-19 briefing goes to New York Governor Andrew Cuomo (Aratani 2020; Dwyer 2020). The prestigious Founders Award from the International Academy of Television Arts and Sciences has been given to individual's like Oprah Winfrey, Steven Spielberg, Shonda Rhimes and Al Gore, whose "work is recognized throughout the world, embodying the vision of the founders of The International Academy of Television Arts and Sciences, crossing cultural boundaries to touch our common humanity" (International Academy of Television Arts and Sciences n.d.).

Cuomo marks the first governor to receive the award and, an even more unprecedented, winning on the basis of pandemic-based briefings that he held throughout the COVID-19 pandemic.

The COVID-19 pandemic has taken the world by storm. The virus was first noted in late-December 2019 in Wuhan City, Hubei China (World Health Organization 2020a; b), with doctors reporting cases of pneumonia from an unknown source, marked by patients having a fever and symptoms similar to several respiratory diseases. By January 30, 2020, with over 9,000 cases worldwide and more than 200 people dead in China, the outbreak was declared a public emergency by WHO Director-General Dr. Tedros Adhanom Ghebreyesu (Gayle et al. 2020; WHO 2020a, b). In the next weeks to follow, cases were reported in almost every continent, including countries like Iran, Italy, Brazil, France, Germany, Australia and Nigeria (Taylor 2020). On March 1, 2020, former United States President Donald Trump declared the COVID-19 outbreak a national emergency and alerted the public that there were local and federal precautions being taken to slow down the spread of the virus (Trump 2020). Less than a month later, the United States became the new epicenter of the virus with more than 500,000 cases and almost 20,000 deaths (Hernandez et al. 2020; Taylor 2020)–nearly half of both were from New York.

The outbreak comes at a time when new media technologies are woven in the daily communications of everyday life (Bromley and Bowles 1995; Huang et al. 2009; Klein and Ford 2003) and where social media has emerged as an optimal platform for receiving news (Gil de Zúñiga et al. 2012; Hermida et al. 2012; Lee and Ma 2012). According to Pew Research data, social media outpaces print newspapers as the preferred medium for American adults to get their news (Shearer 2018) and 72% of adult internet users searched online for health information (Fox 2014; Sherer and Greico 2019). One of the sites that have been studied for news purposes is Twitter. With more than 330 million users worldwide, Twitter has been studied as one of the most prevalent platforms for disseminating and discussing breaking news (Bruns and Burgess 2012), getting information to reach a wide range of audiences (Ju et al. 2013; Kwak et al. 2010) and is used by authoritative figures for communication purposes (Herrera and Requejo 2012).

While the majority of Twitter users are between the ages of 18–29 and Democrat-leaning (Wojcik and Hughes 2019), Twitter was utilized by the former highest official in the United States, Donald Trump, to promulgate information about policies, issues and decisions in real-time. Twitter has also emerged as platform for major health organizations to circulate health-related news, increase public engagement with heath content and to connect to patients (Park et al. 2013, 2015).

Using New York as a single case, this preliminary study examines how Twitter was used by New York Governor Andrew Cuomo to circulate information to New Yorkers and the general public during the different stages of the pandemic. On March 7, 2020, Cuomo announced a state of emergency in New York when there were a total of 76 cases (New York State 2020b).

By March 20, 2020, New York had been declared the United States' outbreak epicenter, with 15,000 people testing positive for COVID (Schumaker 2020). New York was chosen as the area of focus for this study because of its previous epicenter status, and because of Cuomo's record-breaking Emmy win. As defined by Creswell and Poth (2018), a case study is a qualitative research approach that involves thoroughly investigating a contemporary phenomenon by analyzing various forms of data.

Specifically using a critical case rationale, this study will "confirm, challenge or extend" (Yin 2018) the Crisis and Emergency Risk Communication (CERC) model (CDC 2018) as it applies to crisis communication during COVID-19.

7.2 Theoretical Framework

The CERC model was designed by researchers at the United States Center for Disease Control and Prevention (CDC 2018) as a guide for entities looking to spread health-related information on social media (Houston et al. 2014; Reynolds and Seeger 2005). Disaster social media use typically reviews speakers—such as individuals, health authorities, organizations or community leaders—and analyzes the types of information they give during different stages of a disaster (Houston et al. 2014). CERC does the same thing but is specifically designed to analyze and evaluate health communications, with an emphasis on how health authorities use social media (Lwin et al. 2018; Reynolds and Seeger 2005). The CERC model breaks down health communications into five stages: the pre-crisis, initial event, maintenance, revolutions and evaluation. According to the CERC model for best crisis communication practices, stage one criteria should focus on precautions, risk management and overall education to the public; stage two should center information around reduction, reassurance and rapid updates; stage three echoes that of stage two with a particular emphasis on support and positive framing; stage four focuses on updates about the resolutions and new risk; and stage five is an internal reflection where agencies evaluate their responses, what they learned and what can be improved (Reynolds and Seeger 2005). Box 1. shows a working CERC model as guided by Reynolds and Seeger (2005).

Box 1. Working CERC model (Reynolds and Seeger 2005).

I. Precrisis (Risk Messages; Warnings; Preparations) Communication and education campaigns targeted to both the public and the response community to facilitate:

- Monitoring and recognition of emerging risks.
- General public understanding of risk.
- Public preparation for the possibility of an adverse event.
- Changes in behavior to reduce the likelihood of harm (self-efficacy).
- Specific warning messages regarding some eminent threat.
- Alliances and cooperation with agencies, organizations, and groups.
- Development of consensual recommendations by experts and first responders.
- Message development and testing for subsequent stages.

II. Initial Event (Uncertainty Reduction; Self-efficacy; Reassurance) Rapid communication to the general public and to affected groups seeking to establish:

- Empathy, reassurance, and reduction in emotional turmoil.
- Designated crisis/agency spokespersons and formal channels and methods of communication.
- General and broad-based understanding of the crisis circumstances, consequences, and anticipated outcomes based on available information.
- Reduction of crisis-related uncertainty.
- Specific understanding of emergency management and medical community responses.
- Understanding of self-efficacy and personal response activities (how/where to get more information).

III. Maintenance (Ongoing Uncertainty Reduction; Self-efficacy; Reassurance) Communication to the general public and to affected groups seeking to facilitate:

- More accurate public understandings of ongoing risks.
- Understanding of background factors and issues.
- Broad-based support and cooperation with response and recovery efforts.
- Feedback from affected publics and correction of any misunderstandings/rumors.
- Ongoing explanation and reiteration of self-efficacy and personal response activities (how/where to get more information) begun in Stage II.

- Informed decision making by the public based on understanding of risks/benefits.

IV. Resolution (Updates Regarding Resolution; Discussions about Cause and New Risks/New Understandings of Risk) Public communication and campaigns directed toward the general public and affected groups seeking to:

- Inform and persuade about ongoing clean-up, remediation, recovery, and rebuilding efforts.
- Facilitate broad-based, honest, and open discussion and resolution of issues regarding cause, blame, responsibility, and adequacy of response.
- Improve/create public understanding of new risks and new understandings of risk as well as new risk avoidance behaviors and response procedures.
- Promote the activities and capabilities of agencies and organizations to reinforce positive corporate identity and image.

V. Evaluation (Discussions of Adequacy of Response; Consensus About Lessons and New Understandings of Risks) Communication directed toward agencies and the response community to:

- Evaluate and assess responses, including communication effectiveness.
- Document, formalize, and communicate lessons learned.
- Determine specific actions to improve crisis communication and crisis response capability.
- Create linkages to precrisis activities (Stage I).

The goal of the model is to measure if a crisis and risk emergency communication effort achieve the goal to "prevent further illness, injury, or death; restores or maintain calm; and engender confidence in the operational response" (Reynolds 2006, p. 249). The model can help systematically evaluate past and present disaster health communications during its different stages and identify shortcomings or successes, while also helping officials plan their future health communication strategies online.

7.3 Health Crisis and Social Media

There have been many researchers who have looked at social media use during health crises. Walton et al. (2012) looked at the CDC's use of YouTube videos during the 2009 H1N1 epidemic to evaluate if the CDC executed its goal to "reduce transmission and illness severity, and provide information to help health care providers, public health officials and the public address the challenges posed by the new virus" (CDC 2019). The researchers analyzed impressions on social media, page views and video views as the basis of their argument that in times of crisis, government organizations should make resources available on non-traditional platforms, like YouTube. They concluded that the CDC was effective in their social media campaign and that health communications because it was developed "succinctly, appropriately and timely" (Walton et al. 2012). Guidry et al. (2017) also looked at how government organizations use social media during a health crisis, by exploring the use of Instagram and Twitter by the CDC, WHO and Doctors Without Borders during the Ebola outbreak. They found that social media messaging was most effective when the health organizations are well-known and established, and when they follow risk communications strategies like "solution-based messaging, incorporation of visual imagery, and acknowledgment of public fears and concerns" (Guidry et al. 2017).

In a study comparing traditional and social media in crisis communication, Liu et al. (2011) found that organizations benefit from using both old and new media to relay crisis responses and information. They conducted both in-person interviews and an online experiment to understand the different types of media outlets college students use to get health content. While they did find that participants aligned traditional forms of media as being more credible, they also found that where the information came, regardless of the platform, strongly influenced the students' perception of the credibility of it. Kim and Park (2017) corroborated this idea, in their investigation of crisis response strategies by organizations. They found that when crisis information was given by an established organization, a representative from that organization, or other highly influential people, the audience rated that information as more reliable and allowed it to influence their subsequent behaviors or actions. Contributing to scholarship about social media effectiveness, Graham et al. (2015) empirically investigated the social media use of 300 local government officials during crises in Florida and found a positive relationship between its use and the ability to control the emergency. These studies show us that during times of heightened risk or crisis, the public is not only looking to social media to understand what is happening, but they also look for dependable organizations to get trusted information. A similar premise is applied to the current study's examination of Cuomo's Twitter content.

7.3.1 Twitter and Health Communications

Twitter usage during widespread health crises over the past decade, has quickly made it a frequented tool to increase awareness about risk and rapidly update the public. Consequently, Twitter is used by most major health institutions. In an analysis of tweets from the American Heart Association, American Cancer Society, and American Diabetes Association, Park et al. (2015) found that although all the organization used Twitter, each lacked in different sectors relating to messaging, images and retweeting to engage their audiences more effectively. In particular, Twitter users were more likely to react to information that was "health action-based," but the organizations did not capitalize or use their action tweets for community building. The researchers recommend that regular use of Twitter can bolster an organizations reputation, but that they should implement strategic messaging to strengthen public receptiveness to their content (Park et al. 2015). Hagen et al. (2017) looked at how Twitter is used by different organizations and authorities by analyzing tweets during the 2015–2016 Zika virus epidemic. In their study, they focused on the quality of tweets related to Zika and how "twitterers" engage with the content by either refuting it, questioning it, or agreeing with it. The researchers found that politicians, public institutions and scientists can "facilitate the flow of accurate and vital information" just by being active on Twitter because they have the furthest reach and engagement on the platform (Hagen et al. 2017). They also concluded that all public-facing organizations benefit from having social media literacy to be able to relay correct information in times of crisis. Hart et al. (2017) also found that public health professionals and agencies have an immense power on Twitter and can use the platform to debunk myths, advocate for marginalized groups and spread the most pertinent health information at the greatest rate by actively using the platform. The current research will examine these underpinnings, but with the modern health crisis COVID-19.

7.3.2 CERC and Pandemics

The CERC model is a theoretical framework and practical crisis communication model that has been used specifically to study messaging during pandemics. Lwin et al. (2018) used the Crisis and Emergency Risk Communication (CERC) model to analyze the strategic uses of Facebook from Singapore-based health agencies during the Zika epidemic. They conducted a content analysis of Facebook posts by three different agencies during different phases of the Zika epidemic and evaluated it using the model's criteria. In this study, they noted that the organizations followed the CERC model and gave more risk, symptoms and warnings during the pre-outbreak phase; during the outbreak phases, updates were posted on disease case reports and cautions towards highly affected areas, which was coupled with calming messages about government support; and during the disease fade out stages, they focused on individuals' personal responsibility and future preventative measures (Lwin et al. 2018). Their study proved the CERC model as an efficient praxis for evaluating social media use during health crises and it subsequently helped to slow down the spread of Zika, while providing relevant and appropriate information to the public. From the understanding of using social media for health communications, Twitters role in health news, the important of the sender and the best practices identified by the literature, this study will answer these questions:

R1: What different types of information did Governor Cuomo post on Twitter throughout the COVID-19 pandemic as it relates to the CERC model?

R2: What kind of information during a crisis do Twitter users engage with the most?

R3: Was Governor Cuomo's use of Twitter effective in ceasing the spread of COVID-19 in New York?

7.4 Methodology and Research Design

To understand how COVID-19-related information was spread by Governor Cuomo on Twitter, a content analysis was conducted using tweets from his official Twitter account.

According to the official account, Cuomo has been on Twitter since December 2010, just one month before he assumed his seat as governor of New York (NYS 2011). Interactions to his original post, including likes, comments and retweets, will also be included in the evaluation. In addition, press releases and conference videos from the New York State website will be used to identify the timeline for the different phases of the crisis and to establish triangulation, corroborating the contents of his tweets.

The study will replicate Lwin et al. (2018) use of the CERC model, however, it will solely be focusing on Twitter posts from one account. Due to the CERC model of strategic communication during the pre-, active, and post- stages of health crises, tweets will be captured and then clustered into its appropriate category. It is important to note that the COVID-19 pandemic has yet to end in the United States. For the purpose of this preliminary study, the revolution and evaluation phase was determined by the first dip in COVID case numbers. The pre-crisis stage will consist of tweets starting from January 24, 2020, which was his first press release related to COVID-19 (NYS 2020a), to March 6, 2020, the day before the state of emergency was declared (NYS 2020b). The initial event will only include tweets from March 7, 2020, the day the state of emergency was declared (NYS 2020b). The maintenance stage will include tweets from March 8, 2020, to April 7, 2020. April 7 was chosen because that was the day before Cuomo announced that "we are flattening the curve" (NYS 2020c), indicating that the crisis was slowing down. The revolution phase will be represented with tweets from April 8, 2020, the day he said the curve was flattening, to April 11, 2020, giving a few days to ensure cases were still decreasing. The evaluation phase will use tweets from April 12, 2020, as it would offer reflexivity. The press release on April 12, 2020, was also analyzed as it would include reports of cases, deaths, policies and an overall summary of the current status of the disease that may have not been tweeted. Tweets were pulled manually by the researcher on November 1, 2020, and then pulled again on November 7, 2020, to ensure that all tweets were captured and to corroborate a final tweet count. Due to the content being pulled at different time frames, which led to different numbers of interactions per each post, the highest recorded number of interactions was used because it was the most recent. Tweets were captured if it is related to COVID-19 by either explicitly stating "COVID-19" or other indication of the virus in the tweet, which was determined at the researcher's discretion. During week two, the tweets were categorized accordingly into one of the five categories according to the date they were posted.

During week three, an open coding method was used to extract the most prevalent themes and ideas. Open coding helped to establish concepts or phrases that initially emerged from the data (Creswell and Poth 2018; Strauss and Corbin 1994). Next, the researcher utilized second-level coding, pulling concepts together, extracting umbrella themes and condensing the codes. Once the process was done, a final codebook was created for each category and then compared to the CERC model.

7.5 Findings and Discussion

Overall Governor Cuomo sent out a total of 406 COVID-related tweets over the four-month period. A total of 55 tweets were sent out during the pre-crisis phase; eight were sent on the day of the initial event, 305 during the maintenance phase, 32 during the resolution and six on the evaluation day. Cuomo sent out the least number of tweets during the initial event and evolution phase. His tweets accumulated 7,156,597 likes, 1,487,198 retweets and 216,450 comments, most of which were from the maintenance phase. While likes comments and retweets varied per stage, due to the number of tweets that were sent out, the evaluation phase received the largest number of interactions when accounting for tweet proportions, with a ratio of 31,100:1 for likes, 6156:1 for retweets and 726:1 for comments (Table 7.1).

7.5.1 Pre-crisis Stage

During the pre-crisis phase, Governor Cuomo's tweets fell into six categories: risk, precaution, preparedness, COVID-case updates, new policies and health updates, and reassurance. Almost identically following the CERC guidelines, Cuomo's tweets earlier in the year acknowledged the threat of COVID-19 in New York but reassured the public that the state government was prepared to handle the pandemic. Prior to the first case in New York, Cuomo repeatedly mentioned that most people were "low risk," there were no confirmed cases and that citizens should continue taking

Table 7.1 Tweet count and audience response per stage

Stage	Tweet count	Likes	Retweets	Comments
Pre-crisis	55	112.972	35,860	6175
Initial event	8	8888	5,25	1,89
Maintenance	305	6,079,237	1,249,097	180,375
Resolution	32	768,900	152,879	23,429
Evaluation	6	186,600	36,937	4361

precautions as they would for any seasonal cold. However, as time progressed, he introduced more policy and health updates specific to COVID-19, including monetary funds to ensure hospitals were equipped with the necessary tools. After the first case was announced on March 1, 2020, Cuomo's tweets shifted towards updating the public on COVID cases, partnerships with local and national health providers to support the public and best practices to avoid catching the virus. He also announced statewide procedures to reduce risk, including mandatory quarantine for students abroad, reimbursement of planned travel fees and urging for "#PaidSickLeave"—a hashtag he exclusively used several times. More tactics that Cuomo implemented was tweeting several risk precautions translated into Mandarin, with links to the New York coronavirus home page and numbers to the NYS Department of Health. Cuomo tweeted three photos, and one video (unrelated to his daily briefings) showing testing sites, proper elbow bump etiquette and general "happy" looking health workers. Starting on March 1, Cuomo also began his daily COVID-19 briefing in Albany, to which he shared the the link to the Periscope videos on Twitter. Overall engagement with individual tweets stayed in the 100 to 600 range, with likes, comments and retweets combined. The tweet with the most interaction was the March 1, 2020 announcement of the first case and the March 2, 2020 announcement that all health insurers would waive COVID-19 testing fees—both of which accumulated over 100,000 combined interactions. During this time, Cuomo repeatedly used the terms "new yorkers," "update," "coronavirus" and "symptoms" in his tweets (See Fig. 7.1).

Fig. 7.1 A word cloud of the top 50 words in tweets from Governor Cuomo during the pre-crisis phase. The larger the word, the higher the times it was used

7.5.2 Initial Event

Where Cuomo excelled in the pre-crisis stage by following the CERC guidelines exactly, he fell short in the initial event. With only seven tweets sent out on March 7, 2020, the themes were: COVID case number updates, precaution, reassurance and partnership efforts. Shortly after posting the declaration of the State of Emergency in New York State, three subsequent tweets the link to live press conference; a message about how the aforementioned declaration will better help local agencies "by getting them all the tools they need to contain the virus spread"—slightly falling into the CERC's designated "specific understanding of emergency management and media community responses" section (Reynolds and Seeger 2005); and another specifically stating how the declaration can help "expedited procurement, leasing of lab space, hiring, and more," to help local health departments. Cuomo tweeted two COVID case number updates, including the counties where the outbreaks were. In one of the tweets, Cuomo states that cases will be on the rise but that "we" can deal with the situation better. Cuomo reposted the video on proper elbow bump etiquette, that he sent during the pre-crisis stage. Lastly, he sent out another tweet about price gouging of sanitary items like hand sanitizer. Blatantly missing from his tweets from the CERC guidelines was empathetic tweets to reduce emotional turmoil, anticipated outcomes based on the available information, consequences of the circumstance, specific understanding of the medical community response, and generally designated crisis/agency spokesperson and/or formal method of communication (Reynolds and Seeger 2005). Instead, his tweets were brief with general links to the coronavirus homepage, ambiguous messaging about the broad ways on how the declaration of the public emergency will help and updates about the cases. The most substantial tweet he posted that was specific to the CERC model, was the one about price gouging, where he specifically indicated the problem and contact information for the agency in charge of overseeing the problem. The tweets that received the most engagement during this stage was the declaration of the emergency, which had 3400 retweets and 5000 likes. The second most interactive tweet was one on March 7 about COVID case numbers which received 3000 retweets and 3400 likes.

7.5.3 Maintenance Phase

The maintenance phase covered the longest duration of the sample and had the highest number of tweets. Cuomo heavily utilized certain criteria in the CERC guide during the maintenance phase. The main themes of his tweets during this stage were best practices/safety measures; reassurance and gratitude; public restrictions; medical-related content; public health content and case updates. The majority of information that Cuomo tweeted was medical-related content, that included needs of local health providers, solicitation of medical professionals and equipment, and updating the public on the needs that were fulfilled and by what entity. In this group of tweets,

Cuomo was explicit in the exact items that New York State needed, and even directly stating the entities which could provide relief for the needs. He stated things like "I'm calling on @CDCgov to authorize testing at private labs and automated testing to exponentially increase capacity" and "I'll work in full partnership with @POTUS on this—but I ask him to take it seriously," directly calling on the CDC and the former President to help alleviate the burden on New York. These tweets were followed up with fulfillment tweets when donations or needs were met, he specifically thanked airplane company Jet Blue, singer and entrepreneur Rihanna, beauty company Estee Lauder, the National Basketball League and other government entities as donations were made. He also sent a myriad of tweets towards medical professionals, medical students and retired people in medicine to help the state by working at hospitals and health facilities, along with links to surveys and applications. During this stage, a slew of other terms were regularly used like "social distancing," "ventilator" and "spread" amongst other words (See Fig. 7.2).

The next trio of content is reassurance, best practices and emotional-related tweets. Usually, these were given together in the fashion of a tweet like this one on March 26, 2020: We mourn the 385 New Yorkers we have lost to Coronavirus-related illnesses. Tragically, we expect the number to rise as many patients have been on ventilators for weeks. NY will keep fighting flat out to save lives. Help us save lives by staying home. #NewYorkTough.

These tweets received the most interactions in this stage, with one tweet about the indispensability of human life and a roll-out of a public health plan receiving

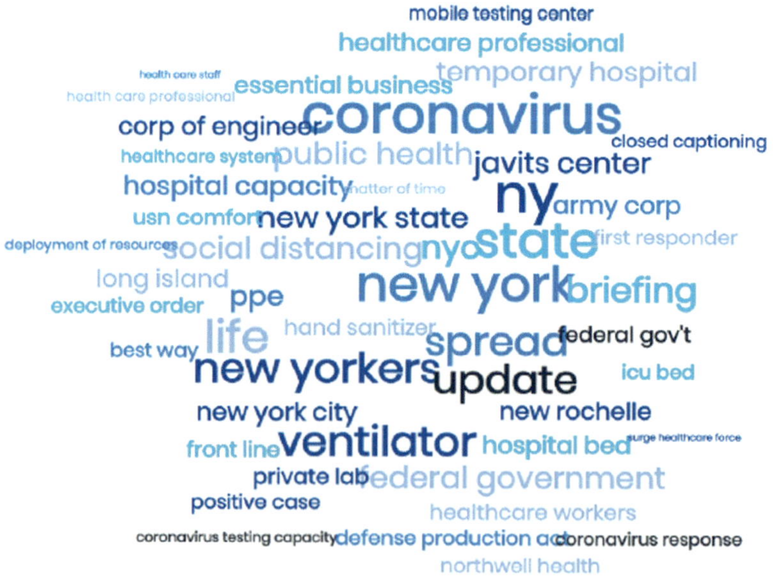

Fig. 7.2 A word cloud of the top 50 words in tweets from Governor Cuomo during the maintenance phase. The larger the word, the higher the times it was used

76,100 retweets, more than 277,000 likes and 8500 comments—nearly double the retweets of the second most interacted tweet during this phase and five times the number of likes. In general, sentimental or positive tweets about a hopeful end were highly circulated and liked by Twitter users more than any other category.

Cuomo also sent out a handful of mournful tweets about New Yorkers who died from COVID-19—almost all ending with a reassurance statement or positive outlook. These were also well circulated and received high engagement from Twitter users.

Throughout this phase, Cuomo utilized the same "best practices" messaging he used during phase one. Usually, it was as simple as "stay at home," other times it was more specific, like slowing down regular use of public services in New York like the subways, parks and grocery stores. He often tweeted restriction updates followed by public health content or coupled them together like this March 19 tweet: "NEW: Today we are mandating that 75% of the non-essential workforce MUST work from home. We are taking this action to further reduce density across the state to slow the spread of #Coronavirus." He almost never tweeted restriction tweets without indicating a public health aspect, reassurance, or a time frame for the restriction.

To a lesser degree, Cuomo tweeted some content that fell into the other CERC categories like understanding factors and issues, correcting misunderstandings and rumors, and feedback from the affected public. However, noticeably missing was the direct objection of misinformation. Instead, Cuomo gave general warnings about the harm of misinformation and told the public to be well-informed, without stating what the misinformation was. One of the most interesting aspects of this time was Cuomo's use of the celebrity to reiterate social distancing tactics, best practices and public health information. He utilized a wide array of celebrity talent like comedians Ben Stiller, Danny Devito; host and actress Lala Anthony; actor Alec Baldwin; and Broadway actress Krystal Joy Brown. All of these videos used the same strategy—an iPhone setup at home advocating for New Yorkers to stay at home to prevent the spread of COVID. All the videos were posted during a two-day span of March 22, 2020, to March 24, 2020, and received generally high engagement, in the thousands for likes, shares and comments.

7.5.4 Resolution Phase

The resolution phase tweets had the second-highest overall engagement from Twitter users, and was almost exclusively best practice content, indicating that New Yorkers should continue to take precautions like staying home and social distancing. On April 8, 2020 Governor Cuomo tweeted that New York's "dramatic actions" are "flattening the curve" followed by self-efficacy actions that New Yorkers can take that mirror the CDC guidelines for reducing the spread of COVID. He also tweeted simple and straight-forward post, like this follow-up tweet on April 8:

> We must keep the curve flat. Keep staying home. Keep practicing social distancing. Keep protecting others. Now is not the time to slack off.

Another theme he utilized was community efforts and public health interest, combining the two to connect the individuals gain of being cautios to statewide community benefits. Content like "New Yorkers will do what we have to do" and using the hashtag "#NewYorkStrong," signaled to the public that a collective effort is the only way to overcome the virus with everyone doing their part. These type of tweets was the most well-received and interactive tweets during this stage, with nearly half of the total interactions coming from tweets like this. Cuomo also continued to provide reassurance and policy/case updates throughout, and of course, his daily coronavirus briefings, which also had a high engagement. The most interacted tweet was an April 8, 2020, update on voting absentee due to COVID. This post had a seventh of the total number of likes, almost a fifth of the retweets and half of all the comments of the post during this stage. Due to the ongoing pandemic status, his tweets during this time differed from much of the tenets in the CERC model—which focus on recovery, having open discussions about the crisis and new understanding of risk—and instead, mirrored more of the content of the maintenance phase. This discrepancy can be addressed in a follow up study after the pandemic has ended to more accurately assess his content when the pandemic is in a resolution phase.

7.5.5 Evaluation

The tweets sent out during the evaluation phase had the most public interaction per tweet than any other phase, although this phase had the least number of tweets overall. Three of the six tweets were dedicated to links for the daily COVID-19 briefing and one commemorated the loss of the 758 New Yorkers who died from COVID-19 on April 11, 2020. The other three tweets were stand-alone tweets. One resonated in the thankful category, with Cuomo tweeting:

> This morning we are returning 35 ventilators lent by Pathways Nursing and Rehabilitation Center to fellow New Yorkers downstate. Their compassion is inspiring. We thank them and salute all the better angels among us.

This tweet also projected that a more hopeful future or resolve was near as previously needed equipment was now being returned.

The next tweet was generally positive that stated that even when things are bad, individuals will "be inspired beyond belief by the goodness of others." He ends the tweet by stating "love wins." While the ambiguity of this tweet is present, it was analyzed as a positive outlook tweet or in the hopeful category.

Lastly, the most interacted tweet was an urgent tweet directed towards the federal government for fair stimulus distribution. This tweet received 20,100 retweets, 78,600 likes and 2800 comments. In this tweet, Cuomo compared the federal stimulus money received by the states Montana and Nebraska, both of which received 25 times the amount of money per COVID case than New York and urged that: "we need a fair federal stimulus bill that is distributed by need." This tweet again highlighted the need for resources for New Yorkers specifically and touched on community cooperation by using "we." In the press release sent out on April 12, 2020, new policies were announced, like an executive order to conduct more antibody tests (NYS 2020d) and statewide efforts to receive more funding. Similar to the revolution phase, these tweets did not follow the CERC guidelines—discussion, consensus, evaluation—because the COVID-19 pandemic is still ongoing in the United States. Furthermore, the press release on this date did not offer any tenets of the evaluation phase either. Further investigation and analysis of the evaluation phase must be determined once the pandemic ends.

7.6 Conclusion

Social media has a multitude of uses in today's modern society, and one of those uses is spreading health-related content. Cuomo's use of Twitter during the COVID-19 pandemic showed that the CERC model can still be implemented in a contemporary health crisis and its criteria are generally well-received by the public. Particularly positive and hopeful reassuring messaging, which is a recommendation in three of the five CERC stages, is the most circulated, shared and interacted types of messaging for Twitter users who interacted with Governor Cuomo's tweets. Cuomo could havegarnered more interactions with the public during the initial phase if he better followed the CERC model and included positive messaging during that period. While Cuomo repeatedly mentioned the preparedness of the state for the pandemic early on, further tweets proved that the need for help was greater than expected. However, his tweets still remained steadfast and optimistic and kept audience reassured, as the model recommends. These preliminary findings indicate that more positive framing and explicitly debunking of false infromation could have been used by Governor Cuomo to better assist New Yorkers, and would have potentially garnered greater engagement. However, his messaging was instrumental and helpful in reducing the spread of COVID-19 and informing the public, as seen in the dip in COVID-19 cases. A subsequent study, focusing on the resolution and evaluation period needs to be conducted to reflect his utilization of Twitter accurately and appropriately.

References

Aratani L (2020, Nov 20) Andrew Cuomo to receive Emmy award for televised covid briefings. The Guardian. https://www.theguardian.com/us-news/2020/nov/20/andrew-cuomo-emmy-award-coronavirus-briefings

Austin L, Fisher Liu B, Jin Y (2012) How audiences seek out crisis information: exploring the social-mediated crisis communication model. Jour of Appl Comm Res 40(2):188–207. https://doi.org/10.1080/00909882.2012.654498

Bromley RV, Bowles D (1995) Impact of internet on use of traditional news media. Newsp Res J 16(2):14–27. https://doi.org/10.1177/073953299501600203

Bruns A, Burgess J (2012) Researching news discussion on twitter. J Stud 13(5–6):801–814. https://doi.org/10.1080/1461670x.2012.664428
CDC (2018) Crisis & emergency risk communication (CERC)|CDC. Cdc.Gov. https://emergency.cdc.gov/cerc/
CDC (2019) CDC 2009 H1N1 Flu. Center for Disease control and prevention. https://www.cdc.gov/h1n1flu/
CDC (2020, Jan 17) Public health screening to begin at 3 U.S. Airports for 2019 novel coronavirus ("2019-nCoV"). Center for Disease Control and Prevention. https://www.cdc.gov/media/releases/2020/p0117-coronavirus-screening.html.
Creswell JW, Poth CN (2018) Qualitative inquiry & research design: choosing among five approaches. Sage, Cop
Dwyer C (2020) Andrew Cuomo to receive international emmy for "Masterful" COVID-19 briefings. NPR.org. https://www.npr.org/sections/coronavirus-live-updates/2020/11/21/937445923/andrew-cu omo-to-receive-international-emmy-for-masterful-covid-19-briefings
Fox S (2014, Jan 15) The social life of health information. Pew Research Center. https://www.pewresearch.org/fact-tank/2014/01/15/the-social-life-of-health-information
Gayle AR, Weaver M, Murray J, Rourke A, Blackall M, Rourke A, Murray J (2020, Jan 31) Virus death toll reaches 213 in China—as it happened. The Guardian. https://www.theguardian.com/world/live/2020/jan/30/coronavirus-live-updates-china-death-toll-wuhan-evacuation-foreign-nationals-citizens-latest-news
Gil de Zúñiga H, Jung N, Valenzuela S (2012) Social media use for news and individuals' social capital, civic engagement and political participation. J Comput-Mediat Commun 17(3):319–336. https://doi.org/10.1111/j.1083-6101.2012.01574.x
Graham MW, Avery EJ, Park S (2015) The role of social media in local government crisis communications. Pub Relat Rev 41(3):386–394. https://doi.org/10.1016/j.pubrev.2015.02.001
Guidry JPD, Jin Y, Orr CA, Messner M, Meganck S (2017) Ebola on Instagram and Twitter: how health organizations address the health crisis in their social media engagement. Pub Relat Rev 43(3):477–486. https://doi.org/10.1016/j.pubrev.2017.04.009
Hagen L, Keller T, Neely S, DePaula N, Robert-Cooperman C (2017) Crisis communications in the age of social media. Soc Sci Comput Rev 36(5):523–541. https://doi.org/10.1177/0894439317721985
Hart M, Stetten N, Islam S, Pizarro K (2017) Twitter and Public Health (Part 2): Qualitative analysis of how individual health professionals outside organizations use microblogging to promote and disseminate health-related information. JMIR Pub Heal and Surv 3(4), e54. https://doi.org/10.2196/publichealth.6796
Hermida A, Fletcher F, Korell D, Logan D (2012) Share, like, recommend. J Stud 13(5–6):815–824. https://doi.org/10.1080/1461670x.2012.664430
Hernandez S, O'Key S, Watts A, Manley B, Pettersson H (2020, Apr 6) Tracking covid-19 cases in the US. CNN. https://www.cnn.com/interactive/2020/health/coronavirus-us-maps-and-cases/
Herrera S, Requejo JL (2012) 10 good practices for news organizations using twitter. J Appl J Media Stud 1(1):79–95. https://doi.org/10.1386/ajms.1.1.79_1
Houston JB, Hawthorne J, Perreault MF, Park EH, Goldstein Hode M, Halliwell MR, Turner McGowen SE, Davis R, Vaid S, McElderry JA, Griffith SA (2014) Social media and disasters: a functional framework for social media use in disaster planning, response, and research. Disasters 39(1):1–22. https://doi.org/10.1111/disa.12092
Huang P, Lurie NH, Mitra S (2009) Searching for experience on the web: An empirical examination of consumer behavior for search and experience goods. J Mark 73(2):55–69. https://doi.org/10.1509/jmkg.73.2.55
International Academy of Television Arts and Sciences. (n.d.). Special awards – international academy of television arts & sciences. *International Academy of Television Arts and Sciences.* https://www.iemmys.tv/international-emmy-awards/special-awards/
Ju A, Jeong SH, Chyi HI (2013) Will social media save newspapers? Journal Pract 8(1):1–17. https://doi.org/10.1080/17512786.2013.794022

Kim Y, Park H (2017) Is there still a PR problem online? Exploring the effects of different sources and crisis response strategies in online crisis communication via social media. Corp Reput Rev 20(1):76–104. https://doi.org/10.1057/s41299-017-0016-5

Klein LR, Ford GT (2003) Consumer search for information in the digital age: an empirical study of prepurchase search for automobiles. J Interact Mark 17(3):29–49. https://doi.org/10.1002/dir.10058

Kwak H, Lee C, Park H, Moon S (2010) What is Twitter, a social network or a news media? In: WWW '10: Proceedings of the 19th international conference on world wide web, pp 591–600. https://doi.org/10.1145/1772690.1772751

Lee CS, Ma L (2012) News sharing in social media: the effect of gratifications and prior experience. Comput Hum Behav 28(2):331–339. https://doi.org/10.1016/j.chb.2011.10.002

Lwin M, Lu J, Sheldenkar A, Schulz P (2018) Strategic uses of Facebook in Zika outbreak communication: implications for the crisis and emergency risk communication model. Int J Environ Res Public Health 15(9):1974. https://doi.org/10.3390/ijerph15091974

NYS (2011) The Official Website of Governor Andrew M. Cuomo. Governor Andrew M. Cuomo. https://www.governor.ny.gov/about

NYS (2020a, Jan 24) Governor Cuomo Outlines State Response to First Two Confirmed Cases of Novel Coronavirus in United States. Governor Andrew M. Cuomo. https://www.governor.ny.gov/news/governor-cuomo-outlines-state-response-first-two-confirmed-cases-novel-coronavirus-united

NYS (2020b, Mar 7) At novel coronavirus briefing, Governor Cuomo Declares State of Emergency to Contain Spread of Virus. Governor Andrew M. Cuomo. https://www.governor.ny.gov/news/novel-coronavirus-briefing-governor-cuomo-declares- state-emergency-contain-spread-virus

NYS (2020c, Apr 8) Amid ongoing COVID-19 pandemic, governor cuomo directs flags to be flown at half-mast in honor of those we have lost to COVID-19. Governor Andrew M. Cuomo. https://www.governor.ny.gov/news/amid-ongoing-covid-19-pandemic-governor-cuomo-directs-flags-be-flown-half-mast-honor-those-we

NYS (2020d, Apr 12) Amid ongoing COVID-19 pandemic, governor cuomo issues executive order directing employers to provide masks to employees who interact with the public. Governor Andrew M. Cuomo. https://www.governor.ny.gov/news/amid-ongoing-covid-19-pandemic-governor-cuomo-issues-executive-order-directing-employers

Park H, Reber BH, Chon M-G (2015) Tweeting as health communication: health organizations' use of twitter for health promotion and public engagement. J Health Commun 21(2):188–198. https://doi.org/10.1080/10810730.2015.1058435

Park H, Rodgers S, Stemmle J (2013) Analyzing health organizations' use of twitter for promoting health literacy. J Health Commun 18(4):410–425. https://doi.org/10.1080/10810730.2012.727956

Reynolds B, Seeger M (2005) Crisis and emergency risk communication as an integrative model. J Health Commun 10(1):43–55. https://doi.org/10.1080/10810730590904571

Reynolds B (2006) Response to best practices. J Appl Commun Res 34(3):249–252. https://doi.org/10.1080/00909880600771593

Schumaker E (2020, Mar 25) Timeline: how coronavirus got started. ABC News. https://abcnews.go.com/Health/timeline-coronavirus-started/story?id=69435165

Shearer E (2018, Dec 10) Social media outpaces print newspapers in the U.S. as a news source. Pew Research Center. https://www.pewresearch.org/fact-tank/2018/12/10/social-media-outpaces-print-newspapers-in-the-u-s-as-a-news-source/

Strauss A, Corbin J (1994) Grounded theory methodology: an overview. In: Denzin NK, Lincoln YS (eds) Handbook of qualitative research. Sage Publications, Inc, pp 273–285

Taylor DB (2020, Mar 4) A timeline of the coronavirus. The New York Times. https://www.nytimes.com/article/coronavirus-timeline.html

Trump DJ (2020, Mar 13) Proclamation on declaring a national emergency concerning the novel coronavirus disease (COVID-19) outbreak. The White House. https://www.whitehouse.gov/pre

sidential-actions/proclamation-declaring-national-emergency-concerning-novel-coronavirus-disease-covid-19-outbreak/

Walton LR, Seitz HH, Ragsdale K (2012) Strategic use of youtube during a national public health crisis: the CDC's response to the 2009 H1N1 Flu epidemic. Case Stud Strateg Commun 1:25–37

Wojcik S, Hughes A (2019, Apr 24) Sizing up Twitter Users. Pew Research Center: Internet, Science & Tech. https://www.pewresearch.org/internet/2019/04/24/sizing-up-twitter-users/

WHO (n.d.) Naming the coronavirus disease (COVID-19) and the virus that causes it. https://www.Who.Int. https://www.who.int/emergencies/diseases/novel-coronavirus-2019/technical-guidance/naming-the-coronavirus-disease-(covid-2019)-and-the-virus-that-causes-it

WHO (2020a, Jan 5) Pneumonia of unknown cause—China. Emergencies preparedness, response; World Health Organization. https://www.who.int/csr/don/05-january-2020-pneumonia-of-unkown-cause-china/en/

WHO (2020b, Jan 30) Statement on the second meeting of the International Health Regulations (2005) Emergency Committee regarding the outbreak of novel coronavirus (2019-nCoV). Who.Int; World Health Organization: WHO. https://www.who.int/news-room/detail/30-01-2020-statement-on-the-second-meeting-of-the-international-health-regulations-(2005)-emergency-committee-regarding-the-outbreak-of-novel-coronavirus-(2019-ncov)

Yin RK (2018) Case study research and applications: design and methods (6th ed). Sage Publications, Inc

Part III
Behavior and Resiliency

Chapter 8
Exploring the Interplay Between Psychological Processes, Affective Responses, Political Identity, and News Avoidance

Nicholas Eng

Abstract With the ever-evolving COVID-19 pandemic, there is a greater need for the public to keep abreast of the news around the novel virus. However, surveys show that there is an increasing proportion of the public that actively avoid news about the pandemic. This chapter elucidates the reasons behind why people avoid COVID-19 news by examining how the current media landscape (both traditional and new media) contributes to COVID-related news avoidance. It argues that psychological processes, affective responses, and political identity can make communicating COVID information through news media an arduous task. This chapter also provides recommendations on how to overcome tendencies to avoid COVID news and motivate individuals to attend to, select, and process this information, through the use of different emotional appeals and news media literacy education.

Keywords News avoidance · News overload · Psychological processing · Emotions · Political identity

> "U.S. Coronavirus Cases Soar as 18 States Set Single-Day Records This Week" – The New York Times.
>
> "Researchers: Covid-19 Death Toll Could Double this Winter to World War 2 Levels" – NBC News.
>
> "COVID-19 is 12 Times Deadlier for Patients with Underlying Health Conditions: CDC" – Fox News.

As the novel COVID-19 virus became more widespread, with over 55 million cases worldwide and more than a million deaths in a span of less than one year, news coverage of the pandemic increased dramatically as well. Naturally, as the situation became more dire, news coverage too was filled with doom and gloom with increasing instances of reports emphasizing the gravity of the issue. In times of uncertainty and crisis, such as during a pandemic, people are expected to rely more heavily on the media (Ball-Rokeach and DeFleur 1976), yet we see the reverse occurring, with an increasing proportion of the public actively avoiding news about the pandemic.

N. Eng (✉)
The Pennsylvania State University, State College, USA
e-mail: nje49@psu.edu

D. M. Berube (ed.), *Pandemic Communication and Resilience*, Risk, Systems and Decisions, https://doi.org/10.1007/978-3-030-77344-1_8

In fact, according to a Pew Research Center (2020) survey, the percentage of U.S. adults who follow news about the COVID-19 outbreak very closely dropped from 51% in March to 35% in September of 2020. This drop is even more apparent when looking at political identity, with a 9% drop for Democrats, but a 22% drop among Republicans during the same time period (Pew 2020).

Since the average public does not generally seek out information from primary sources (e.g., research published in academic journals), the news becomes an even more essential source of information and news avoidance evolves into a public health issue. News consumption is vital in helping the public to stay up to date with health threats, prevention efforts, and treatments (Myrick 2014). Especially during times of a pandemic where new scientific research is being published about the virus, how it spreads, and how to protect oneself and those around them, keeping up with the news becomes particularly essential. Furthermore, new research influences public health guidance, and has policy implications that require public attention. Because the implications of news avoidance are so grave, the focus of this chapter is to elucidate why people tend to avoid news during times of a pandemic and provide recommendations on how to overcome this. Centering this discussion around the COVID-19 pandemic, it will first define news avoidance before exploring the interplay between psychological processes, affective responses, and political identity as shaped by the current media landscape, on why people tend to avoid the news. It will then conclude with recommendations on alternative media forms and emotional appeals that may overcome tendencies to avoid COVID-19 news and nudge (Thaler and Sunstein 2008) people in the right direction to select, attend to, and process information from the news.

8.1 What is News Avoidance?

News avoidance has been conceptualized and operationalized by scholars in a multitude of ways. For example, a review conducted by Skovsgaard and Andersen (2020) identified four approaches scholars have taken to operationalize news avoidance. While some apply cluster analysis or latent class analysis to define groups according to their news exposure, others use a cut-off point based on amount of news consumption, a cut-off point for news avoidance, or relying on self-reports of whether people identify as actively avoiding the news (Skovsgaard and Andersen 2020). Regardless of methodology, there are generally four different types of news viewers: (1) intentional news selectors, (2) unintentional news selectors, (3) unintentional news avoiders, and (4) intentional news avoiders (Van den Bulck 2006). This chapter is concerned with the fourth group—those who intentionally avoid the news during the COVID-19 pandemic.

To provide a definition of news avoidance, this chapter follows Toff and Kalogeropoulos (2020) in defining this as "an intermittent practice that may occur at differing rates among the public" (p. 368). The use of the adjective *intermittent* suggests that people who habitually "resist" or abstain from news entirely (Edgerly

2017; Toff and Nielsen 2018; Trilling and Schoenbach 2013; Woodstock 2014) are not included. This chapter deals not with people who are "chronic" news avoiders since they have a different set of reasons as to why they choose to avoid the news completely. What this chapter is interested in exploring are people who in general may not avoid the news, but actively choose to avoid particular types of news, during a specific time period. For example, during the COVID-19 pandemic, people may choose to avoid all virus-related updates, but may still keep up with news about politics and the economy.

8.2 Why and How Do People Choose News Media?

Before diving into reasons why people avoid COVID-19 news, it is imperative to look at the reasons behind why and how people select media messages. A long tradition of media effects and mass communication research sheds some light on this. As one of the earliest lines of communication research, Lazarsfeld et al. (1948) found that people tend to select media messages that will help them to achieve their needs and goals, as well as to confirm their beliefs. Media, therefore, has very specific functions for an individual and are not just used in a futile manner. Communication theories like uses and gratifications (U&G; Katz et al. 1973) and selective exposure (Knobloch-Westerwick 2014) speak to this assumption. The key proposition of the U&G approach is that media users are cognizant and aware of what motivates them to select media messages (Katz and Blumler 1974). For example, people may be motivated to use the news for escapism, entertainment, information-seeking, or to converse with others (Diddi and LaRose 2006; Lee 2013; Rubin and Perse 1987; Rubin et al. 2006). Selective exposure refers to "any systematic bias in audience composition for a given medium or message, as well as any systematic bias in selected messages that diverges from the composition of accessible messages" (Knobloch-Westerwick 2015, p. 3). That is to say that people choose to select specific mediated messages, rather than to attend to all the media messages that they can be potentially exposed to. These theories developed the idea that in a sea of media messages, people can only attend to a limited number of messages and they choose them based on what serves their needs and desires best. Additionally, there are a number of dispositional (e.g., individual needs), situational (e.g., mood and emotion), social (e.g., group identity), and psychological factors (e.g., information seeking that confirms one's beliefs) that guide message selection (Rosengren 1974). In the context of COVID-19 news, people may choose to attend to news that satiates their need to keep abreast of COVID-19 updates (dispositional) that can reduce their anxiety about the pandemic (situational), and that appeal to their political beliefs (social and psychological). With a basic understanding of why and how people choose different media, the next section explores the psychological and affective responses, as well as the role of group identity to explain news avoidance in the context of the present media landscape.

8.3 Current Media Landscape and the Psychological Processes of News Overload

In the twenty-first century, people have an abundance of choice when it comes to receiving news. There has been an extensive rise in the mediums in which people are able to receive news. Not only can people receive news from the radio, newspaper, and television, there is access to 24-h news channels. This abundance of traditional news media is supplemented by news that is distributed on social media. In fact, with the widespread penetration of social media, news is being distributed expeditiously and extensively more so than ever before (Fedeli and Matsa 2018). News from both traditional and new media is so ubiquitous and omnipresent that it has been termed "ambient news" to describe how it is almost impossible to shy away from news coverage (Gil de Zúñiga et al. 2017; Hermida 2010a, b). What is unique about the COVID-19 pandemic is that it has the potential to be as deadly as the 1918 influenza pandemic (Faust et al. 2020) and needless to say the media landscape then was dramatically different from how it is now with the advent of social media and an increase in news channels on traditional media platforms. With the plethora of news choices both on traditional and new media, increasingly, people are feeling a sense of information and news overload.

Information overload refers to the concept of receiving too much information (Eppler and Mengis 2004). As a matter of fact, the World Health Organization (WHO 2020) has described the COVID-19 pandemic with the term "infodemic" to reference the "overabundance of information, both online and offline." While "infodemic" was used by the WHO to also acknowledge the widespread mis- and disinformation regarding the virus, the term in itself indicates the volume of news and information that are available at one's fingertips. Information overload has implication for information processing. With an overload of information, people are unable to cognitively process that information (Eppler and Mengis 2004), which follows the limited capacity model (Lang 2000) that posits that processing mediated messages requires cognitive effort. Since there is a limited capacity in everyone, people's ability to process messages that they receive are also limited (Lang 2000). As the saying goes, too much of anything is bad, so when there is an overwhelming amount of information, people's processing is reduced and is now divided among different sources. Furthermore, perceived information overload has been associated with feelings of stress and anxiety (Eppler and Mengis 2004), which not just requires emotional effort to regulate these emotions, but also increases the level of cognitive effort needed to attend to incoming media messages.

With respect to news overload, similar definitions and outcomes hold. To be clear, news overload occurs when one is exposed to too much news (Holton and Chiyi 2012) whereas information overload can be invoked from a variety of sources not specific to the news, such as advertising and promotions (Edmunds and Morris 2000). With the advent of social media and a saturation of offline news programs, news has become so ubiquitous that is now seen as a task or chore (Nordenson 2008). Empirical studies show that perceived news overload is associated with news fatigue and analysis paralysis (Song et al. 2017). As such, when people feel that there is too much news, they feel tired of news consumption and become incapable of cognitively processing news stories. In addition, in the context of social media, Park (2019) found that people who perceive news overload feel less confident about their ability to seek out and understand news on social media, which increased their avoidance of news consumption on social media platforms. News avoidance can then be seen as a coping strategy for individuals who perceive high levels of news overload because of psychological factors such as fatigue and the lack of self-efficacy to cognitively process the news (Islam et al. 2018; Park, 2019; Song et al., 2017). Interestingly, a study in 2012 found that exposure to news through computers and Facebook were significant and positive predictors of news overload but exposure to news from the television had a negative association with perceptions of overload (Holton and Chiyi 2012). Almost a decade since the publication of Holton and Chiyi's (2012) study, it is highly plausible that exposure to news from both traditional and new media are now associated with news overload.

With the increased news coverage of COVID-19, people are feeling a sense of news overload and are opting to avoid news coverage of this topic as a coping mechanism. The impact of news overload impairs not just one's cognitive and psychological processing but also has influence on one's mental health. Research on how media exposure to COVID-19 affects mental health and well-being is growing (e.g., Cummins 2020; Fiorillo and Gorwood 2020; Gao et al. 2020; Taylor et al. 2020) with stress, anxiety, and depression being some of the negative effects of being constantly bombarded by COVID-19 news. After all, research on previous pandemic periods show that following news related to the disease can increase people's anxiety levels (Taylor 2019). It is therefore a natural transition to explore the nature of COVID-19 media coverage and the emotional and affective responses that follow to help explain active news avoidance.

8.4 Emotions and Affective Responses to COVID-19 News

COVID-19 news can arouse a number of negative affective and emotional responses. Looking at news headlines around the pandemic, such as the one from NBC News comparing the deaths of COVID-19 to that of World War 2, can be very anxiety-inducing and rather disconcerting. Together with the "ambient" nature of news (Gil de Zúñiga et al. 2017; Hermida 2010a, b), social distancing, lockdowns, and isolation, it seems almost impossible to feel positively about the situation. While hardly the only issue, media exposure does contribute to how overwhelming COVID-19 can be emotionally. In fact, a sentiment analysis of 141,208 COVID-19 related news headlines found that 52% of the news headlines evoked negative sentiments (compared to 30% evoking positive sentiments, and 18% being neutral), with 20% of the headlines evoking emotions of fear, 15% anticipation, and 14% sadness (Aslam et al. 2020).

Emotions are very powerful drivers of action selection in that they can guide information processing, and influences attention, memory and judgment (Lang 2000;

Lazarus 1991; Nabi 2003). But more importantly, emotions can drive people to select (or avoid) certain types of news. The emotion-as-frames model (EFM; Nabi 2003, 2007) posits that when ideas or events are consistently appraised with certain emotions, it will subsequently shape the way in which individuals respond to similar circumstances. In the EFM, emotions are conceptualized as frames in which individuals interpret stimuli (Nabi 2003, 2007) whereby existing schemas are activated that guide one's response (Fiske and Taylor 1991). For example, if COVID-19 is often framed in the media in ways that arouse fear, then people will tend to associate fear with any COVID-19 news that they might potentially be exposed to. In addition, Lazarus (1991) asserts that discrete emotions are associated with a core relational theme that is associated with an action tendency. Going back to Aslam et al.'s (2020) study, the three commonly evoked emotions in headlines are fear, anticipation, and sadness. Fear and anxiety (or anticipation) are associated with the correlational theme of an imminent threat, and when faced with fear-inducing information, people's action tendency is to avoid the threat (Lazarus 1991). Additionally, sadness is characterized by feelings of irrevocable loss and pessimism about whether things will get better (Lazarus 1991). Sadness is accompanied by the action tendency to change one's circumstances, which might explain why people may turn to alternative, more uplifting media forms rather than to attend to the overly depressing news coverage.

It is, however, perfectly reasonable that news headlines and its content convey a certain level of fear and threat in order to make salient the threat of COVID-19. However, fear can be very debilitating. Fear appeal theories such as the fear-as-acquired drive model (Hovland et al. 1953), parallel process model (PPM; Leventhal 1970, 1971), protection motivation theory (PMT; Rogers 1975, 1983), and the extended parallel process model (EPPM; Witte 1992, 1994) show the persuasiveness of fear appeals. The fear-as-acquired drive model (Hovland et al. 1953), for example, posits that people learn through their experiences with threats. When faced with a threat, individuals will actively try to reduce the unpleasant emotion of fear. If the action taken was successful in achieving their goal of fear reduction, similar actions will be taken when faced with a related threat. Extending this line of research, the PPM (Leventhal 1970, 1971), PMT (Rogers 1975, 1983), and EPPM (Witte 1992) contend that in response to a threat, maladaptive and/or adaptive outcomes may arise. Depending on people's appraisal of the threat severity, threat susceptibility, response efficacy, and self-efficacy, they may engage in fear control (a maladaptive outcome to reduce fear) or danger control (an adaptive outcome to take action to reduce the threat) (Rogers 1975, 1983; Witte 1992, 1994). If threat perceptions are high, but perceived efficacy is low, then people will be fearful and frightened, but will not know how to control the threat. As such, they may opt instead to control their feelings of fear—through avoidance (Lazarus 1991). Looking at the headlines at the beginning of the chapter, it is evident that these headlines heighten people's threat perceptions. They show just how severe COVID-19 is through the number of deaths reported, and how susceptible people are to the virus, especially if they are residing in the U.S. What is missing is an indication on how to control the threat. Of course, some of these COVID-19 news articles will provide recommendations on protection (e.g., wearing a mask or social distancing) in the content itself, but considering that this is a novel virus, there was a lot of uncertainty about how COVID-19 spreads and how to protect oneself in the early stages of the outbreak. To add another layer of complexity into the equation, even if news articles do provide recommendations of protection, psychological reactance may be aroused (Brehm 1966, 1972). According to psychological reactance theory (Brehm 1966, 1972), when people perceive that their behavioral freedoms are threatened, restricted, or removed (e.g., not being able to gather with one's friends), they try to restore their freedoms through strategies such as source derogation (Brehm 1966; Worchel 1974; Worchel and Brehm 1970). In derogating the source, individuals may think to themselves that the journalist or the scientist that conducted the research are not credible sources of information, and as such, avoid COVID-19 news in general.

Put together, research on emotions and affective responses paint a clearer picture as to why people may choose to avoid COVID-19 news. The negative emotions that are potentially aroused from constant exposure to news content may motivate people to attend to other forms of media. After all, mood management theory (MMT; Zillmann, 1988, 2000) posits that people are driven by their hedonistic desires and want to maintain and prolong positive moods while changing their negative moods. As such, when it comes to media choices, people will strive to avoid media content that perpetuates their negative moods, while seeking out content that will put them in a positive mood. This is supported by a study conducted in 2019, across 40 countries, which found that 58% of people tend to avoid the news because they feel that it negatively affects their moods (Newman 2019). This makes sense considering that people generally want to be in a positive affective state (Bless and Fiedler 2006). On top of emotional and affective reasons why people may avoid the news, one's political identity can also play a crucial role in explaining people's decision to avoid specific sources of COVID-19 news.

8.5 Partisanship, Selective Exposure and Avoidance, and (Dis)trust

Political identity may motivate someone to seek media messages that are aligned with one's existing political beliefs. In turn, they may avoid exposure to counter-attitudinal news media. Politically motivated partisan selective exposure has been confirmed by a number of studies (e.g., Iyengar and Hahn 2009; Mukerjee and Yang 2020; Stroud 2010). Selective exposure (Knobloch-Westerwick 2014) to certain news media can be explained by one's subjective need to reaffirm how correct they are (Garrett et al. 2013). That is, people actively look for news to collect evidence that their beliefs are correct. Similarly, confirmation bias (Knobloch-Westerwick and Kleinman 2012) may also be at play where people seek and interpret information that supports their prior beliefs. These concepts can be traced back to Festinger's (1957) theory of cognitive dissonance that people are partial to cognitively consistent information, and when faced with dissonance, they aim to reduce feelings of discomfort, and will actively avoid potentially dissonant information. Selective avoidance on the other hand is to actively avoid information that contracts one's prior political beliefs (Garrett et al. 2013; Jang 2014; Song 2017). Iyengar and Hahn (2009) found that depending on one's political ideology, individuals tend to avoid news from *the other side*. For example, conservatives and Republicans tend to avoid news from MSNBC (left-leaning media outlet) while liberals and Democrats actively avoid news from Fox News (right-leaning media outlet). The psychological mechanisms are similar in that people do not like to feel that they are incorrect, and this cognitive dissonance is something that they want to avoid.

This selective exposure and avoidance based on political identity is also an artifact of the media environment and how COVID-19 is being reported by news media. In a content analysis conducted by Hart et al. (2020), they find evidence that both newspaper and network news coverage are highly polarizing in their reporting of COVID-19, with Republican coverage using language that is associated with federal responses to COVID-19, the need for a vaccine, China, and the potential for hydroxychloroquine as a cure for the coronavirus. Democratic coverage on the other hand used language regarding responses from Democratic governors (as opposed to the federal government) and the impact of the virus on hospitals and residents. In fact, it was also found that politicians appeared in newspaper coverage more frequently than scientists did, showing evidence of high degrees of politicization. This discrepancy in media coverage, increases perceptions of bias, which also influences levels of trust and how the pandemic is being reported by media outlets.

Trust in news from mainstream media is greatly divided among partisans. The concept of trust is extremely complex and defined in a plethora of ways by different scholars, but in general consists of perceptions of benevolence, integrity, competence, and predictability (McKnight and Chervany 2001). In the context of trust in news, benevolence refers to perceptions of the media outlet and journalist in being motivated to act in the interest of the public; integrity is whether they are committed to telling the truth; competence is the ability to report accurately, and predictability refers to their ability to report in a consistent manner. Distrust on the other hand focuses on perceived value incongruence that is intrinsically linked to expectancy violation (McKnight and Chervany 2001). This is where expectations of the news outlet or journalist's benevolence, integrity, competence, and predictability are violated resulting in the perception that there is a mismatch in values. While distrust was previously treated as the absence of trust and exists on the same continuum (Bigley and Pearce 1998), more scholars are acknowledging that trust and distrust are distinct concepts and may even co-occur (e.g., Liu and Wang 2010). Perceptions of trust and distrust in news media is especially polarizing. For example, Democrats trust, more than distrust, 22 out of 30 news sources, while Republicans trust only six (Jurkowitz et al. 2020). Conversely, Democrats distrust, more than trust, eight of 30 news sources, while Republicans distrust an overwhelming 20 out of 30 news sources (Jurkowitz et al. 2020). Jurkowitz et al. (2020) also find that over the years, Republicans are increasingly distrusting of a larger number of news outlets, while Democrats' trust in news sources has remained stable, even strengthening in certain cases. This partisan divide in terms of trust and distrust in news media is also translated to people's perceptions of media coverage of COVID-19. While 66% of Democrats believe that media coverage of COVID-19 is accurate, only 31% of Republicans share this sentiment (Gottfried et al. 2020). This same stark contrast is seen in levels of agreement of whether media coverage of COVID-19 is working to benefit the public (Democrats: 66%, Republicans, 31%), and giving people the information that they need (Democrats: 73%, Republicans: 44%) (Gottfried et al. 2020). This also has implications for attention to COVID-19 news as a Pew Research Center (2020) survey finds that 43% of Republicans actively try and tune out COVID-19 news compared to 20% of Democrats. Clearly, the data show that based on one's political beliefs and identity, people have varying levels of trust in news media, which may translate into differences in levels of news avoidance in this COVID-19 pandemic.

This political polarization is further exacerbated by social media. Not only do social media algorithms learn which posts tend not to be clicked on or liked, and halt promoting such news stories (e.g., COVID-19 news) on people's feeds (Thorson 2020), partisans are also unlikely to interact with those who differ from their own political views. While some scholars contend that with social media, the probability of coming across diverse information and viewpoints is higher (Papacharissi 2002), research shows that ideological strength is predictive of practices of unfriending on social media platforms (Bode 2016). That is to say that those who identify as being very conservative or very liberal, tend to unfriend others for political reasons like disagreeing with their views, or for posting too much (Bode 2016). A caveat here is that while one might be tempted to conclude that people on social media are in echo chambers, this is not the case. There is in fact heterogeneity in social networks on social media (Lee et al. 2014) and unfriending on social media due to political disagreements tends to be a very rare occurrence (Bode 2016). What is concerning is that people may already have preconceived notions about the news content that is being shared on social media because of the political identity of the source creating or sharing that news content. In what has been called a "hostile media phenomenon" (Vallone et al. 1985) or "hostile media effect" (Arpan and Raney 2003; Giner-Sorolla and Chaiken 1994; Gunther et al. 2009; Lee et al. 2018), people tend to perceive neutral media content to be biased against their own viewpoint. More specifically, Lee et al. (2018) found support that partisans will perceive more bias in news content that are shared by those of an opposing party compared to those shared by their own party. Notably, Lee et al. (2018) show that Republicans who are exposed to a news story shared by Democrats on Twitter, tend to perceive more bias in the news story, and the same trend is seen in Democrats (who saw a news story shared by a Republican) albeit to a smaller extent. As such, even if people in general are not unfriending those in their social media networks from an opposing party, this perception of bias may lead them to actively avoid such content. In addition, even if there is heterogeneity in one's social network, one's social media feed generally tends toward political agreement (Peterson et al. 2019) possibly because algorithms stop promoting news stories on people's feeds if they are not interacted with (e.g., liked or clicked) (Thorson 2020), making news avoidance by partisans much easier.

In summary, this chapter identified four reasons that explain why people may actively avoid COVID-19 news: (1) perceptions of news overload, (2) negative emotions evoked from the news, and (3) politically driven selective exposure/avoidance and relatedly, (4) distrust and perceptions of bias in news by partisans. It is also important to note that these reasons are not mutually exclusive. For example, it is entirely plausible that perceptions of news overload can lead to politically-driven selective avoidance because people are cognitive misers and do not have unlimited cognitive abilities (Lang 2000), as such, they may rely on their political identity to filter what news to avoid (and select). Alternatively, the negative emotions that people feel from reading the news may also be a consequence of news overload or at least intensified by the abundance of news. Further, the more one identifies with a political party, there is also an increase in feelings of anger (Mason 2015) which could then also increase feelings of distrust and bias in news media. As such, this is an interplay between these four reasons and news avoidance. With these potential explanations for why people avoid COVID-19 news, the next section proposes possible solutions which could apply to a number of overlapping reasons, as well as suggestions for future research directions.

8.6 Overcoming News Avoidance

There are a number of directions that can be taken to overcome news avoidance, one of which includes changing the tonality of messaging through the use of humor and awe appeals to elicit different emotional responses. Humor has been found to be a more positive alternative to typical threat-based messages (often seen in COVID-19 news), and is extremely successful at attracting attention (Eisend 2009). Humorous Internet memes have been found to be a positive emotional boost that can help people to better cope with the stresses of a pandemic (Folkman and Moskowitz 2000). In fact, in an experiment conducted by Myrick et al. (2021), it was found that participants exposed to memes reported greater empathy for people with COVID-19 and greater willingness to consume COVID-19 news compared to participants in the non-meme control condition. Another emotion that could be leveraged to reduce news avoidance is awe, extending Skovsgaard and Andersen (2020) suggestion that presenting news in a hopeful, other-oriented manner can help alleviate news avoidance. Awe is a positive emotion (Gordon et al. 2017; Shiota et al. 2007) that is indicated by perceptions of vastness and a need to accommodate new information (Keltner and Haidt 2003). Awe has the ability to encourage individuals to think not just about themselves, but to also consider the people and the environment around them—building a stronger connection with others—while also diminishing their self-importance (Bai et al. 2017; Shiota et al. 2007; Stellar et al. 2017). Awe appeals could reduce perceptions of news overload and/or reduce the negative emotions aroused by the news, shifting perceptions of fear, anxiety, and stress, to curiosity. Furthermore, as Stellar et al. (2017) observes, "awe helps individuals fold into cohesive collectives by leading to a reduced estimation of one's individual importance" (p. 203). In other words, awe has the potential to bypass people's partisanship and to feel connected to others who may not share their political beliefs. Future research should probe how certain message features in news could evoke awe and humor responses, or if exposure to awe-inspiring or other humorous media prior to COVID-19 news, can be viable solutions to news avoidance.

Apart from emotional appeals, a promising approach to reduce news avoidance is through news media literacy education. Such forms of education can be effective in reestablishing people's loss of self-efficacy from news overload, while also building trust between citizens and news organizations. According to Park (2019), "if politicians and educators make an effort to boost citizens' confidence in their news finding and handling (news efficacy), news overload may not remain as an obstacle to informing citizens" (p. 9). This suggestion comes from research that shows that news overload is associated with analysis paralysis (Song et al. 2017) which may reduce people's confidence that they can comprehend news effectively (Islam et al. 2018; Park 2019; Song et al. 2017). One way to increase people's news efficacy is through news media literacy education. This is akin to research on media literacy, which provide media consumers with the skills to access, analyze, evaluate, and create media content (Livingstone 2004). News media literacy education emphasizes the same skills but tailored to apply to news consumption (Fleming 2013; Kahne et al. 2012; Maksl et al. 2015; Vraga and Tully 2015). In educating the public on the skills needed to select and process news, it is possible that perceptions of news overload will be reduced, and consequently, reduces news avoidance as well. Furthermore, these news media literacy education programs also emphasize the relationships between journalists and citizens, which may also help build trust between citizens and the news content that they are exposed to. Indeed, exposure to news media literacy messages has been found to decrease hostile media perceptions for neutral content, by conservatives (Vraga and Tully 2015), and reduces liberals' perceptions of bias in news stories (Vraga et al. 2009). Overall, these studies show the potential for news media literacy education but do not show a direct link between such forms of education and reductions in news avoidance. In one study conducted by Vraga and Tully (2019) exposure to a news media literacy message prior did not reduce selective exposure or avoidance. However, the study was conducted a week prior to the 2016 election, which the authors call an examination of "explicitly partisan news stories during a particularly combative political season" (Vraga and Tully 2019, p. 81). Future research should continue studying whether the use of news media literacy education can reduce news avoidance, as well as increase trust in news media and news efficacy, during times of crisis like a pandemic.

8.7 Conclusion

Avoiding news in times of a pandemic is not only a tremendous communication challenge but has implications for public health. In times of crisis, citizens need to be kept updated on the evolving health risks and treatments and news avoidance especially with the novel COVID-19 pandemic is detrimental. In this chapter, the psychological processes of news overload, negative affective and emotional responses to the news, as well as the role of political identity were proposed as reasons why people may actively avoid COVID-19 news. These reasons are shaped by the present media landscape and should be seen as overlapping and not mutually exclusive reasons. They each have an influence on one another and news avoidance should be seen as an interplay of these reasons. News media literacy education could be potential solution to news avoidance which can motivate individuals to use and comprehend COVID-19 news. On top of providing reasons for news avoidance and potential solutions, this chapter also opens up additional research opportunities in fields related to public health, media, and communication.

References

Arpan LM, Raney AA (2003) An experimental investigation of news source and the hostile media effect. J Mass Commun Q 80(2):265–281. https://doi.org/10.1177/107769900308000203

Aslam F, Awan TM, Syed Jabir H, Kashif A, Mahwish P (2020) Sentiments and emotions evoked by news headlines of coronavirus disease (COVID-19) outbreak. Hum Soc Sci Commun 7(1). https://doi.org/10.1057/s41599-020-0523-3

Bai Y, Maruskin LA, Chen S, Gordon AM, Stellar JE, McNeil GD, Peng K, Keltner D (2017) Awe, the diminished self, and collective engagement: universals and cultural variations in the small self. J Pers Soc Psychol 113(2):185–209. https://doi.org/10.1037/pspa0000087

Ball-Rokeach SJ, DeFleur ML (1976) A dependency model of mass-media effects. Commun Res 3(1):3–21. https://doi.org/10.1177/009365027600300101

Bigley GA, Pearce JL (1998) Straining for shared meaning in organization science: problems of trust and distrust. Acad Manag Rev 23(3):405–421. https://doi.org/10.2307/259286

Bless H, Fiedler K (2006) Mood and the regulation of information processing and behavior. In: Forgas JP (ed) Affect in social thinking and behavior. Psychology Press, pp 65–84

Bode L (2016) Pruning the news feed: unfriending and unfollowing political content on social media. Res Polit 3(3):2053168016661873. https://doi.org/10.1177/2053168016661873

Brehm JW (1966) A theory of psychological reactance. Academic Press

Brehm JW (1972) Responses to loss of freedom: a theory of psychological reactance. General Learning Press

Cummins E (2020) How to fight fear and anxiety when quarantine ends. Vox. Accessed 16 Dec 2020. https://www.vox.com/the-highlight/2020/5/13/21248281/coronavirus-mental-health-reopen-reopening-covid-anxiety-bars-restaurants-office

Diddi A, LaRose R (2006) Getting hooked on news: uses and gratifications and the formation of news habits among college students in an internet environment. J Broadcast Electron Media 50(2):193–210. https://doi.org/10.1207/s15506878jobem5002_2

Edgerly S (2017) Seeking out and avoiding the news media: young adults' proposed strategies for obtaining current events information. Mass Commun Soc 20(3):358–377. https://doi.org/10.1080/15205436.2016.1262424

Edmunds A, Morris A (2000) The problem of information overload in business organisations: a review of the literature. Int J Inf Manage 20(1):17–28. https://doi.org/10.1016/S0268-4012(99)00051-1

Eisend M (2009) A meta-analysis of humor in advertising. J Acad Mark Sci 37:191–203. https://doi.org/10.1007/s11747-008-0096-y

Eppler MJ, Mengis J (2004) The concept of information overload: a review of literature from organization science, accounting, marketing, MIS, and related disciplines. Inf Soc 20(5):325–344. https://doi.org/10.1080/01972240490507974
Faust JS, Lin Z, del Rio C (2020) Comparison of estimated excess deaths in New York City during the COVID-19 and 1918 influenza pandemics. JAMA Netw Open 3(8):e2017527–e2017527. https://doi.org/10.1001/jamanetworkopen.2020.17527
Fedeli S, Matsa KE (2018) Use of mobile devices for news continues to grow, outpacing desktops and laptops. Pew Research Center. Accessed 20 Dec 2020. https://www.pewresearch.org/fact-tank/2019/11/19/americans-favor-mobile-devices-over-desktops-and-laptops-for-getting-news/
Festinger L (1957) A theory of cognitive dissonance. Stanford University Press
Fiorillo A, Gorwood P (2020) The consequences of the COVID-19 pandemic on mental health and implications for clinical practice. Eur Psychiatry J Assoc Eur Psychiatrists 63(1):e32–e32. https://doi.org/10.1192/j.eurpsy.2020.35
Fiske ST, Taylor SE (1991) Social cognition. Mcgraw-Hill Book Company
Fleming J (2013) Media Literacy, News literacy, or news appreciation? a case study of the news literacy program at Stony Brook University. Journalism & Mass Communication Educator 69(2):146–165. https://doi.org/10.1177/1077695813517885
Folkman S, Moskowitz JT (2000) Stress, positive emotion, and coping. Curr Dir Psychol Sci 9(4):115–118. https://doi.org/10.1111/1467-8721.00073
Gao J, Zheng P, Jia Y, Chen H, Mao Y, Chen S, Wang Y, Fu H, Dai J (2020) Mental health problems and social media exposure during COVID-19 outbreak. PLoS ONE 15(4):e0231924. https://doi.org/10.1371/journal.pone.0231924
Garrett RK, Carnahan D, Lynch EK (2013) A turn toward avoidance? selective exposure to online political information, 2004–2008. Polit Behav 35(1):113–134. https://doi.org/10.1007/s11109-011-9185-6
Gil de Zúñiga H, Weeks B, Ardèvol-Abreu A (2017) Effects of the news-finds-me perception in communication: social media use implications for news seeking and learning about politics. J Comput-Mediat Commun 22(3):105–123. https://doi.org/10.1111/jcc4.12185
Giner-Sorolla R, Chaiken S (1994) The causes of hostile media judgments. J Exp Soc Psychol 30(2):165–180. https://doi.org/10.1006/jesp.1994.1008
Gordon AM, Stellar JE, Anderson CL, McNeil GD, Loew D, Keltner D (2017) The dark side of the sublime: distinguishing a threat-based variant of awe. J Pers Soc Psychol 113(2):310–328. https://doi.org/10.1037/pspp0000120
Gottfried J, Walker M, Mitchell A (2020) Republicans and Democrats sharply divided in views of news media's coronavirus coverage. Pew Research Center. Accessed 30 Dec 2020. https://www.journalism.org/2020/05/08/americans-are-more-likely-than-not-to-think-the-news-media-are-fulfilling-key-roles-during-the-coronavirus-outbreak-but-partisans-are-starkly-divided/
Gunther AC, Miller N, Liebhart JL (2009) Assimilation and contrast in a test of the hostile media effect. Commun Res 36(6):747–764. https://doi.org/10.1177/0093650209346804
Hart PS, Chinn S, Soroka S (2020) Politicization and polarization in COVID-19 news coverage. Sci Commun 42(5):679–697. https://doi.org/10.1177/1075547020950735
Hermida A (2010a) From TV to Twitter: Hhow ambient news became ambient journalism. M/C J 13(2). https://doi.org/10.5204/mcj.220
Hermida A (2010b) Twittering the news. J Pract 4(3):297–308. https://doi.org/10.1080/17512781003640703
Holton A, Chiyi HI (2012) News and the overloaded consumer: factors influencing information overload among news consumers. Cyberpsychol Behav Soc Netw 15(11):619–624. https://doi.org/10.1089/cyber.2011.0610
Hovland CI, Janis IL, Kelley HH (1953) Communication and persuasion. Yale University Press
Islam AKM, Whelan E, Brooks S (2018) Social media overload and fatigue: the moderating role of multitasking computer self-efficacy. In: 24th American conference on information systems, New Orleans, LA. Accessed 30 Dec 2020. https://aisel.aisnet.org/amcis2018/SocialComputing/Presentations/1/

Iyengar S, Hahn KS (2009) Red media, blue media: evidence of ideological selectivity in media use. J Commun 59(1):19–39. https://doi.org/10.1111/j.1460-2466.2008.01402.x
Jang SM (2014) Challenges to selective exposure: selective seeking and avoidance in a multitasking media environment. Mass Commun Soc 17(5):665–688. https://doi.org/10.1080/15205436.2013.835425
Jurkowitz M, Mitchell A, Shearer E, Walker M (2020) U.S. media polarization and the 2020 election: a nation divided. Accessed 30 Dec 2020. https://www.journalism.org/2020/01/24/u-s-media-polarization-and-the-2020-election-a-nation-divided/
Kahne J, Lee N-J, Feezell JT (2012) Digital media literacy education and online civic and political participation. Int J Commun 6:24. https://ijoc.org/index.php/ijoc/article/view/999
Katz E, Blumler JG (1974) The uses of mass communications: current perspectives on gratifications research. Sage
Katz E, Blumler JG, Gurevitch M (1973) Uses and gratifications research. Pub Opin Q 37(4):509–523. http://www.jstor.org/stable/2747854
Keltner D, Haidt J (2003) Approaching awe, a moral, spiritual, and aesthetic emotion. Cogn Emot 17(2):297–314. https://doi.org/10.1080/02699930302297
Knobloch-Westerwick S (2014) The selective exposure self- and affect-aanagement (SESAM) model: applications in the realms of race, politics, and health. Commun Res 42(7):959–985. https://doi.org/10.1177/0093650214539173
Knobloch-Westerwick S (2015) Choice and preference in media use: advances in selective exposure theory and research. Routledge
Knobloch-Westerwick S, Kleinman SB (2012) Preelection selective exposure: confirmation bias versus informational utility. Commun Res 39(2):170–193. https://doi.org/10.1177/0093650211400597
Lang A (2000) The limited capacity model of mediated message processing. J Commun 50(1):46–70. https://doi.org/10.1111/j.1460-2466.2000.tb02833.x
Lazarsfeld PF, Berelson B, Gaudet H (1948) The people's choice: how the voter makes up his mind in a presidential campaign. Columbia University Press
Lazarus RS (1991) Emotion and adaptation. Oxford University Press, Oxford
Lee AM (2013) News audiences revisited: theorizing the link between audience motivations and news consumption. J Broadcast Electron Media 57(3):300–317. https://doi.org/10.1080/08838151.2013.816712
Lee JK, Choi J, Kim C, Kim Y (2014) Social media, network heterogeneity, and opinion polarization. J Commun 64(4):702–722. https://doi.org/10.1111/jcom.12077
Lee TK, Kim Y, Coe K (2018) When social media become hostile media: an experimental examination of news sharing, partisanship, and follower count. Mass Commun Soc 21(4):450–472. https://doi.org/10.1080/15205436.2018.1429635
Leventhal H (1970) Findings and theory in the study of fear communications. In Berkowitz L (ed) Advances in experimental social psychology. Academic Press, pp 119–186
Leventhal H (1971) Fear appeals and persuasion: the differentiation of a motivational construct. Am J Public Health 61:1208–1224. https://doi.org/10.2105/AJPH.61.6.1208
Liu M, Wang C (2010) Explaining the influence of anger and compassion on negotiators' interaction goals: an assessment of trust and distrust as two distinct mediators. Commun Res 37(4):443–472. https://doi.org/10.1177/0093650210362681
Livingstone S (2004) Media literacy and the challenge of new information and communication technologies. Commun Rev 7(1):3–14. https://doi.org/10.1080/10714420490280152
Maksl A, Ashley S, Craft S (2015) Measuring news media literacy. J Media Literacy Educ 6(3):29–45. Accessed 3 Jan 2021. https://digitalcommons.uri.edu/cgi/viewcontent.cgi?article=1174&context=jmle
Mason L (2015) "I disrespectfully agree": The differential effects of partisan sorting on social and issue polarization. Am J Polit Sci 59(1):128–145. https://doi.org/10.1111/ajps.12089
McKnight DH, Chervany NL (2001) Trust and distrust definitions: one bite at a time. In: Falcone R, Singh M, Tan YH (eds) Trust in cyber-societies. Springer, pp 27–54

Mukerjee S, Yang T (2020) Choosing to avoid? A conjoint experimental study to understand selective exposure and avoidance on social media. Polit Commun, pp 1–19. https://doi.org/10.1080/10584609.2020.1763531
Myrick JG (2014) Journalism and health. In: Thompson TL (ed) Encyclopedia of health communication. Sage, pp 605–608
Myrick JG, Nabi RL, Eng N (2021) COVID memes, COVID news, and coping: Can a good mood and good citizenship coexist? [Unpublished manuscript]. College of Communications, Pennsylvania State University
Nabi RL (2003) Exploring the framing effects of emotion: do discrete emotions differentially influence information accessibility, information seeking, and policy preference? Commun Res 30(2):224–247. https://doi.org/10.1177/0093650202250881
Nabi RL (2007) Emotion and persuasion: a social-cognitive perspective. In: Monahan R-EDRJ (ed). Erlbaum
Newman N (2019) Reuters institute digital news report 2019. Digital News Report. https://www.digitalnewsreport.org/survey/2019/overview-key-findings-2019/
Nordenson B (2008) Overload! Journalism's battle for relevance in an age of too much information. Columbia J Rev. https://archives.cjr.org/feature/overload_1.php
Papacharissi Z (2002) The virtual sphere: the internet as a public sphere. New Media Soc 4(1):9–27. https://doi.org/10.1177/14614440222226244
Park CS (2019) Does too much news on social media discourage news seeking? Mediating role of news efficacy between perceived news overload and news avoidance on social media. Soc Media + Soc 5(3):2056305119872956. https://doi.org/10.1177/2056305119872956
Peterson E, Goel S, Iyengar S (2019) Partisan selective exposure in online news consumption: evidence from the 2016 presidential campaign. Polit Sci Res Methods, pp 1–17. https://doi.org/10.1017/psrm.2019.55
Pew RC (2020) Before trump tested positive for coronavirus, republicans' attention to pandemic had sharply declined. Accessed 3 Jan 2021. https://www.journalism.org/wp-content/uploads/sites/8/2020/10/PJ_2020.10.07_Attention-To-COVID-19-News_FINAL.pdf
Rogers RW (1975) A protection motivation theory of fear appeals and attitude change1. J Psychol 91(1):93–114. https://doi.org/10.1080/00223980.1975.9915803
Rogers RW (1983). Cognitive and psychological processes in fear appeals and attitude change: a revised theory of protection motivation. In B. L. Cacioppo & L. L. Petty (Eds.), *Social psychophysiology: A sourcebook* (pp. 153–176). Guilford.
Rosengren KE (1974) Uses and gratifications: A paradigm outlined. In: Blumler JG, Katz E (eds) The uses of mass communications: Current perspectives on gratifications research. Sage, pp 269–286
Rubin AM, Perse EM (1987) Audience activity and television news gratifications. Commun Res 14(1):58–84. https://doi.org/10.1177/009365087014001004
Rubin AM, Perse EM, Powell RA (2006) Loneliness, parasocial interaction, and local television news viewing. Hum Commun Res 12(2):155–180. https://doi.org/10.1111/j.1468-2958.1985.tb00071.x
Shiota MN, Keltner D, Mossman A (2007) The nature of awe: elicitors, appraisals, and effects on self-concept. Cogn Emot 21(5):944–963. https://doi.org/10.1080/02699930600923668
Skovsgaard M, Andersen K (2020) Conceptualizing news avoidance: towards a shared understanding of different causes and potential Solutions. J Stud 21(4):459–476. https://doi.org/10.1080/1461670X.2019.1686410
Song H (2017) Why do people (sometimes) become selective about news? The role of emotions and partisan differences in selective approach and avoidance. Mass Commun Soc 20(1):47–67. https://doi.org/10.1080/15205436.2016.1187755
Song H, Jung J, Kim Y (2017) Perceived news overload and its cognitive and attitudinal consequences for news usage in South Korea. J Mass Commun Q 94(4):1172–1190. https://doi.org/10.1177/1077699016679975

Stellar JE, Gordon AM, Piff PK, Cordaro D, Anderson CL, Bai Y, Maruskin LA, Keltner D (2017) Self-transcendent emotions and their social functions: compassion, gratitude, and awe bind us to others through prosociality. Emot Rev 9(3):200–207. https://doi.org/10.1177/1754073916684557

Stroud NJ (2010) Polarization and partisan selective exposure. J Commun 60(3):556–576. https://doi.org/10.1111/j.1460-2466.2010.01497.x

Taylor S (2019) The psychology of pandemics: preparing for the next global outbreak of infectious disease. Cambridge Scholars Publishing.

Taylor S, Landry CA, Paluszek MM, Rachor GS, Asmundson GJG (2020) Worry, avoidance, and coping during the COVID-19 pandemic: a comprehensive network analysis. J AnxiEty Disord 76:102327. https://doi.org/10.1016/j.janxdis.2020.102327

Thaler RH, Sunstein CR (2008) Nudge: improving decisions about health, wealth, and happiness. Yale University Press

Thorson K (2020) Attracting the news: Algorithms, platforms, and reframing incidental exposure. Journalism 21(8):1067–1082. https://doi.org/10.1177/1464884920915352

Toff B, Kalogeropoulos A (2020) All the news that's fit to ignore: How the information environment does and does not shape news avoidance. Public Opin Q 84(S1):366–390. https://doi.org/10.1093/poq/nfaa016

Toff B, Nielsen RK (2018) "I just Google it": folk theories of distributed discovery. J Commun 68(3):636–657. https://doi.org/10.1093/joc/jqy009

Trilling D, Schoenbach K (2013) Skipping current affairs: the non-users of online and offline news. Eur J Commun 28(1):35–51. https://doi.org/10.1177/0267323112453671

Vallone RP, Ross L, Lepper MR (1985) The hostile media phenomenon: biased perception and perceptions of media bias in coverage of the Beirut massacre. J Pers Soc Psychol 49(3):577–585. https://doi.org/10.1037//0022-3514.49.3.577

Van den Bulck J (2006) Television news avoidance: exploratory results from a one-year follow-up study. J Broadcast Electron Media 50(2):231–252. https://doi.org/10.1207/s15506878jobem5002_4

Vraga EK, Tully M (2015) Media literacy messages and hostile media perceptions: processing of nonpartisan versus partisan political information. Mass Commun Soc 18(4):422–448. https://doi.org/10.1080/15205436.2014.1001910

Vraga EK, Tully M (2019) Engaging with the other side: using news media literacy messages to reduce selective exposure and avoidance. J Inform Tech Polit 16(1):77–86. https://doi.org/10.1080/19331681.2019.1572565

Vraga EK, Tully M, Rojas H (2009) Media literacy training reduces perception of bias. Newsp Res J 30(4):68–81. https://doi.org/10.1177/073953290903000406

WHO (2020) Managing the COVID-19 infodemic: promoting healthy behaviours and mitigating the harm from misinformation and disinformation. Accessed 20 Dec 2020. https://www.who.int/news/item/23-09-2020-managing-the-covid-19-infodemic-promoting-healthy-behaviours-and-mitigating-the-harm-from-misinformation-and-disinformation

Witte K (1992) Putting the fear back into fear appeals: the extended parallel process model. Commun Monogr 59(4):329–349. https://doi.org/10.1080/03637759209376276

Witte K (1994) Fear control and danger control: a test of the extended parallel process model (EPPM). Commun Monogr 61(2):113–134. https://doi.org/10.1080/03637759409376328

Woodstock L (2014) The news-democracy narrative and the unexpected benefits of limited news consumption: the case of news resisters. Journalism 15(7):834–849. https://doi.org/10.1177/1464884913504260

Worchel S (1974) The effect of three types of arbitrary thwarting on the instigation to aggression. J Pers 42(2):300–318. https://doi.org/10.1111/j.1467-6494.1974.tb00676.x

Worchel S, Brehm JW (1970) Effect of threats to attitudinal freedom as a function of agreement with the communicator. J Pers Soc Psychol 14(1):18–22. https://doi.org/10.1037/h0028620

Zillmann D (1988) Mood management: using entertainment to full advantage. In: Communication, social cognition, and affect. Lawrence Erlbaum Associates, Inc., pp 147–171

Zillmann D (2000) Mood management in the context of selective exposure theory. Ann Int Commun Assoc 23(1):103–123. https://doi.org/10.1080/23808985.2000.11678971

Chapter 9
A Story About Toilet Paper: Pandemic Panic-Buying and Public Resilience

David M. Berube

Abstract This chapter examines panic-buying (PB) as byproduct of pandemics. Panic buying is associated with disasters including pandemics. If and when disasters become more prominent and prevalent, understanding phenomena like PB becomes an important part of disaster communication and disaster management. Panic buying of toilet paper was examined during the international Covid-19 crisis of 2019–2020 in an effort to study and explain the behavior of PB, who panic buyers are, which drivers motivate PB as well as the socio-psychological explanations of panic-buying found in the existing literature. The chapter concludes with some recommendations to reduce PB in the future. This is a snapshot of a behavior intrinsic to disasters of all sorts and should serve as both explicative and proscriptive warnings.

Keywords Covid-19 · Pandemic · Panic-buying · Stockpiling · Toilet paper

9.1 Introduction

As this chapter is completed in early 2021, a cold snap hit Houston, Texas in the United States. Water pipes froze, burst, and leaked. Water treatment collapsed and water where it could be found needed to be boiled. Panic ensued and thousands went shopping for a host of things. This behavior is called panic-buying (PB) and it the subject of what follows.

A pandemic is neither an accidental nor a random event. Humankind has helped prepare ecological niches for all sorts of viruses. With 8 billion people in densely packed cities, animal habitats are squeezed and they need to live somewhere or go extinct, so they move where they can, much closer to humans. Also, climate change appears to be reducing opportunities for human and animal habitation. We are easily able to travel anywhere on the planet, so small outbreaks can easily become epidemics

D. M. Berube (✉)
North Carolina State University, Raleigh, USA
e-mail: dmberube@ncsu.edu

D. M. Berube (ed.), *Pandemic Communication and Resilience*, Risk, Systems and Decisions, https://doi.org/10.1007/978-3-030-77344-1_9

and in turn, pandemics. "Particularly relevant is the multiplication of contacts with bats, which are the natural reservoir of innumerable viruses capable of crossing the species barrier and spilling over to humans" (Rose 2020, p. 3).

9.2 Introduction to Panic Buying

PB is serious though PB for toilet paper may be less so. Disasters recur and under the current threats of climate change and associated disasters, we may expect an increase in its frequency as well as severity. In addition, pandemics will recur due to weak international efforts to control animal to human-to-human transmission, and dozens of other phenomena such as wet marketing, bushmeat consumption, and so forth.

Specifically, PB behaviors have been frequently observed when there is supply disruption risk, any seemingly major disaster. For example, when severe snowstorms are forecasted, consumers who worried about the regular supply of goods rushed to stores and buy unusually large amounts of food (Zheng et al. 2020). As such, the literature on PB also has been called stockpiling.

This happened many times before: during the oil shock in October 1973, during the infamous Y2K Millennium Bug (ask your parents), after the election of President Obama (guns sales increased fearing gun purchases would be regulated), and so forth. It happened in East Asia amidst rumors that imports from China would soon collapse (Boxall 2020; Otsuki 2020). It occurs during snowpocalypse(s) when Winter storms arrive below the Mason-Dixon line and triggered when Storm Watchers interrupt your favorite TV shows (definitely a Southern thing). It happens before and after natural disasters, especially hurricanes and typhoons, (e.g., superstorms Sandy and Haiyan) (Sneath et al. 2009, p. 46) though not exclusively.

Covid-19, a pandemic event described throughout this book, is a "hot crisis". "A hot crisis is defined as a real-world event that creates acute fear among the public (Ungar 1998). "In a hot crisis scenario, we expect a heavy reliance on news that includes panic-intensifying devices (the emphasis of risks, blame, and speculation) rather than news with panic-reducing devices (the emphasis of solutions and praise" (Kilgo et al. 2019, p. 813). These crises tend to produce panic, anxiety and fear and they, in turn, produce PB much like what happened in Texas during the ice storm of February 2021.

PB is not "hoarding disorder" thought both involve stockpiling. "Hoarding disorder" affects approximately 2–6% of the population (Mathes et al. 2017). Generally, when we think of "hoarding" we visualize rooms piled with cardboard boxes collected over years by people with serious psychological issues.

> As many as half of those suffering from hoarding disorder will also suffer from depression, and 30% or more will have an anxiety disorder. Hoarding-related clutter in homes increases the risk of falls, pest or vermin infestation, unstable or unsafe living conditions, and difficulty with self-care. It may stun you to know that up to 25% of deaths by house fire are due to hoarding.... Instead, stockpiling is a normal behavior that many people practice in

preparation for a known or anticipated shortage. The goal of stockpiling is to create a reserve in case there's a future need (Matthews 2020).

As well, "PB and hoarding of food, fuel, and medicines are common during wartime and periods of civil unrest, in regions where goods are chronically scarce (e.g., Russia and eastern Europe under communism), and in times of fear and uncertainty (e.g., the 1979 US gasoline crisis…the 2001 anthrax attacks)… Rationing of flu vaccine in the fall of 2004 led to PB, followed within months by surplus stocks, and a fear of avian flu triggered hoarding of the antiviral drug oseltamivir (Tamiflu)" (Sterman and Dogan 2015, p. 9).

The scarce products subject to PB generally include essential medicines (especially cold and flu medicines), vitamins, thermometers, masks, hand sanitizers (antiseptic spray cleaners, antiseptic hand sanitizing gels, and a host of toiletries that carry the label antiseptic), rubber and latex gloves, condoms, tampons, groceries, esp. milk, bread and eggs, bottled water, fuels as in propane, kerosene and automobile grade gasoline, tinned and preserved foods, food grains such as rice and refined flour, dried pasta, soap, lotion, and paper products, facial tissues, and paper towels. Bikes, trampolines, swimming pools, exercise equipment, hair clippers, freezers, and even webcams. Still, this chapter is about none of these products nor about those who crammed these articles into towering piles onto shopping carts.

Finally, PB is a relatively new and unexplored area in consumer behavior research… It is now a frequent occurrence in many countries, leading to stock-outs and supply chain disruption (Yuen et al. 2020). "Rushing to stock up on toilet paper before it vanished from the supermarket isle, stashing cash under the mattress, purchasing a puppy or perhaps planting a vegetable patch—the Covid-19 pandemic has triggered some interesting and unusual changes in our behavior" (UTS 2020). All behavior worth noting.

9.3 Why Toilet Paper?

Why toilet paper? Matthew Loeb put it best: "I like toilet paper. When I run out of it, there is that frantic millisecond when I am desperately scrambling for the nearest paper substitute. It is a helpless feeling… (Loeb 2016). Monica Heese from *The Washington Post* was much more poetic. "When apocalypses seem imminent, rolls of toilet paper unspool into the yardage by which Americans measure panic. We can imagine no scenario worse than the world ending while our pants are around our ankles" (2020).

The first mention of toilet paper in the Western world comes from the sixteenth century, with a short description by the French novelist (and physician) François Rabelais arguing its ineffectiveness. China, however, had toilet paper in the second century BCE and the Japanese used chuugi (20–25 cm wooden sticks) during the Nara period (eighth century CE) for both external and internal cleaning of the anal canal. (Charlier et al. 2012).

Before paper was invented, or readily available, people used leaves, seashells, fur pelts and corn cobs. The ancient Greeks and Romans used small ceramic disks and also sponges on the ends of sticks, which were then plunged into a bucket of vinegar or salt water for the next person to use" (Murphy 2020) (Fig. 9.1).

Toilet paper has value beyond the obvious and its very sensitive to PB. For example, "a government press release warning of a potential shortage in toilet paper led to a lot of press coverage but no outright PB until Johnny Carson, a famous late night television host, joked about it during one his opening monologue on *The*

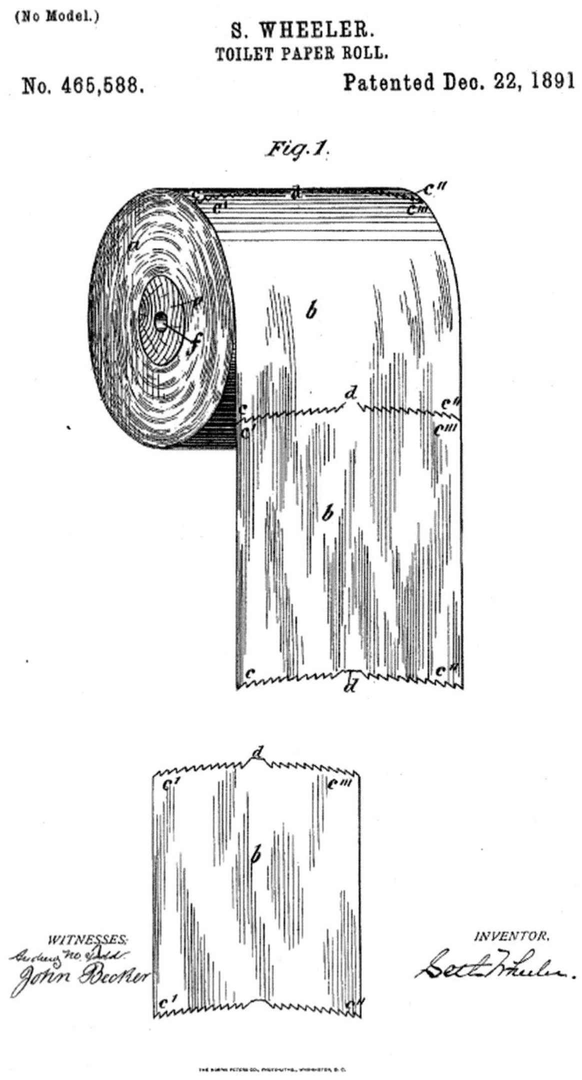

Fig. 9.1 Wheeler S (1891 cited in Murphy 2020) toilet paper roll. US patent 465,588, 22 Dec 1891

Tonight Show… In 1973, U.S. consumers cleared store shelves of the rolls for a month based on little more than rumors, fears and a joke." (Zogorsky 2020).

"Most toilet paper historians (there are more than you would think) credit Seth Wheeler with inventing modern toilet paper, perforated and on a roll, an idea he patented in 1891. The diagram on the patent application should put to rest any arguments about how to load the roll: The flap comes over the top and down the front. While manufacturers might have added dyes, prints, perfumes and soothing aloe, toilet paper has remained pretty much the same ever since" (Murphy 2020).

The following description helped organize this chapter. There is something about toilet paper that is comforting and hints our lives are manageable.

As Bussel puts it: "More than any other belongings in my home, opening my storage closet and seeing rolls upon rolls of toilet paper, neatly stacked, fluffy, white and seemingly endless, offers me a sense of calm. It reminds me that even when the present feels utterly out of control—which is often—there will be a future where the most pressing issue will be whether to bring another roll into the bathroom with me (2020).

9.4 PB of Toilet Paper During the Covid-19 Pandemic

"During the Covid-19 pandemic, hoarding toilet paper, was reported nearly across the globe. Despite no signs of imminent shortages, the highest reported rise in PB for toilet papers seemed to come from Argentina, Australia, Italy, Japan, the United Kingdom, Singapore, Spain, and the USA (Anastasiadou et al. 2020; Keane and Neil 2020; Sim et al. 2020).

In March 2020, Statista found that toilet paper sales exceeded estimates by 140% in Italy, 98% in Australia, 82% in Spain and 80% in the UK (Buchholz 2020). The *Statista Consumer Market Outlook* compared data and calculated estimates for 16 countries to show that revenues had risen most in Italy, followed by Vietnam and Australia. "In other countries hit hard by the virus, for example Spain and France, the sale of toilet paper rose by 82% and 30%, respectively" (Buchholz 2020).

9.4.1 Australia

Panic buying and stockpiling commenced in early March 2020 and increased sharply towards the end of the month when the lockdown scale was extended, and social distancing was tightened. The ANZ reported that supermarket spending was up 40% from the same week ending March 16 the previous year… There was a 35% increase in retail food spending, a 60% jump in the pharmacy and toiletry spend, and a 22% rise in electronic purchases… Among the items stockpiled, toilet paper was the common and greatest stockpiled item… However, PB appeared to slow in April. This may be accounted for by repeated assurances from government and the interventions practiced by retailers or supermarkets (Prentice et al. 2020).

Some illustrations: New South Wales Police were called to a supermarket in Western Sydney after an argument broke out between two women over toilet paper. The argument was captured on video, which has been viewed more than 210,000 times on Twitter (ABC 2020). "In Victoria, Woolworths Managing Director Claire Peters said her supermarket is selling seven weeks' worth of toilet paper a day. She also added it is now working on getting more onto shelves. "The demand is just so much significantly higher than supply," she said (Swain 2020).

The New Yorker reported a case where a café in Australia accepted toilet paper for currency (three rolls for a coffee, 36 rolls for a kilo of beans) while a newspaper in Australia ran eight mostly blank pages for its readers. "Run out of loo paper?" the tabloid asked (Alford 2020). Rolls were being flogged for hundreds of dollars online in Australia, while listeners were calling into radio stations to win packs of 3-ply loo roll (Mao 2020).

9.4.2 United States

"Following similar trends of previous PB scenarios, the US saw a 32.5% increase year-on-year in grocery spending in the month of March and a 12.8% reduction in April (Trading Economics 2020). This suggests that PB is either a relatively short-term occurrence and/or that over-spending in March meant consumers had sufficient supplies to last them through April" (Loxton et al. 2020).

"At some Seattle stores on Sunday, in a throwback to earlier days of the pandemic, people were already buying up stacks of bathroom tissue, which seems to turn to spun gold when things look grim" (Brodeur 2020).

Some illustrations: "A worker at a Seattle WinCo Food market told *The Daily Beast* that when she arrived for work at 4 A.M., the store was fully stocked with toilet paper and paper towels. But by 1 P.M., Deirdre, who declined to give her last name, said the store decided to limit the number of such items each customer could buy. An hour later, she said, "We were already down to where you could see the back walls." In Minnesota, meanwhile, a grocery store worker said people are hoarding toilet paper in particular. "People are stockpiling now not because they're afraid of being stuck at home, but because they've seen everyone else buying it up and are afraid, they won't be able to get any when they need it later," Bennett told The *Daily Beast* without giving his last name (Steinbuch 2020).

"In Virginia, ABC8 News reports some large chains, including Kroger and Walmart, were already experiencing the second wave of PB in mid-summer. One local store owner Norm Gold told the network: 'There will be a second rush. I am very confident in my store and what we have now. They learned from what happened six months ago and they brought product in early.' Target told the network they will enforce purchase limits if necessary. Earlier this year grocers were forced to limit purchases of products like Purell sanitizers, Lysol cleaning spray and canned soup. Companies like Walmart, Target and Wegman's curbed store hours for the public in order to give workers time to restock shelves" (Fruen 2020).

9.5 Why Study PB of Toilet Paper?

Toilet paper seems emblematic to PB (Yoshizaki et al. 2020). Here's what we've learned. Some PB makes sense, pharmaceuticals, fuel, potable water, and food. Disaster events reduce supplies of materials and scarcity results in less materials, brand unavailability, and/or much higher prices (Tsao et al. 2019). Some PB makes less sense, such as the case of toilet paper and that is why it was selected for this chapter. Focus needed to be set on PB not the subject of the buying.

There is plenty of toilet paper. There was no "real" toilet paper shortage in 2020 (DePaulo 2020; Hughes 1988). It is produced domestically (Ohnsman 2020). Stores only store limited amounts because it is bulky and takes up a lot of storage space (Chadwick 2020). Commercial sales can be controlled simply by limiting the amount allowable for purchase at any one time (Bachelor 2020; Besson 2020).

Without very good reasons at all, people across the world panicked and decided they needed to PB toilet paper and this instantiation will let us learn why without being overly troubled by deep and well-established demand and supply issues. "Instead of scoffing at people for buying all the toilet paper, let's get to the root of behavior in people we're close to. Let's help make people feel safe. Given someone's history, maybe toilet paper doesn't make them feel safe right now. The question is, what does" (Wood 2020)?

Before lurching forward, please consider that some have argued PB of toilet paper may have its own undesirable consequences. Stockpiling of commodities, including toilet paper, can trigger shortages and stoke price rises. (For example, in the early days of the Covid-19 crisis, toilet rolls were listed on eBay in Australia at prices up to AU$1,000,000 for 600 rolls—AU$1,667 per roll" (Baddeley 2020). Singh and Rakshit (2020) found toilet paper for sale in black markets in India in response to PB.

In some instances, people least able to afford supplies find themselves priced out of the market as prices increase reflecting demand. Although this may be temporary, it disproportionately affects vulnerable people such as those in rural areas and those with low incomes more than others" (Arafat et al. 2020c). In addition, "...(t)here is a significant positive correlation between average income per capita and PB.... Panic buying happens in every income class, including low-income ones and contribute to enhancing the understanding of demand behavior during periods of crisis" (Yoshizaki et al. 2020).

Another implication, alternatives for toilet paper, like facial tissues and paper towels could seriously overtax water remediation plants. "The resulting scarcity of toilet paper in some households has led to problematic consequences such as the clogging of outfall pipes after people started using products other than toilet paper" (Garbe et al. 2020).

9.6 Who Are Panic Buyers?

There seems to be positive association between PB and age. "Older people are more prone to a severe course of the disease and, thus, may be more eager to prepare for strict self-isolation. In addition, in some countries, older people were asked to self-isolate before more comprehensive lockdowns were put in place which might partly account for the age effect" (Garbe et al. 2020).

They found: "older participants shopped more frequently, bought more packages of toilet paper and had more toilet papers rolls in stock as compared to younger participants. Participants residing in Europe shopped toilet paper more frequently than North American residents ($p = 0.039$) but had less toilet paper in stock ($p < 0.001$)" (Garbe et al. 2020).

Both (Colombus 2020; Garbe et al. 2020) did not find the 'Honesty-Humility' characteristic from HEXACO[1] to be a significant predictor of toilet paper PB. Toilet paper stockpiling might not be resulting from a lack of solidarity and, as such, moral appeals by public authorities asking people to refrain from stockpiling might be less fruitful than expected.... More conscientious people tend to stockpile more toilet paper. This finding is in line with the expectation that longsighted and more orderly individuals engage in more stockpiling and does not support the counternarrative that conscientious individuals refrain from impulsive PB due to increased self-control" (Garbe et al. 2020).

Just to be fair, it's not just those who exhibit mildly neurotic tendencies that are prone to stockpiling. "At one end of the spectrum for 'Conscientiousness' is the need for order and being prepared. 'Conscientiousness' is positively associated with toilet paper consumption. In particular, participants high on 'Conscientiousness' tended to shop more frequently ($p = 0.065$), shopped more toilet paper ($p = 0.045$), and stocked more toilet paper ($p = 0.048$)... (Garbe et al. 2020).

Models for toilet paper shopping intensity and the amount of stocked toilet rolls seemed to have an indirect effect of 'Emotionality' through perceived threat was significant (c' = 0.016; 95% CI = [0.002; 0.031] for toilet paper shopping intensity; c' = 0.019; 95% CI = [−0.006; 0.036] for stocked toilet rolls. The indirect effects for toilet paper frequency was marginally significant (c' = 0.014; 95% CI

[1] The HEXACO model (see Lee and Ashton 2008) is rooted in lexical studies of personality descriptors across various languages and organizes individual differences along six broad personality domains: Honesty-Humility (characterized by the facets sincerity, fairness, greed avoidance, and modesty), Emotionality (fearfulness, anxiety, dependence, sentimentality), eXtraversion (sic) (social self-esteem, social boldness, sociability, liveliness), Agreeableness (forgiveness, gentleness, flexibility, patience), Conscientiousness (organization, diligence, perfectionism, prudence), and Openness to Experience (aesthetic appreciation, inquisitiveness, creativity, unconventionality).

= [0.001; 0.029]. These results suggest that Emotionality may fuel the feelings of being threatened by the Covid-19 pandemic which may consequently foster toilet paper stockpiling (Garbe et al. 2020).An individual with this disposition, when faced with scenes of PB, might feel the need to be ready for 'worst-case scenarios' and avoiding the temptation to hoard could be very difficult" (Mead 2020).

Garbe, Rau and Toppe data was mostly corroborated by McCrae's results (2020). Firstly, those qualifying as emotional in the personality domain—meaning they experience more anxiety and feel a greater deal of empathy and sentimental attachment to others—were more likely to stockpile. Lastly, being conscientious correlated with the behavior of purchasing big on toilet rolls (McCrae 2020).

All these effects held across North American and European countries and were robust across different indicators of toilet paper stockpiling (i.e., shopping frequency, shopping intensity, and stocked toilet rolls) (Garbe et al. 2020). McCrae (2020) added that older people were slightly more likely to purchase more rolls too (see above), and found Americans shopped less frequently than European customers, but bought big when they did. However, this may be explained away by packaging protocols in the USA where products are often sold in bulk.

9.7 Why Do We Panic Buy?

In a meta-analysis of 27 journal articles, Yuen et al (2020) summarized that PB is influenced by themes: social psychological, perceived threat and perceived scarcity, fear of the unknown and issues of control (Yuen et al. 2020). There are probably five main categories of reasons consumers during a crisis, like a pandemic, engage in PB. They appear below in increasing degree of prominence in the literature and have all been discussed in articles about PB of toilet paper.

9.7.1 Psychological Factors

Least discussed in the literature is the Freudian spin. Andrea Greenman, the president of the Contemporary Freudian Society, said: "The fact that now we are all presumably losing control creates a regressive push to a very early time. So, I guess that translates in the unconscious to 'If I have a lifelong supply of toilet paper, I'll never be out of control, never be a helpless, dirty child again (Alford 2020)".

"Psychologists say it's more than a little Freudian, what with the anal personality being tied to a need for order, hoarding and fear of contamination and once these characteristics align with obsessive compulsive tendencies triggered when people feel threatened." Nick Haslam, a professor of psychology at the University of Melbourne in Australia and the author of "Psychology in the Bathroom" claims this may explain some PB behavior involving toilet paper (Murphy 2020).

There are a series of other sociopathic disorders that may help explain PB behavior as well. Lins and Aquino summarized these very well.

> "Personal disruptive events could be also related to PB, and to psychological disorders, which panic and fear are related symptoms (e.g., post-traumatic stress disorder, anxiety disorders, panic disorder, social phobia, agoraphobia), and other types of buying behavior disorder (e.g., compulsive buying). Furthermore, it is probable that the link between psychological disorders and PB can be stronger during stressful periods" (2020).

9.7.2 Perceived Threats and Shortages

Sometimes PB is driven by simple subjective utility calculations: on balance, more toilet paper than needed is better than not enough or none at all. "A risk expert in Australia explained it this way: "Stocking up on toilet paper is… a relatively cheap action, and people like to think that they are 'doing something' when they feel at risk." This is an example of "zero risk bias," in which people prefer to try to eliminate one type of possibly superficial risk entirely rather than do something that would reduce their total risk by a greater amount" (Zogorsky 2020).

Sheu and Kuo (2020) drew from cognitive psychology and added a human's perception and expectation of post-disaster supply shortage reflect a mental process of assessment of a person's ability to cope with a potential supply shortage by speculative hoarding after a disaster, as well as induced gains and losses. They reported "hoarding behavior prior to or during a disaster is a form of self-protection behavior, planned behavior in an attempt to minimize risk using stores of supplies conferring a sense of safety and well-being."

Additional factors include a minimal amount of harm in purchasing products that last for an extended period of time and are needed in any case" especially so when the shelf half-life for a roll of toilet paper is a very long time (Laato et al. 2020).

Garbe et al. (2020) reported participants who felt more threatened shopped for toilet paper more frequently, bought more packages of toilet paper, and had more toilet paper in stock. McCrae's results concurred (2020).

9.7.3 Misinformation and Unknowns

This section may deserve its own chapter, a subject for a future publication. Covid-19 has been hailed as the first social media amplified pandemic (Cinelli et al. 2020).

Within weeks of the emergence of the novel coronavirus disease 2019 (Covid-19) in China, misleading rumors and conspiracy theories about the origin circulated the globe paired with fearmongering, racism and mass purchase of face masks, all closely linked to social media" (Depoux et al. 2020).

Mentions by former US President Donald Trump may comprise by far the largest single component of the infodemic. Trump mentions comprised 37.9% of the overall infodemic (see below), well ahead of "miracle cures", which comprised 26.4%. However, a substantial proportion—possibly even the majority—of the "miracle cures" topic was also driven by the president's comments, so a substantial overlap can be expected between these topics. Evanega et al (2020) conclude therefore that the President of the United States was likely the largest driver of the Covid-19 misinformation "infodemic".

Uscinski and Enders from *The Atlantic* report "nearly a third of the people we polled believe that the virus was manufactured on purpose… the late Rush Limbaugh suggested that our public-health officials are deep-state operatives and might not even be health experts. Other conservative commentators claimed hospitals aren't actually treating any Covid-19 patients… (2020).

Only 16.4% of the misinformation conversation was "fact-checked" in nature, suggesting that the majority of COVID misinformation is conveyed by the media without question or correction (Uscinski and Enders 2020).

9.7.3.1 Infodemic

COVID has been called the first infodemic though that distinction seems to be a matter of perspective. On February 2, the WHO dubbed the new coronavirus "a massive 'infodemic,'" referring to"an overabundance of information—some accurate and some not—that makes it hard for people to find trustworthy sources and reliable guidance when they need it." It's a distinction that sets the coronavirus apart from previous viral outbreaks" (see Ball & Maxman 2020) and WHO (2020) said WHO Director-General Tedros Adhanom Ghebreyesus. He reaffirmed the term at the Munich Security Conference on February 15 (Zarocostas 2020).

"We're not just fighting an epidemic; we're fighting an infodemic. Fake news spreads faster and more easily than this virus and is just as dangerous." Ghebreyesus "called on all governments, companies and news organizations to work with us to sound the appropriate level of alarm, without fanning the flames of hysteria" (Ghebreyesus, WHO 2020).

9.7.3.2 Social Media

"During (COVID), the increasing popularity of social media has made health information about coronavirus spread more rapidly and become widely accessible on the Internet" (Cheng and Lo 2020).

"During a pandemic situation such as Covid-19, it can be difficult for individuals to organize all online information clearly and accurately" (Laato et al. 2020)…. "While SARS, MERS, and Zika all caused global panic, fears around the coronavirus have been especially amplified by social media. It has allowed disinformation to spread and flourish at unprecedented speeds, creating an environment of heightened uncertainty that has fueled anxiety … online" (Hao and Basu 2020).

Furthermore, many media companies and others creating Covid-19 news online might be in a hurry to publish stories. As information is unclearly available even for news creators, this rush increases the presence of inaccurate information…" (Laato et al. 2020).

"Allison Chase, a psychologist at Insight Behavioral Health in Austin, Texas, says panicking can be contagious and spread faster than the actual virus—especially with social media" (Hernandez 2020). Many people shared pictures, videos, and posts regarding the unavailability of necessities of life in their local markets. These pictures, videos, and posts created an extreme panic situation as more people rushed to markets to stockpile and to avoid future uncertainties (Naeem 2021).

9.7.4 Anxiety and Coping

"Anxiety spreads faster than the virus," Catherine Belling at Northwestern's Feinberg School of Medicine, told ABC News… According to a fairly recent ABC News/Ipsos poll (Karson 2020), "two-thirds of Americans are anxious that they or someone they know could get infected with the virus—with increasing fears of being quarantined" (Muwahed 2020).

Anxiety and fear is intensified by poorly built communication messages from government entities, both national and international, print and broadcast media (live and recorded then streamed), and social media (such as Facebook and Twitter). The amplification of anxiety associated with pandemics is done by many of the same entities as they report on the PB triggering a fear of missing out when consumers learn what other consumers and engage in a herding effect sometimes referred to a FOLO (fear of losing out).

Consequently, "PB can be a response to the perceived lack of control regarding the future and social demands… (T)his sense of loss of control has a big impact on stress levels (see Sim et al. 2020; Sneath et al. 2009). An ambiguous and intense situation, such as the unprecedented Covid-19 restrictions, can induce feelings of distress and helplessness (Kennett-Hensel et al. 2012). Thus, consumers are more likely to participate in activities that offer a sense of security and comfort, regardless of long-term or broader societal implications (see Rapoport and Chammah 1965; Loxton et al. 2020).

Buying things as a way of coping to deal with stressful events is not new. According to Terror Management Theory (Routledge and Vess 2018), people have an internal psychological conflict resulting from a desire for self-preservation contrasted with knowledge and the certainty that death is inevitable (See Harmon-Jones et al. 1997).

In turn, this coping behavior can be driven by "herding". In the context of purchasing behavior during a crisis, collective societal anxieties are perpetuated and consumers are more likely to pay attention to the purchasing behaviors of their peers…." (Loxton et al. 2020).

Loxton linked herding to many other aforementioned variables. "Herd mentality and hoarding behaviors are driven by innate vulnerability factors, including intolerance of uncertainty (see Mathes et al. 2017), and loss of control, whereby possession of a threatened consumer good restores a sense of security, comfort and preparedness (see Frost and Hartl 1996; Loxton et al. 2020).

This behavior is predicated on fear, both an antecedent and a product of stress. "Fear has played a particularly vital role in coverage of the coronavirus outbreak. Since reports first started circulating about the new mystery illness on January 12, and up until February 13 2020, Wahl-Jorgensen, a journalism professor from Cardiff, tracked reporting in major English-language newspapers around the world, using the LexisNexis UK database. This included almost 100 high-circulation newspapers from around the world, which have collectively published 9,387 stories about the outbreak. Of these, 1,066 articles mention "fear" or related words, including "afraid". Such stories often used other frightening language—for example, 50 articles even used the phrase "killer virus" (Wahl-Jorgensen 2020).

In this regard, PB can be viewed as an outlet to regain control over the situation, which compensates for the psychological losses experienced by individuals.... According to compensatory control theory (Kay et al. 2009), when the source of discomfort (i.e., disease outbreak) is not amenable to control, the individual will turn to increase control over other domains. Individuals can exert control over the environment through problem solving" (Yuen et al. 2020).

9.7.5 Positive Feedback Loops

A positive feedback (exacerbating feedback, self-reinforcing feedback) is a process that occurs in a feedback loop which exacerbates the effects of a disturbance and there are two disturbing positive feedback loops involving PB of toilet paper. One is called herding and the other involves media amplification sometimes of herding behavior.

Anxiety and fear is intensified by poorly built communication messages from government entities, both national and international, print and broadcast media (live and recorded then streamed), and social media (such as Facebook and Twitter). The amplification of anxiety associated with pandemics is done by many of the same entities as they report on the PB triggering a fear of missing out when consumers learn what other consumers and engage in a herding effect.

9.7.5.1 Herding Begets More Herding

"Some social scientists have pinned PB on a herd instinct that is triggered by fear and spread through social media".... (INSEAD 2020). "Herd mentality also adds to this behavior. The experts say the fact that PB is happening at all can prompt people to participate, according to says David Savage, professor of behavioral economics at the University of Newcastle in Australia" (Lufkin 2020).

In addition, more herding reduces the individual capacity to resolve stress and fear drivers. "Shortages created by PB also force consumers to devote extra time and effort to shopping, diverting time from welfare-improving activities like work, leisure, and sleep, and generating psychic costs of anxiety and stress. Furthermore, the shortages produced by PB may heighten anxiety about the pandemic—and the government's response—among the general population" (Keane and Neal 2021).

9.7.5.2 Media Amplification

Highly associated with herding is media amplification (Kasperson 1988). Most consumer psychology experts say the aforementioned PB behavior is "obviously irrational", and a clear example of herd mentality whipped up by social media and news coverage (Mao 2020).

Two quantitative studies on amplification and mediation from Asia (Vietnam and China) have reported similar effects. (Nguyen and Nguyen 2020; Ning et al. 2020).

Media amplification of Covid-19.

Media, analogue and digital, have been associated with Covid-19 and heightened anxiety (Sigurvinsdottir et al. 2020), emotional contagion (Valenzano et al. 2020), and nocebo effects (a negative placebo effect, for example how negative thinking may affect recovery) (Amanzio et al. 2020).

Negative information which is spread through mass media repetitively can affect public health negatively in the form of nocebo effects and mass hysteria. Mass and digital media in connection with the state may have had adverse consequences during the Covid-19 crisis… [M]ass hysteria can be exacerbated and self-reinforcing when the negative information comes from an authoritative source, when the media are politicized, and social networks make the negative information omnipresent (Bagus et al. 2021).

"On the news, we see hazmat suits, masks and hastily assembled hospitals that look like shipping containers. So, it is natural that this frightening imagery, combined with the fact that this is a foreign disease with a mysterious (and supposedly animal) origin, has caused the circulation of panic" (Garrett 2020).

Media amplification of herding behavior.

"The data show that small changes in the habits of a minority of shoppers prompted lurid headlines about empty shelves—which then made others, quite rationally, change their behavior." (Lewis 2020). "…(R)eading about PB or visuals of empty shopping shelves may lead to more people indulging in the same behavior: social learning theory (Bandura 1971) suggests as much" (Arafat et al. 2020a; Sim et al. 2020). "The more images of stockpiling that emerged on social media, the more panicky buying that ensued" (Corkery and Maheshwari 2020).

Visualizations accentuate the loop. If there's one image that captures the panic seeping through the United States it is empty store shelves where toilet paper sits no longer and... "(w)e're seeing pictures and videos of people shopping and empty shelves," said Andy Yap, professor of organizational behavior at INSEAD Business school in Singapore. "Seeing things makes it more salient, weighted more heavily in your head and it creates the echo chamber of people buying these products. (Devlin 2020; Yap 2020). "Media can display the photos of empty shelves indicating the scarcity of goods, specify the goods and increase tension, anxiety, and fear among the general population, resulting in a further increase in PB" (Arafat et al. 2020b).

9.8 Recommendations

There are many options. We can shut down information channels to make certain only the "correct" information is communicated. "The extent of false rumors and news depends on the rapid response of governments and the local city councils (sic), the imposition of sanctions and laws to deter people and unofficial parties from circulating and disseminating false information and creating a state of panic among citizens and creating a gap and a state of distrust between the official authorities and the general public...." (Almomani and Al-Qu'ran 2020). Obviously, this has drawbacks in a free society.

We can nudge people along using what we have learned from behavioral economics (Thaler and Sunstein 2009). Nudging will need certain characteristics. As Chris Stiff from Keele University hypothesized, "Social sanctions—such as naming and shaming or ostracizing those who hoard—can often be effective but relies on the person caring about being shamed. Direct punishment—such as fines—can work, but only if it is strong enough to be a deterrent to others. Too weak a punishment can actually increase selfishness if individuals realize the cost/benefit analysis will work in their favor" (Stiff 2020). Again, you can expect pushback.

What follows are the some of the prevalent and palatable options supported in the literature.

9.8.1 Communicate Security

Based on their understanding of which variables may be fluid and optimal for messaging, Garbe, Rau and Toppe added that communicators may wish to stress the functioning of supply chains and the long-term availability of vital commodities. Such rational appeals might exploit people's long-sightedness and effectively counter the dysfunctional intuition that commodities may become scarce in the near future" (Garbe et al. 2020).

"PB is a compensatory consumption behavior. Individuals purchasing products as a way to restore deficits triggered by perceived needs and desires that can only be fulfilled indirectly (Koles et al. 2018). In this regard, "the deficit refers to the loss of control over the situation, and this can be compensated through problem solving such as panic PB" (Yuen et al. 2020). Hence, it might be prudent to give the public some efficacy. "Just telling people to stop is not going to stop them. People are PB because of the need to feel they are in control. They need to be told or given something positive to do, such as helping out their elderly neighbors in isolation or donating to food banks, so they feel they are doing something to help" (Ellson 2020).

9.8.2 Debunk Misinformation

It is commonly assumed that misinformation is largely a phenomenon of social media which has led to calls for stricter regulation of the content on platforms such as Facebook and Twitter. However, misinformation appears in traditional media as well.

Here, it typically takes two forms: amplification of false claims through widespread coverage of prominent persons whose views and comments are considered newsworthy; and to a lesser degree, active fact-checking and debunking of false claims and misinformation (Evanega et al. 2020).

Faced with a fast-moving threat, people are hungry for information, which creates an opportunity for authorities to reach people who might ordinarily be unreceptive to public health advice. "It is up to officials to be agile and nimble and making the science as understandable as possible" (Devlin 2020).

"SteelFisher et al. (2015) recommend that for future outbreaks, public health officials should establish relationships with independent health professional associations that are trusted by the public and work with them to shape policies and messages with the public.

Debunking campaigns must be carefully crafted. While Monica Schoch-Spana from the Johns Hopkins Center for Health Security may be correct when she framed pandemics in a certain way: "People don't listen to health authorities on sunny days, but under extreme circumstances people are all ears" (Devlin 2020). On the flip side, some online debunking campaigns have been shown to create a reinforcement effect in usual consumers of conspiracy stories" (Bessi et al. 2015).

> Communication strategies (including specific messages, media vehicles, spokespeople and images targeted at different audiences) can be developed and pre-tested for use by government, medical authorities, non-government organizations and other relevant organizations in an attempt to increase the public's understanding of the risk. Such strategies will minimize fear, refute misinformation that the public may encounter (e.g., from co-workers or media sources) and enhance the likelihood of the public taking the recommended preventive and remedial actions should these become necessary (Jones et al. 2010).

9.8.3 Build Trust

"In times of crisis, public authorities tend to focus their concern on avoiding panic and filtering the information they provide to the public. But trust is a crucial support to public health systems. It is during crises such as the Covid-19 outbreak that this trust is put to the test" (Atlani-Dualt et al. 2020).

Trust is one of the cornerstones of outbreak communication (Seeger and Reynolds 2007) and the media is the primary communication outlet for updating the public, public health officials and the media must develop concrete communication strategies to build Americans' trust and confidence in order to be credible and prepared for future outbreaks" (Kelly et al. 2015).

Almost all of US adults (over 90%) reported trusting information about medical topics from doctors and other health care professionals in 2019. In contrast, a majority of respondents to a poll conducted on March 13–14, 2020 reported not trusting information about Covid-19 from President Trump, and about half did not trust information from the news media (Earnshaw and Katz 2020).

Much too often PB represents a lack of trust in the officials responsible for controlling the consequences of a disaster event and this is worsened as media and other sources amplify a non-existent risk. The public attitude is simple: if government can't help then we are on our own!

Garbe, Rau and Toppe's study indicates there is some opportunity to impact the public with messaging maybe with some trust-building.

> Around 20% of the differences in toilet paper consumption that were explained by feelings of threat were based on people's dispositional tendency to worry a lot and generally feel anxious. At the same time, the remaining 80% of this effect were not found to be rooted in personality differences. This suggests that how much people feel personally threatened by Covid-19 also depends on psychological factors not accounted for in our study or on malleable external factors such as the risk management by and trust in local authorities. Hence, these findings highlight the potential of public communication to address individuals' perceptions of threat and thereby alter their shopping behavior (Garbe et al. 2020).

Ostenhoff and Palmer (2020) drew a similar conclusion in their study of adolescents. "…[A]dolescents' beliefs about the severity of the virus, the extent to which they value social responsibility, their social trust, and their prioritization of their own self-interest over others were independently associated with their news monitoring, social distancing, disinfecting, and hoarding behavior in the days after the United States declared Covid-19 a national emergency."

Yuen pulled it all together. "…[A] high level of social trust would indicate that individuals would be more cooperative and considerate by not hoarding and sharing limited supplies with others. Conversely, a high level of social distrust could cause the public to act individualistically, fearing others to buy more than their share and leaving none for others. This triggers PB" (Yuen et al. 2020).

9.9 Conclusion

"Most people engaging in these behaviors would agree that toilet paper offers no real immunity against Covid-19. Over time, these seemingly protective and innocuous behaviors have the capacity to keep individuals in a cycle of continuous stress and anxiety and do compromise individual health," says clinical psychologist Sara Houshmand" (Whitehead 2020).

Disasters of all sorts will continue to plague humanity as will pandemics (Hotez 2021). Each of these events provoke consumer behavior which is undesirable. While this chapter focused on toilet paper, it might easily have been about N95 masks and other PPE (personal protective equipment) that are desperately needed by first responders and health professionals or pharmaceuticals and over-the-counter therapies exhausting supplies desperately needed by others. Also, and most disheartening, it might make it more difficult for many people with the least of means to provide for the general well-being of themselves and their families.

PB needs to be understood in order for us to be able to build strong messaging that not only discourages PB per se but also communicates the challenges of responding to a disaster, especially a pandemic with its extensive set of challenges such as behavioral changes, wearing masks properly, stay-at-home orders, and significantly abbreviated public behavior, and vaccine hesitancy.

Acknowledgements My gratitude to the Genetic Engineering in Society Center and the Public Communication of Science and Technology (PCOST) program for continuing support and the Research Triangle Nanotechnology Network for its ongoing sponsorship. All comments represent, my own and do not necessarily reflect the views of the National Science Foundation (ECCS 1542015), the RTNN and Duke University, University of North Carolina at Chapel Hill and North Carolina State University.

References

ABC News (2020) Toilet paper shortage from coronavirus panic buying sees argument break out at Woolworths in Sydney. ABC News. March 7. https://www.abc.net.au/news/2020-03-07/two-women-brawl-in-supermarket-over-toilet-paper/12036218. Accessed 22 Dec 2020

Alford H (2020) What would Freud make of the toilet-paper panic? The New Yorker. March 30

Almomani H, Al-Qu'ran W (2020) The extent of people's response to rumors and false news in light of the crisis of the corona virus. Ann Méd-Psycho 178:684–689

Amanzio M et al (2020) How do nocebo phenomena provide a theoretical framework for the COVID-19 Pandemic? Front Psychol. March 30.https://doi.org/10.3389/fpsyg.2020.589884.

Anastasiadou E et al (2020) The coronavirus' effects on consumer behavior and supermarket activities: insights from Greece and Sweden. Int J Sociol Soc Policy 40(9/10):893–907

Arafat SMY, Kar SK, Marthoenis M, Sharma P, Apu EH, Kabir R (2020a) Psychological underpinning of panic buying during pandemic (COVID-19). Psychiatry Res 289:103061

Arafat SMY, Kar SK, Menon V, Kaliamoorthy C, Mukherjee S, Alradie-Mohamed A, Sharma P, Marthoenis M, Kabir R (2020c) Panic buying: an insight from the content analysis of media reports during COVID-19 pandemic. Neurol Psychiatry Brain Res 37:101–102

Arafat SMY, Kar SK, Menon V, Marthoenis M, Sharma P, Alreadie-Mohamed A, Mukherjee S, Kaliamoorthy C, Kabir R et al (2020b) Media portrayal of panic buying: a content analysis of online news portals. Global Psychiatry 3(2):1–6. https://doi.org/10.2478/gp-2020-0022

Atlani-Dualt L, Ward JK, Roy M, Morin C, Wilson A (2020) Tracking online heroisation and blame in epidemics. Lancet 5:e137–e138

Bachelor L (2020) MPs in plea to government over UK's Covid19 stockpiling. Guardian. March 21. https://www.theguardian.com/world/2020/mar/21/mps-plea-government-uk-covid-19-stockpiling-coronavirus. Accessed 23 Dec 2020

Baddeley MC (2020). Toilet paper mania how behavioral economics can explain why people are stockpiling toilet paper. Psychology Today. March 8. https://www.psychologytoday.com/us/blog/copycats-and-contrarians/202003/toilet-paper-mania. Accessed 23 Dec 2020

Bagus P, Peña-Ramos JA, Sánchez-Bayón A (2021) COVID-19 and the political economy of mass hysteria. Int J Environ Res Public Health 18:1376. https://doi.org/10.3390/ijerph18041376

Ball P, Maxman A (2020) The epic battle against coronavirus misinformation and conspiracy theories. Nature. May 27. https://www.nature.com/articles/d41586-020-01452-z. Accessed 29 Dec 2020

Bandura A (1971) Social learning theory. General Learning Press, NY

Bessi A, Coletto M, Davidescu GA, Scala A, Caldarelli G, Quattrociocchi W (2015) Science versus conspiracy: collective narratives in the age of misinformation. PLoS ONE 10(2):e0118093. https://doi.org/10.1371/journal.pone.0118093

Besson EK (2020) COVID-19 (coronavirus): panic buying and its impact on global health supply chains. World Bank Blogs. April 28. https://blogs.worldbank.org/health/covid-19-coronavirus-panic-buying-and-its-impact-global-health-supply-chains. Accessed 29 Dec 2020

Boxall E (2020) Lessons from history: 'Don't panic' buying. Investors Chronicle. November 13. https://www.invetorschronicle.co.uk/education/2020/11/12/lessons-from-history-don-t-panic-buying/. Accessed 23 Dec 2020

Brodeur N (2020) Toilet paper shelves again left bare, as grocery store shoppers worry about Washington restrictions. The Seattle Times. November 16. https://www.seattletimes.com/seattle-news/toilet-paper-shelves-again-left-bare-as-grocery-store-shoppers-worry-about-washington-restrictions/. Accessed 22 Dec 2020

Buchholz K (2020) Toilet paper producers roll'ing in the dough. Statista. April 2. https://www.statista.com/chart/21327/rise-in-revenue-toilet-paper-selected-countries/#:~:text=Because%20not%20only%20food%2C%20but,the%20same%20month%20last%20year. Accessed 22 Dec 2020

Bussel RK (2020) Want to know why people stockpile toilet paper? I'm a hoarder and I have a few ideas. Huffington Post Personal. March 30. https://www.huffpost.com/entry/coronavirus-toilet-paper-hoarding_n_5e7a2320c5b62f90bc51e1db. Accessed December 29, 2020.

Chadwick J (2020) Why people are panic-buying toilet paper during the coronavirus crisis: pandemic expert says a heightened sense of DISGUST at dirt and germs during outbreaks fuels the behaviour. Mail Online. November 2. https://www.dailymail.co.uk/sciencetech/article-8160043/Pandemic-psychologist-explains-people-panic-buy-toilet-roll.html. Accessed 28 Dec 2020

Charlier P, Brun L, Huynh-Charlier. (2012). Toilet hygiene in the classical era. Br Med J. December 17

Cheng Y, Luo Y (2020) The presumed influence of digital misinformation: examining US public's support for governmental restrictions versus corrective action in the COVID-19 pandemic. Online Inf Rev. December 2

Cinelli M et al (2020) The COVID-19 social media infodemic. Sci Rep 10:16598. https://doi.org/10.1038/s41598-020-73510-5

Corkery, M. & Maheshwari, S. (2020). Is there really a toilet paper shortage? The New York Times. December 2. https://www.nytimes.com/2020/03/13/business/toilet-paper-shortage.html. Accessed 23 Dec 2020

DePaulo B (2020) Why are people hoarding toilet paper? Psych Central. https://psychcentral.com/blog/why-are-people-hoarding-toilet-paper/. Accessed 23 Dec 2020

Depoux A, Karafillakis E, Wilder-Smith A (2020) The pandemic of social media panic travels faster than the COVID-19 outbreak. J Travel Med. March 3. https://doi.org/10.1093/jtm/taaa031
Devlin, H. (2020). Don't let coronavirus tip society into panic, say psychologists. Guardian. March 6. https://www.theguardian.com/science/2020/mar/06/dont-let-coronavirus-tip-society-into-panic-say-psychologists. Accessed 26 Dec 2020
Earnshaw VA, Katz IT (2020) Educate, amplify, and focus to address COVID. JAMA Network. April 17. https://jamanetwork.com/channels/health-forum/fullarticle/2764847. Accessed 11 Feb 2021
Ellson, A. (2020). Pandemic psychologist explains lavatory roll panic. The Times. March 27. 14. https://www.thetimes.co.uk/article/coronavirus-pandemic-psychologist-explains-lavatory-roll-panic-tjddffqbd. Accessed 9 Jan 2021
Evanega, S., Lynas, M., Adams, A. & Smolenyak, K. (2020). Coronavirus misinformation: quantifying sources and themes in the COVID-19 'infodemic'. Alliance for science. https://allianceforscience.cornell.edu/wp-content/uploads/2020/10/Evanega-et-al-Coronavirus-misinformation-submitted_07_23_20-1.pdf. Accessed 11 Feb 2021
Frost RO, Hartl TL (1996) A cognitive-behavioral model of compulsive hoarding. Behav Res Ther 34(4):341–350. https://doi.org/10.1016/0005-7967(95)00071-2
Fruen L (2020) Here we go again: panic buyers across U.S. empty shelves of toilet paper, food and disinfectant as 40 states see record daily increases in cases and 20 states report all-time highs in deaths. The Daily Mail. November 17. https://www.dailymail.co.uk/news/article-8957785/Panic-buying-2-0-Shoppers-hoard-essentials-states-impose-stricter-COVID-lockdowns.html. Accessed 3 Jan 2021
Garbe L, Rau R, Toppe T (2020) Influence of perceived threat of Covid-19 and HEXACO personality traits on toilet paper stockpiling. PLoS ONE 15(6):e0234232
Garrett O (2020) A case of modern mass hysteria? The coronavirus. Exeposé. February 13. https://exepose.com/2020/02/13/a-case-of-modern-mass-hysteria-the-coronavirus/. Accessed 23 Dec 2020
Ghebreyesus TA WHO (World Health Organization) (2020) In: Munich security conference. February 15. https://www.who.int/director-general/speeches/detail/munich-security-conference. Accessed 29 Dec 2020
Hao K, Basu T (2020) The coronavirus is the first true social-media "infodemic". MIT Technol Rev. February 12https://www.technologyreview.com/2020/02/12/844851/the-coronavirus-is-the-first-true-social-media-infodemic/. Accessed 9 Jan 2021
Harmon-Jones E, Simon L, Greenberg J, Pyszczynski T, Solomon S, McGregor H (1997) Terror management theory and self-esteem: evidence that increased self-esteem reduced mortality salience effects. J Pers Soc Psychol 72(1):24–36
Heese M (2020) Pandemic, panic and toilet paper math. The Washington Post. March 14. https://www.washingtonpost.com/lifestyle/style/pandemic-panic-and-toilet-paper-math/2020/03/14/8945312e-657b-11ea-b3fc-7841686c5c57_story.html. Accessed 26 Dec 2020
Hernandez S (2020) Panic spreads fear especially on social media, local doctor says. KXAN NEWS. https://www.kxan.com/news/stop-hoarding-toilet-paper-says-local-psychologist-youre-spreading-fear-faster-than-covid-19/. Accessed 29 Dec 2020
Hotez PJ (2021) Preventing the next pandemic: vaccine diplomacy in a time and anti-science. Johns Hopkins University Press, Baltimore, MD
Hughes W (1988) A tribute to toilet paper. Rev Infect Dis 10(1):218–222
INSEAD (2020) The psychology behind coronavirus panic buying. https://knowledge.insead.edu/economics-finance/thepsychology-behind-coronavirus-panic-buying-13451. Accessed 29 Dec 2020
Jones SC, Waters L, Holland O, Bevins J, Iverson D (2010) Developing pandemic communication strategies: Preparation without panic. J Bus Res 63:126–132
Karson K (2020) Two-thirds of Americans concerned about contracting coronavirus, as country grapples with growing crisis: POLL. ABC NEWS. March 13. https://abcnews.go.com/Politics/

thirds-americans-concerned-contracting-coronavirus-country-grapples-growing/story?id=69557806. Accessed 24 Dec 2020
Kasperson RE et al (1988) The social amplification of risk: a conceptual framework. Risk Anal 8(2):177–187
Kay AC et al (2009) Compensatory control achieving order through the mind, our institutions, and the heavens. Curr Dir Psychol Sci 18:264–268. https://doi.org/10.1111/j.1467-8721.2009.01649.x
Keane M, Neal T (2021) Consumer panic in the COVID-19 pandemic. J Econometrics 220(1):86–105. https://doi.org/10.1016/j.jeconom.2020.07.045
Kelly B, Squiers L, Bann C, Stine A, Hansen H, Lynch M (2015) Perceptions and plans for prevention of ebola: results from a national survey. BMC Public Health 15:1136. https://doi.org/10.1186/s12889-015-2441-7
Kennett-Hensel P, Sneath J, Lacey R (2012) Liminality and consumption in the aftermath of a natural disaster. J Consum Mark 29:52–63
Kilgo DK, Yoo J, Johnson TJ (2019) Spreading ebola panic: newspaper and social media coverage of the 2014 ebola health crisis. Health Commun 34(8):811–817. https://doi.org/10.1080/10410236.2018.1437524
Koles B, Wells V, Tadjewski M (2018) Compensatory consumption and consumer compromises: a state-of-the-art review. J Mark Manage 34(1–2):96–133
Laato S, Islam AKMN, Farooq A, Dhir A (2020) Unusual purchasing behavior during the early stages of the COVID-19 pandemic: the stimulus-organism-response approach. J Retail Consum Serv 57:102224
Lee K, Ashton MC (2008) The HEXACO personality factors in the indigenous personality Lexicons of english and 11 other languages. J Pers 76(5):1001–1054
Lewis H (2020) How panic-buying revealed the problem with the modern world. The Atlantic. March 20. https://www.theatlantic.com/international/archive/2020/03/coronavirus-panic-buying-britain-us-shopping/608731/. Accessed 28 Dec 2020
Lins S, Aquino S (2020) Development and initial psychometric properties of a panic buying scale during COVID-19 pandemic. Heliyon 6(9):E04746. https://doi.org/10.1016/j.heliyon.2020.e04746
Loeb M (2016) The rich life measured in toilet paper. Psychother Consultants. March 20. http://psychotherapyconsultants.blogspot.com/2016/11/the-rich-life-measured-in-toilet-paper.html. Accessed 17 Feb 2021
Loxton M et al (2020) Consumer behaviour during crises: preliminary research on how coronavirus has manifested consumer panic buying, herd mentality, changing discretionary spending and the role of the media in influencing behaviour. J Risk Financ Manage 13(8):166. https://doi.org/10.3390/jrfm13080166
Lufkin B (2020) Coronavirus: the psychology of panic buying. March 4. BBC. https://www.bbc.com/worklife/article/20200304-coronavirus-covid-19-update-why-people-are-stockpiling. Accessed 29 Dec 2020
Mao F (2020) Coronavirus panic: why are people stockpiling toilet paper? BBC—Australia. March 4. https://www.bbc.com/news/world-australia-51731422. Accessed 22 Dec 2020
Mathes BM, Oglesby ME, Short NA, Portero AK, Raines AM, Schmidt NB (2017) An examination of the role of intolerance of distress and uncertainty in hoarding symptoms. Compr Psychiatry 72:121–129
Matthews C (2020) Hoarding, stockpiling, panic buying: what's normal behavior in an abnormal time? Conversation. November 17. https://theconversation.com/hoarding-stockpiling-panic-buying-whats-normal-behavior-in-an-abnormal-time-149422. Accessed 22 Dec 2020
McCrae M (2020). Study reveals which people bought all that toilet paper during the pandemic. Sci Alert. June 12. https://www.sciencealert.com/study-finds-just-the-kind-of-people-who-were-panic-buying-all-of-the-toilet-paper. Accessed 24 Dec 2020
Mead E (2020). How personality type explains why we panic-buy toilet paper. Psych Central. April 7. https://www.pinterest.com/pin/147492956535432391/. Accessed 29 Dec 2020

Murphy K (2020). Stop using toilet paper. The New York Times. April 3. https://www.nytimes.com/2020/04/03/opinion/toilet-paper-hoarding-bidets.html. Accessed 24 Dec 2020
Muwahed J (2020) Coronavirus pandemic goes viral in the age of social media, sparking anxiety. ABC NEWS. March 13. https://abcnews.go.com/Politics/coronavirus-pandemic-viral-age-social-media-sparking-anxiety/story?id=69580796. Accessed 24 Dec 2020
Naeem M (2021) Do social media platforms develop consumer panic buying during the fear of covid-19 pandemic. J Retail Consum Serv 58. https://doi.org/10.1016/j.jretconser.2020.102226
Nguyen H, Nguyen A (2020) Covid-19 misinformation and the social (media) amplification of risk: a Vietnamese perspective. Media Commun 8(2):444–447. https://doi.org/10.17645/mac.v8i2.3227
Ning L et al (2020) The impacts of knowledge, risk perception, emotion and information on citizens' protective behaviors during the outbreak of COVID-19: a cross-sectional study in China. BMC Public Health 20:1751. https://doi.org/10.1186/s12889-020-09892-y
Ohnsman A (2020) The national toilet paper scare: why our supplies are safe. Forbes. March 16. https://www.forbes.com/sites/alanohnsman/2020/03/16/the-national-toilet-paper-sca.re-why-our-supplies-are-safe/?sh=51c704569a47. Accessed 23 Dec 2020
Ostenhoff B, Palmer CA (2020) Attitudes and psychological factors associated with news monitoring, social distancing, disinfecting, and hoarding behaviors among US adolescents during the coronavirus disease 2019 pandemic. JAMA Pediatr June 29. https://doi.org/10.1001/jamapediatrics.2020.1876.
Otsuki GJ (2020) Shit's getting real: a cultural analysis of toilet paper. Blog: anthropology of science, technology and Japan. March 11. https://www.gjotsuki.net/shits-getting-real-a-cultural-analysis-of-toilet-paper/. Accessed 23 Dec 2020
Prentice C, Chen J, Stantic B (2020) Timed intervention in COVID-19 and panic buying. J Retail Consum Serv 57. https://doi.org/10.1016/j.jretconser.2020.102203. Accessed 11 Feb 2021
Rapoport A, Chammah A (1965) Sex differences in factors contributing to the level of cooperation in the prisoner's dilemma game. J Pers Soc Psychol 2(6):831–838
Reynolds B, Seeger MW (2007) Crisis and emergency risk communication as an integrative model. J Health Commun 10(1):43–55
Rose PI (2020) Book review: Frank M. Snowden, epidemics and society: from the black death to the present. Society. July 20. 1-3.https://doi.org/10.1007/s12115-020-00517-z. Accessed 15 Feb 2021
Routledge C, Vess M (eds) (2018) Handbook of terror management theory. Elsevier Science and Technology Book, Oxford
Sheu JB, Kuo HT (2020) Dual speculative hoarding: a wholesaler-retailer channel behavioral phenomenon behind potential natural hazard threats. Int J Disaster Risk Reduction 44:101430
Sigurvinsdottir R, Thorisdottir IE, Gylfason HF (2020) The impact of COVID-19 on mental health: the role of locus on control and internet use. Int J Environ Res Public Health 17:6985
Sim K, Chua HC, Vieta E, Fernandez G (2020) The anatomy of panic buying related to the current COVID-19 pandemic. Psychiatry Res 288:113015. https://doi.org/10.1016/j.psychres.2020.113015
Singh CK, Rakshit P (2020) A critical analysis to comprehend panic buying behaviour of Mumbaikar's in COVID-19 era. Stud Indian Place Names (UCG Care J) 40–60
Sneath JZ, Lacey R, Kenneth-Hensel PA (2009) Coping with a natural disaster: losses, emotions, and impulsive and compulsive buying. Mark Lett 20:45–60
Steelfisher GK, Blendon RJ, Lasala-Blanco N (2015) Ebola in the United States—public reactions and implications. New England J Med 373(9):789–791
Steinbuch Y (2020) COVID-19 panic buying: toilet paper, essentials fly off shelves again. New York Post. November 17. https://nypost.com/2020/11/17/covid-19-panic-buying-toilet-paper-essentials-fly-off-shelves/. Accessed 22 Dec 2020
Sterman JD, Dogan G (2015) "I'm not hoarding, I'm just stocking up before the hoarders get here". Behavioral causes of phantom ordering in supply chains. J Oper Manage 39–40:6–22

Stiff C (2020) The game theory of panic-buying—and how to reduce it. Conversation. March 13. https://theconversation.com/the-game-theory-of-panic-buying-and-how-to-reduce-it-134107. Accessed 28 Dec 2020

Swain S (2020) Woolworths halts 'click and collect' shopping, home delivery in Victoria amid 'exceptional' sales spike. 9News. March 16. https://www.9news.com.au/national/coronavirus-woolworths-to-introduce-elderly-only-shopping-hour/1704e12a-cbd3-4a16-879e-1b0b8dabca6b. Accessed June 2, 2021

Thaler R, Sunstein C (2009) Nudge: improving decisions about health, wealth, and happiness. Penguin Books, NY

Trading Economics (2020) U.S. Retail sales 1992–2020 Data. https://tradingeconomics.com/united-states/retail-sales. Accessed 11 Jan 2021

Tsao YC, Raj PVRP, Yu V (2019) Product substitution in different weights and brands considering customer segmentation and panic buying behavior. Ind Mark Manage 77:209–220. https://doi.org/10.1016/j.indmarman.2018.09.004

UTS (University of Technology—Sydney) Press Release (2020) Hoarding and herding during the COVID-19 pandemic. Sci Daily. September 10. https://www.sciencedaily.com/releases/2020/09/200910100608.htm. Accessed 26 Dec 2020

Ungar, S. (1998). Hot crises and media reassurance: a comparison of emerging diseases and ebola Zaire. Br J Sociol 49(1):36–65. March. https://doi.org/10.2307/591262. Accessed 15 Feb 2021

Uscinski JE, Enders AM (2020) The coronavirus conspiracy boom. Atlantic. April 30. https://www.theatlantic.com/health/archive/2020/04/what-can-coronavirus-tell-us-about-conspiracy-theories/610894/. Accessed 3 Jan 2021

Valenzano A et al (2020) The social brain and emotional contagion: COVID-19 effects. Medicina 56:640

WHO (World Health Organization) (2020) Infodemic management. https://www.who.int/teams/risk-communication/infodemic-management. Accessed 11 Jan 2021

Wahl-Jorgensen K (2020). Coronavirus: how media coverage of epidemics often stokes fear and panic. Conversation. February 14. https://theconversation.com/coronavirus-how-media-coverage-of-epidemics-often-stokes-fear-and-panic-131844. Accessed 14 Feb 2021

Whitehead K (2020) Why Hong Kong panic buying happened: herd mentality, the media, over-reaction and distrust. South China morning post. February 25. https://www.scmp.com/lifestyle/health-wellness/article/3052101/why-hong-kong-panic-buying-happened-herd-mentality-media. Accessed 3 Jan 2021

Wood KA (2020) A psychoanalysis of toilet paper and freudian wisdom in the corona-pocalypse. Medium. March 23. https://medium.com/@kristiewood/a-psychoanalysis-of-toilet-paper-freudian-wisdom-in-the-corona-pocalypse-e664f2c2e3c6. Accessed 29 Dec 2020

Yap A (2020) Coronavirus: the psychology behind panic-buying. DW (Deutsche Welle). March 11. https://www.dw.com/en/coronavirus-the-psychology-behind-panic-buying/a-52721736. Accessed 23 Dec 2020

Yoshizaki HTY, de Brito Dr, I, Hino CM, Aguiar LL, Pinheiro MCR (2020) Relationship between panic buying and per capita income during COVID-19. Sustainability 12. https://ideas.repec.org/a/gam/jsusta/v12y2020i23p9968-d452933.html. Accessed 7 Jan 2021

Yuen KF, Wang X, Ma F, Li KX (2020) The psychological causes of panic buying following a health crisis. Int J Environ Res Public Health 17:3513. https://doi.org/10.3390/ijerph17103513

Zarocostas J (2020) How to fight an infodemic. The Lancet. February 29. https://doi.org/10.1016/S0140-6736(20)30461-X. Accessed June 2, 2021

Zheng R, Shou B, Jang J (2020) Supply disruption management under consumer panic buying and social learning effects. Omega. https://doi.org/10.1016/j.omega.2020.102238. Accessed 14 Feb 2021

Zogorsky J (2020) There's plenty of toilet paper—so why are people hoarding it? Conversation. March 23

Chapter 10
Celebrity, Resilience, and Communication: The Role of *Some Good News* During the Covid-19 Pandemic

Elizabeth Fish Hatfield

Abstract This chapter considers the role of celebrities and entertainment personalities in the first months of the Covid-19 pandemic. As the United States began quarantining and social distancing, the usual coverage of celebrities "just like us" shifted. Not only did the activities change so did the voice, as paparazzi could no longer snap celebrities eating out or meeting with friends. Celebrity pandemic communication operated in many ways: by not communicating, by promoting safety-oriented behaviors, acting as virtual entertainers, and sharing cautionary tales. For this paper, we will analyze the online show Some Good News that emerged in March 2020. Hosted by John Krasinski, the show's purpose was clear: as the news focused almost entirely on the global Covid-19 numbers, SGN would work to promote positive news meant to be a bright alternative for audiences. This chapter will explore how SGN offered an outlet that ultimately reinforced the role of celebrity culture as influential role models, aspirational motivators, and behavioral leaders even as it served an important need for audiences during the early months of 2020.

Keywords Celebrity culture · Covid-19 · "Some Good News" · Positive news · Krasinski

10.1 Introduction

In times of crisis, celebrities often communicate messages meant to promote resilience, optimism, and hope when audiences might otherwise feel quite the opposite. Star-studded fundraisers bring in necessary support in moments of utter devastation such as hurricanes, tsunamis, and fires. The celebrity impact for these causes is undeniable, as Brown argues that celebrities remain "powerful agents of social change" (2004, p. 97) with significant social capital. During 2020, the global coronavirus pandemic presented a new form of crisis. Unlike typical disasters, Covid-19 was not geographically limited. Instead, the health crisis impacted worldwide populations, including celebrities, with serious, fatal potential. As United States (US)

E. Fish Hatfield (✉)
University of Houston-Downtown, Houston, TX, USA

D. M. Berube (ed.), *Pandemic Communication and Resilience*, Risk, Systems and Decisions, https://doi.org/10.1007/978-3-030-77344-1_10

citizens watched the pandemic ravage nations like China and Italy in early 2020, the local presence of Covid-19 steadily emerged on both the east and west US coastlines. While some local leaders gave clear instructions for preventing the disease from spreading (Landin 2020), the state-to-state approach varied significantly. However, in the earliest days of the Covid-19 pandemic, one message for affected areas was generally clear: stay home (The Coronavirus Outbreak March 20, 2020).

This chapter considers the role of celebrities and entertainment personalities in the first months of the Covid-19 pandemic when they, too, were asked to stay home, social distance, and follow health recommendations meant to curb the spread of the virus. Celebrity pandemic communication operated in many ways: reducing content for celebrity news publications, individually promoting safety-oriented behaviors, creating personal content as virtual entertainers, and sharing cautionary tales. For this chapter, we will analyze the online show *Some Good News* (SGN) that emerged in March 2020. Hosted by John Krasinski, star of television shows *The Office* and *Jack Ryan*, SGN's purpose differed from other media messages. As traditional news focused almost entirely on global Covid-19 statistics, SGN would work to promote positive news meant to be a bright alternative for audiences. This chapter explores how SGN ultimately reinforced the role of celebrities as influential role models, aspirational motivators, and behavioral leaders even as it served an important need for audiences during the early months of 2020.

10.2 The Role of News During the Covid-19 Pandemic

Prior to the Covid-19 pandemic, the news environment in the United States had become highly polarized (Jurkowitz and Mitchell 2020). Though liberal viewers were more likely to trust the news in general as compared to conservative viewers, each group placed greater levels of trust in opposing news outlets (Jurkowitz and Mitchell 2020). Regardless of the source, when the coronavirus began spreading in the United States audiences paid attention. A March 2020 Pew Research Study found 89% of US adults were watching Covid-19 news either very closely or fairly closely (Mitchell and Oliphant 2020). However, early news and messaging about the pandemic varied based on the chosen news source (Jurkowitz and Mitchell 2020), resulting in debate about its severity (Jurkowitz and Mitchell 2020). As news stations presented up to the minute counts of both current and recovered US coronavirus case numbers, audiences could not avoid coverage of the growing pandemic regardless of the news source they engaged.

Celebrity news also changed abruptly in March 2020 because, as Carras wrote, "the stars are not out these days" (2020). When the nation slowed down, so did celebrities. Like all other non-pandemic news, celebrity media coverage vanished seemingly overnight. As Rottenberg (2020) noted:

> With the majority of the nation under lockdown, celebrities, like the rest of us, are stuck in their homes. The beaches where they might otherwise be spotted sunbathing are empty. The cafes, restaurants, and nightclubs they normally frequent are closed. The film and TV

> sets where they work are shut down. The gyms and yoga studios where they exercise are shuttered. The red carpets they strut during film premieres have been rolled up and stowed away indefinitely.

"Just like us," celebrities were not immune to quarantining and social distancing. Harnessing their social power, celebrities quickly began documenting their quarantine experience on social media with messages promoting social distancing (Carras 2020). Amidst this environment, *Some Good News* emerged. Created by John Krasinkski from his home office, the show differed from other celebrity communications as a more organized media message and way for Krasinski himself to be productive during quarantine.

10.3 Some Good News

In many ways, SGN had a familiarity to it. It was celebrity-driven media content in response to a crisis with a goal of entertainment. However, one big difference characterized this event: it was not a one-time fundraiser. Instead, the program gave Krasinski and his peers an ongoing focus during quarantine while meeting an unprecedented demand for good news that went beyond *SGN* (Lorenz 2020). Within a week of posting its first episode to Youtube, SGN had over 1.5 million subscribers (Lorenz 2020). In total, Krasinski produced eight episodes between March and May 2020. Eventually, he returned to work and had his own good news: selling the show's rights to Viacom CBS (Goldberg 2020). While some audiences disagreed with the sale, Krasinski noted the show "wouldn't be sustainable with my prior commitments" (Henderson 2020).

The concept of good news has not been widely conceptualized in academic scholarship. McIntyre (2016) defines good news as:

> Stories with particularly positive overtones, such as rescues, cures, acts of heroism, economic growth, reunifications, or love. These stories often do not have conflict. Think of a good news story as one in which the majority of the site's readers/viewers would be satisfied or pleased that the event happened or happened as it did. (p. 224)

Good news stories generally emphasize "entertainment and emotional impact," (McIntyre 2016, p. 228) and audiences feel good after watching them (McIntyre and Gibson 2016). Many of the stories presented on SGN fell under a category labeled in McIntyre and Gibson's work as the silver-lining story (2016). A silver-lining story includes positive outcomes that result from a negative situation (McIntyre and Gibson 2016). With Covid-19 as a framework for most of SGN's narratives, the silver-lining story as a form of good news allowed Krasinski to report on the current pandemic, while adhering to the goals of the show.

SGN presented a unique message of resilience and positivity, working to make sense of a new, relatively unprecedented health crisis. Hosted by a popular celebrity figure and frequented by a diverse array of other media stars from all industries, SGN's tone struck the right balance of sensitivity without being saccharine. As Park wrote: "Krasinski's uplifting humanity-first attitude feels perfectly matched for the era; he doesn't shy away from the crisis, but he doesn't wallow in it either" (2020). For this project, the role of celebrity communication on SGN will be analyzed using qualitative thematic analysis as described by Terry et al. (2017). Project results made clear that SGN was more than a news show. It was an important tool for shaping audience understanding and behavior at a time when critical health outcomes were at stake.

10.4 The Role of Celebrity Communication and Influence on Audiences

The celebrity persona remains both ordinary and extraordinary—making a beloved star simultaneously likeable and special (Brown and Fraser 2004). As audiences identify with celebrities, research indicates they are more likely to modify behavior because of a celebrity's educational-entertainment messages as compared to other forms of information (Brown and Fraser 2004). Educational messages in narrative form may successfully avoid issues such as "reactance, counterarguing, and perceived invulnerability" that impact how a message is perceived (Moyer-Gusé and Nabi 2010, p. 26). Indeed, the "intrinsically persuasive" (Dahlstrom 2014, p. 13,614) nature of narratives frequently found in educational-entertainment can be useful for sharing health information with non-expert audiences. As long as persuasive intent is not obvious, the message may successfully influence audiences (Dahlstrom 2014). However, sensing persuasive goals leaves audiences feeling manipulated and therefore, negates the usefulness of the message (Dahlstrom 2014). SGN, while presented as a news program, arguably serves instead as entertainment media with its focus on feel-good stories, as the narrative format "has long been known to influence perceptions about the real world" (Dahlstrom 2014, p. 13,616). For those at home, SGN offered fresh content when little else was available and viewer time was in abundance.

Educational-entertainment programs can influence audience behavior because within their content, the celebrity acts as a social model. Social models "serve as transmitters of knowledge, values, cognitive skills, and new styles of behavior" (Bandura 2003, p. 78). Audiences may look to celebrities to confirm their own ideas, finding reinforcement when a celebrity acts similarly, or alternately, gaining new information that may shift how one makes sense of a situation or event. Learning does not have to be intentional to occur, and a "vast amount of information about human values, styles of thinking, and behavior patterns is gained from the extensive modeling in the symbolic environment of the mass media" (Bandura 2002, p. 126). As people watch and learn from the media, behavioral change hinges on self-efficacy. If a person does not believe they are capable of mirroring the celebrity's behavior or taking on a new style of behavior, that person may not find the effort worth attempting. As Bandura writes, "people must have a strong belief in their own efficacy in order to sustain the perseverant effort needed to succeed" (2003, p. 79). Watching a social model prevail builds self-efficacy, as well as working towards a goal in manageable steps versus overwhelmingly large tasks (Bandura 2003). While self-efficacy is critical, Bandura identifies that collective efficacy also matters. "The strength of families, communities, school systems, business organizations, social institutions, and even nations lies partly in people's sense of collective efficacy that they can solve the problems they face and improve their lives through unified effort," writes Bandura (2003, p. 80). While media messages may or may not change audience behavior, the celebrity as a social model broadens the experience of viewers with the potential to demonstrate ways to build the skills needed to navigate new environments or expectations at the individual and communal level.

10.5 Some Good News During a Period of No Good News

SGN's episodes focused on life during quarantine in the early days of the 2020 pandemic. Two themes emerge from the messages communicated via SGN. First, the celebrity-hosted show repeatedly reinforces the value and necessity of personal sacrifice. Second, the show presents creativity as a form of resilience. Krasinski shares ways to survive and thrive amidst a pandemic by utilizing silver-lining narratives about everyday people and encouraging efficacy and adherence to new "bizarre" norms for everyday behavior (Some good news 2020f). While the show set out to entertain during the pandemic, it ultimately communicates important information about health behaviors and navigating new social norms.

10.5.1 The Value of Sacrifice

In the spring of 2020, people around the globe were asked to stay home, social distance, and wear masks to protect vulnerable populations and slow the onslaught of cases facing medical professionals. The US cultural value of individual freedom often contradicts the idea of prioritizing the greater good. However, pandemic orders to quarantine at home did just that. The personal sacrifice staying home requires is first modeled by Krasinski. Unlike the mainstream news, which presented quarantine as a directive, Krasinski does not instruct audience members, treating quarantine as normative versus debatable. Krasinski models health-conscious behavior for his audience by filming from his home office wearing the stereotyped quarantine virtual meeting outfit of a jacket, tie, and something unexpected on the bottom. In the first episode, Krasinski speaks as if talking to an audience, but quickly corrects himself for viewers, "I am alone in this room" (Some good news 2020a). Throughout the series, Krasinski reports that his family has remained home and not seen friends or family. He commiserates with parents schooling at home and connects with friends via Zoom. In the final episode, viewers see a behind-the-scenes video demonstrating how Krasinski used Zoom to work with the show's producers (both also working from home). Rather than mope about the missed opportunities in his own life, Krasinski takes pleasure from recognizing others and encourages audiences to do the same. He tells them "this level of joy is contagious" (Some good news 2020d) and regularly reminds them that "no matter how tough life can get, there's always good in the world" (Some good news 2020a). The positive outlook Krasinski presents is also modeled by the many celebrities he brings onto the show. Regardless of the weather, celebrity observers note: "It looks, ah, pretty good" (Some good news 2020d).

As a popular celebrity, Krasinski utilizes his platform to encourage health-conscious behaviors recommended by the Center for Disease Control by presenting them as typical. When discussing the cultural environment, Krasinski speaks inclusively, always using the term "we" to describe the experience. The viewers of SGN are assumed to be quarantining at home, a behavior Krasinski communicates to be both expected and challenging. In the second episode, he tells a young guest: "This corona thing is a real bummer, isn't it? But the social distancing thing is very important" (Some good news 2020a). In this moment, Krasinski empathizes as someone who, just like many others, is home with his family waiting out the initial first wave of the coronavirus in the United States. He also recognizes the behavior of others, rewarding them with praise. During a story on health care workers in Boston, Krasinski tells them, "I love that you are social distancing!" (Some good news 2020c). Krasinski models and rewards adherence to recommended health behaviors.

On SGN, Krasinski recognizes big and small acts as equally important. Krasinski's interaction with a hospital worker unable to see her family highlights his appreciation: "We are all missing something, but you guys are missing something on an extreme scale that we can never repay you for" (Some good news 2020c). Indeed, while healthcare "heroes" put their "lives on the line" (Some good news 2020d) for everyone else, others sacrifice by staying home. Krasinski presents sacrifice as something everyone is capable of, even if that sacrifice may look and feel different. In Episode 1 Krasinski interviews Coco, a young girl welcomed home by a socially distant car parade after completing her final cancer treatment. Video shows joy and support for the girl, who tells Krasinski:

> There are a lot of people like me going through things that have low immune systems, or everyone that isn't going to be extremely affected by it, to protect people that will be. It's like everyone came together to help the people that will be affected by it (Some good news 2020a).

While Coco's story shares the special moment of her homecoming, this silver-lining story results from her underlying narrative of severe illness. Coco's story represents one of the many reasons sacrifice is worth doing and demonstrates how one community came together to maintain safe health practices while supporting a loved member. Presenting Coco as optimistic and grateful, SGN reminds audiences of more challenging scenarios (such as a child with cancer) that frame the simplicity of staying home as easier in comparison. Similarly, a video chat with astronauts getting ready to return home after months on the International Space Station reminds audiences that many have gone through periods of relative social isolation in pursuit of a greater goal.

As the host, Krasinski presents an empathetic voice in urging quarantine, relating to the feeling of fatigue, and desiring normal activity just as audience members might. He urges viewers to be patient, relating to their emotional state in commenting: "You know what, it's okay if you have asked yourself every now and again, what is this all for?" (Some good news 2020c). During these moments, Krasinski works to build audience efficacy. Treating each week as a small step to complete, audience members are encouraged to continue their success staying home. Though he notes, "we are by no means out of the woods," he shares stories of patients recovering and leaving the hospital as signs that "there's light at the end of the tunnel" (Some good news 2020c). Indeed, throughout the series, Krasinski highlights the sickest cases of Covid-19 as a reminder of the seriousness presented by the disease even as he demonstrates health workers' success fighting it. Turning these health narratives into his typical upbeat story, Krasinski jokes that survivors are competing for the best one-liners as they exit the hospital. Stories of collective efficacy, such as Coco's parade or medical teams' victories, reinforce individual self-efficacy by communicating the value of teams working together.

SGN treats health care workers as possessing the efficacy to do their jobs, a distinction from the communicated need to build and encourage ongoing self-efficacy to stay home. As Bandura identified, social learning through models is particularly useful for new behaviors and staying home presented a drastic behavioral change for many Americans. Krasinski recognizes different forms of sacrifice as equally important, thanking both front line workers and audiences at home for their willingness to sacrifice what they can, while also presenting other less common forms of sacrifice. A mailman offers to run errands for anyone on his local route, a bar owner removes years of decorative signed dollar bills so her employees can eat, Trader Joe's employees envision themselves as superheroes, and a taxi driver in Spain gives free rides to and from the hospital each day. Each story presents personal sacrifices with positive attitudes. These narratives model individual ways people can go above and beyond depending on their personal ability.

Krasinski regularly refers to the "weird world of isolation," (Some good news 2020h) but makes clear that quarantine offers an important way ordinary people can help out during the global pandemic. Indeed, the commitment to sacrifice emerges as a shared value for the SGN community. Throughout the stories shared, with visuals of Krasinski, other celebrities, and regular families in their personal homes, SGN offers audiences a narrative demonstrating what sacrifice should look like and how audience members can make behavioral changes that impact greater society. The sacrifices of being an essential worker, staying home, and acts of kindness and generosity are presented as equally important ways people work toward a shared goal of returning to normal—an important message that instills efficacy. As Krasinski reminds viewers, "when this thing ends, and it will come to an end," (Some good news 2020e) he, too, will appreciate uneventful trips out of the house.

10.5.2 The Need for Creativity

While presenting sacrifice as a testament to the human spirit, SGN describes early 2020 as a challenging and "weird" time (Some good news 2020h). As people in the United States slowed down and stayed home, SGN emerged as a unique voice of encouragement during the pandemic. In the early weeks of the pandemic, the American public was hungry for distracting entertainment as they made sense of confusing media messages surrounding Covid-19 (Park 2020). With each new episode shared, creativity as a form of resilience emerges as a narrative theme. The show itself offers a representation of creativity, taking the sour news of Covid-19 and finding the silver-lining each week. Krasinski includes quirky news, highlights of the pandemic trends, and content created by regular people. With many cherished moments lost to quarantine and the Covid-19 pandemic, SGN shows how ongoing creativity allows people to persist, remain optimistic, and find sources of entertainment.

Creativity is first modeled by Krasinski through his homemade logo, the celebrity encounters he creates, online celebrations, and even his development of the show itself as a way to pass time during quarantine. But audiences cannot access impressive celebrity networks or expensive gifts meant to cheer people up, so while we see Krasinski being creative, his enactment does not necessarily communicate replicable actions that lead to efficacy. Instead, he works to build viewer efficacy through the activities and video clips of at-home viewers, which make up the bulk of the show. He works to present creativity as accessible—noting that he has "absolutely no idea what I am doing" and regularly reminding viewers "you are the good news" (Some good news 2020a). Krasinski highlights a wide variety of ways people creatively fight boredom and loss. As a result, he communicates that everyone can be creative—and the result is a good thing. As one product of pandemic-inspired creativity, SGN's format inspired viewers who utilized its concept to create artwork, introductions, and home versions from around the world. While Krasinski joked about copyright infringement, the show's interactive element emphasizes the shared cultural experience and ways everyone's input is valued. With the potential for your own creative endeavors to be featured each week on the program, the show directly promotes creativity as a successful antidote to challenges experienced during life's temporary disruption.

By recognizing the originality of viewers, Krasinski rewards the actions he refers to as a triumph of the "human spirit" (Some good news 2020a). Many stories included on SGN solve problems created by the pandemic's distance. One teacher brings a white board to teach outside a child's glass door. Other parents stage at-home Disney rides to entertain their children. People turn ordinary activities like laundry and dishes into competitions. In all of these moments, Krasinski urges that instead of finding quarantine to be a restriction, the resulting creativity allows people to develop new ways to learn and explore the world around them. The range of ordinary to extraordinary examples of creativity make clear that worthy ideas can take on many forms.

While Krasinski models creativity and encourages it in others to build efficacy, he also presents creativity as a way to cope with loss during the pandemic. Audiences see clips from weddings with friends and family watching via Zoom as the couple cuts the cake, just as they might in a more traditional venue. New mothers sit in their front window to safely introduce a new baby to friends and family. The palpable loss many express in this series is buffered by an appreciation for the bigger picture. Towards the end of the series, one father finally gets to hold his child and both end up in tears. By dreaming up alternatives, the people highlighted in this series serve as role models for other audience members by accepting the reality of the situation and making the best of it. Creativity is more than a coping mechanism, but an important process of resilience and efficacy.

Missed moments on SGN represent individual experiences as well as societal milestones. Krasinski frequently uses his celebrity connections to create unique and powerful experiences for audience members meant to mitigate lost moments. SGN plays on the glamour of celebrity presence to substitute for audience members' celebrations because, as Krasinski reiterates, these events cannot happen as they typically would. One little girl, whose dream of seeing *Hamilton* was cancelled, is treated to a montage of the original cast singing the musical's lead song via Zoom. Another couple's *The Office*-themed proposal offers inspiration for an on-the-spot home wedding replicating the one experienced by characters Jim and Pam on the show, complete with the entire *Office* cast and a new song from country singer Zac Brown. As the couple greets their parents via Zoom, the bride wipes tears from her eyes. The challenges of pandemic are temporarily replaced with celebrity-infused moments that take advantage of the uniqueness of being in a celebrity's presence—even if it is online. Inspiring words from Oprah, Steven Spielberg, and Malala help new graduates to process the experience of graduating without the formal ceremony. At the SGN prom, the Jonas Brothers join the celebration since, like the class of 2020, they did not get to attend prom. In these episodes, messages of resiliency through creativity also remind audiences that celebrities are at home and willing to participate in 2020s Zoom culture. Each *Hamilton* cast member films from home, each *Office* actor dances to the wedding from their living room or backyard, and just like John Krasinski's homemade logo, perfection is substituted with good-enough to make these performances happen.

10.6 Celebrity Communication and Resiliency

This analysis identifies how sacrifice and creativity work together to promote resilience on SGN. Krasinski invited viewers into his home and family life on SGN, creating a parasocial interaction that made him feel more like a friend than a celebrity. His self-deprecating demeanor bypassed the challenges other media faced communicating to a polarized US audience and instead, offered an important voice sharing expectations for behavior during the 2020 pandemic. As a form of communication, SGN was more entertainment than news, and as entertainment, the show made salient what quarantine should look and feel like for audiences. With a focus on positive news, SGN ultimately finds a functional space within the cultural environment to reinforce messages about health safety and behavior without the issues of politics facing other media outlets.

Each week, Krasinski shows viewers that staying home during quarantine does not have to be a time of sadness. He empathizes as someone also staying home and models what that behavior should look like. Interactions with other celebrities and ordinary people reinforce the message that sacrifice is normative. While this looks different for different people, SGN portrays all acts as valuable and even necessary. Krasinski shares individual stories to highlight how each person can make a difference. By modeling appropriate behavior, reinforcing audience members' good behavior, and highlighting creativity, Krasinski encourages viewers to believe they too can make a difference. Messages of self-efficacy integrate the show as Krasinski motivates those at home to do what they can to mitigate the effects of Covid-19.

On SGN, the shared experience of the pandemic results in an additional silver-lining: the emergence of new forms of community and connection. The Covid-19 pandemic presented a shared challenge at the global and local levels. Managing the illness and its impact became an important goal for US communities during 2020. Bandura outlined the challenges presented by shared goals (2003): "Goals have little impact unless they are translated into explicit plans and strategies for realizing them," and argued that programs like SGN "model how to translate a vision of a desired future into a set of achievable subgoals" (p. 81). By showing audiences how coworkers, neighbors, and even strangers came together during the first months of the Covid-19 pandemic, SGN offers both strategy and a vision of the future for viewers. As Krasinski ends the show, he notes, "Every single week, if you can look past the goofy guy wearing half a suit, you'd see what resilience really looks like. What unbroken really means. What the true definition of good really is" (Some good news 2020h). Though the global pandemic was far from over when SGN ended in May 2020, the show communicates a path toward a brighter future while allowing audience members to believe in its possibility.

References

Bandura A (2002) Social cognitive theory of mass communication. In Bryant J, Zillmann D (eds) Media effects: advances in theory and research, pp 121–154. Lawrence Erlbaum Associates

Bandura A (2003) Social cognitive theory for personal and social change by enabling media. In: Singhal A, Rogers EM, Cody MJ, Sabido M (eds) Entertainment-education and social change, pp 97–118. Routledge

Brown WJ, Fraser BP (2004) Celebrity identification in entertainment-education. In: Singhal A, Rogers EM, Cody MJ, Sabido M (eds) Entertainment-education and social change: history, research, and practice, pp 97–115. Taylor & Francis

Carras C (2020) 'It's okay to cry': celebrities keep their distance to cope with coronavirus. The Los Angeles Times. https://www.latimes.com/entertainment-arts/story/2020-03-16/coronavirus-celebrities-social-distancing

Dahlstrom MF (2014) Using narratives and storytelling to communicate science with nonexpert audiences. Proc Natl Acad Sci 111(Supplement 4):13614–13620

Goldberg (2020) John Krasinski's 'Some Good News' sells to Viacom CBS following massive bidding war. Hollywood Reporter. https://www.hollywoodreporter.com/live-feed/john-krasinskis-some-good-news-sells-viacomcbs-massive-bidding-war-1295491

Henderson C (2020) John Krasinski explains why he sold 'Some Good News' series to CBS after backlash from fans. USA Today. https://www.usatoday.com/story/entertainment/celebrities/2020/05/27/john-krasinski-explains-why-he-sold-some-good-news-after-backlash/5272133002/

Jurkowitz M, Mitchell A (2020) Cable TV and covid-19: how Americans perceive the outbreak and view media coverage differ by main news source. Pew Research Center. https://www.journalism.org/2020/04/01/cable-tv-and-covid-19-how-americans-perceive-the-outbreak-and-view-media-coverage-differ-by-main-news-source/

Landin T (2020) Covid-19 public policy update: March 20, 2020. Greater Houston Partnership. https://www.houston.org/news/covid-19-public-policy-update-march-20-2020

Lorenz T (2020) The news is making people anxious. You'll never believe what they're reading instead. The New York Times. https://www.nytimes.com/2020/04/14/style/good-news-coronavirus.html?searchResultPosition=3

McIntyre K (2016) What makes "good" news newsworthy? Commun Res Rep 33(3):223–230

McIntyre KE, Gibson R (2016) Positive news makes readers feel good: a "Silver-Lining" approach to negative news can attract audiences. South Commun J 81(5):304–315

Mitchell A, Jurkowitz M, Oliphant JB, Shearer E (2020) Three months in, many Americans see exaggeration, conspiracy theories and partisanship in Covid-19 news. Pew Research Center. https://www.journalism.org/2020/06/29/three-months-in-many-americans-see-exaggeration-conspiracy-theories-and-partisanship-in-covid-19-news/

Mitchell A, Oliphant JB (2020) Americans immersed in covid-19 news; most think media Are doing fairly well covering it. Pew Research Center. https://www.journalism.org/2020/03/18/americans-immersed-in-covid-19-news-most-think-media-are-doing-fairly-well-covering-it/

Moyer-Gusé E, Nabi RL (2010) Explaining the effects of narrative in an entertainment television program: overcoming resistance to persuasion. Hum Commun Res 36(1):26–52

Park D (2020) Sharing good news: the dos and don'ts. Turner. https://turnerpr.com/spin-kitchen/sharing-good-news

Rottenberg J (2020) As celebs shelter at home, paparazzi hustle to find new angles. The Los Angeles Times. https://www.latimes.com/entertainment-arts/story/2020-04-03/coronavirus-quarantine-celebrities-paparazzi

Some good news (2020a) Some good news with John Krasinski Ep. 1 [Video]. Youtube. https://www.youtube.com/watch?v=F5pgG1M_h_U

Some good news (2020b) Hamilton cast zoom surprise: Some good news with John Krasinski (Ep. 2) [Video]. Youtube. https://www.youtube.com/watch?v=oilZ1hNZPRM

Some good news (2020c) Baseball is back: Some good news with John Krasinski (Ep. 3) [Video]. Youtube. https://www.youtube.com/watch?v=Eg08rJGKjtA

Some good news (2020d) SGN prom with Billie Eilish, Jonas brothers & Chance the rapper (Ep. 4) [Video]. Youtube. https://www.youtube.com/watch?v=VQLi2GYVULc

Some good news (2020e) SGN potluck with Martha Stewart, Guy Fieri, David Chang, & Stanley Tucci (Ep. 5) [Video]. https://www.youtube.com/watch?v=o1zIgTwENPg

Some good news (2020f). SGN graduation with Oprah, Steven Spielberg, Jon Stewart, and Malala (Ep. 6) [Video]. https://www.youtube.com/watch?v=IweS2CPSnbI

Some good news (2020g) The office cast reunites for zoom wedding: Some good news with John Krasinski (Ep. 7) [Video]. https://www.youtube.com/watch?v=NDjNX3nEfYo

Some good news (2020h) Some good news with John Krasinski: the SGN community episode! (Ep. 8) [Video]. https://www.youtube.com/watch?v=TXdKrtmexWU

Terry G, Hayfield N, Clarke V, Braun V (2017) Thematic analysis. In: Willig C, Rogers W (eds) The Sage handbook of qualitative research in psychology, 2nd edn, pp 17–37. Sage

The Coronavirus Outbreak (2020) The New York Times. https://www.nytimes.com/2020/03/20/world/coronavirus-news-usa-world.html?searchRes

Chapter 11
Economic Feedback Loops: Crisis Communication Methods and Exhibited by the Travel and Tourism Industry During the COVID19 Pandemic

Jennifer Edwards

Abstract This chapter focuses on the resiliency exhibited by the travel and tourism industries during the COVID19 pandemic. Through quick responses focused on risk management and crisis communication, the industries have become more poised to focus on their post-COVID19 future. The travel industry was greatly affected by the COVID19 pandemic. From significant decreases in check-ins and ticket sales to expensive enhanced cleaning procedures, the transportation and hotel industries continue to experience economic hardships. Since the beginning of the COVID19 pandemic, the World Travel and Tourism Council (World Travel and Tourism Council (Recovery scenarios 2020 and economic impact from COVID19: America's recovery scenarios updates. Retrieved from https://wttc.org/Research/Economic-Impact/Recovery-Scenarios/moduleId/1902/) indicates over 10.8 million travel and tourism jobs were lost in the United States. In addition to the job losses, the travel and tourism industry has lost 907 billion dollars in GDP in 2020. Despite the financial and job losses experienced by the travel and tourism industries, they have had to quickly develop communication policies, procedures, and marketing techniques to prevent further losses. This chapter focuses on the external crisis communication and pandemic communication techniques adopted by U.S. based hotel chains and airlines during the COVID19 pandemic. Best practices for pandemic-based crisis communication for the travel and tourism industries will also be presented during this chapter.

Keyword COVID-19 · Travel · Tourism · Airlines · Hotels

J. Edwards (✉)
Tarleton State University, Stephenville, TX, USA

D. M. Berube (ed.), *Pandemic Communication and Resilience*, Risk, Systems and Decisions, https://doi.org/10.1007/978-3-030-77344-1_11

11.1 The Travel and Tourism Industry During COVID-19

The travel and tourism industry in the United States continues to be impacted by the COVID19 crisis. From decreased ticket sales and the rapid development of policies regarding COVID19 testing by airports and governmental agencies to a delayed distribution of the COVID19 vaccine, airlines and hotels have encountered many travel-related barriers during the COVID19 time period. As a result of continued budget cuts, airlines have had to furlough thousands of airline attendants and other personnel (Sider 2020). From early retirement offers to buyout packages, these airlines had to continue to communicate with their personnel during the pandemic crisis. Two issues of airline communication during a pandemic are timely health and safety policy and the effectiveness of corporate messaging.

COVID19 pandemic communication began generally in the United States with increased messaging from The White House. This communication was distributed in the form of Presidential proclamations and the first such proclamation restricted travel from certain countries (Alward 2020). A writer for the Journal of the Air Mobility Command indicated: Travel restrictions due to the coronavirus are regularly updated based on presidential proclamations and, as of March 16, 2020, included certain travel restrictions for entry to and from China, Iran, South Korea, United Kingdom, Ireland, and the European Schengen area. The CDC raised England, Scotland, Wales, Northern Ireland, and the Republic of Ireland to level 3 warning to avoid nonessential travel due to widespread, ongoing transmission (p. 25).

After this proclamation, the airline and hotel industry was deeply impacted by the COVID19 crisis. From decreased travel reservations to increased COVID19 cases, individuals in the United States had to adapt to the new normal.

Mandatory flying regulations like receiving a negative COVID19 test prior to boarding the aircraft and temperature checks have become the norm across the world. Inflight dining has also been modified to aid in the safe consumption of beverages and food. Most airlines, hotels, and restaurants have adopted rapid and mass disinfection techniques to keep their employees and customers safe while traveling. In addition to the aforementioned techniques, other deployed solutions include "mobility tracing apps, robotised-AI touchless service delivery, digital health passports and identity controls, social distancing and crowding control technologies, big data for fast and real time decision-making" (Sigala 2020, p. 314). During the pandemic, robots have been utilized to deliver materials, clean public spaces, measure body temperature, and provide security. "Current travel practices will certainly not be sustained … Therefore, we must reconsider our strategies to protect our health to prevent travel-related diseases" (Felkai 2021, p. 1).

Health is an important factor of traveling to a destination and it is also essential when arriving at a hotel. Hotels are posting signs to remind customers to wash their hands and to wear a face mask when entering buildings (Dial 2020). Some hotel chains are requiring their employees complete a COVID19 questionnaire before entering the doors of the workplace. These questionnaires usually include questions focused on the employee's potential prior interactions with individuals who may have had COVID19. In addition to providing services for employees and guests, some hotels were utilized by the government. Some countries chose to use hotels that were under capacity as quarantine hotels for individuals who recently traveled outside of the country and other individuals with possible exposure to COVID19. The U.K. required all arrivals from COVID19 hot spots to quarantine inside of designated hotels (Ross 2021).

11.2 Pandemic Communication, COVID19, and Consumer Confidence

A study published in August 2020, investigated how 66 major airline brands communicated with the public during a pandemic (Research and Markets 2020). This study organized airline pandemic communication into four overarching categories: policy, communication, credibility, and citizenship.

During COVID19, airlines have experienced less customers purchasing flights ("A4A reports states with air service and travel hardest hit by COVID-19 pandemic" 2020). New York received the largest decrease in departure bookings (70% less) and New Jersey received the second largest decrease in departure bookings (67% less). Across the nation, states experienced an average of 50% less flights booked (McCartney 2020). In November 2020, the number of ticketed passengers at New York's three airports has decreased 95% since mid-April.

The Medical Advisory Group of the International Air Transport Association (IATA) provided COVID19 safety recommendations for the aviation industry (IATA Medical Advisory Group 2020). These recommendations include: screening for symptoms, integrating personal protective equipment (PPE) and masks, distancing among passengers, cleaning procedures, COVID-19 testing (and vaccinations), antibody testing, and additional level of safety for crew members and passengers.

Prior to COVID19, the airline industry expected growth in the areas of passenger load and aircraft fleet growth. Unfortunately, the industry had to adjust their expectations after numerous border closures, country lockdowns, and increased levels of uncertainty among current and potential customers regarding the transmission of COVID19. The Global Airport and Airline Industry: Post-Pandemic (COVID19) Growth Opportunity Analysis, 2020 and Beyond (2020) report stated:

> The uncertain nature of this pandemic has severely marred passenger confidence, which may take time to rebuild post-crisis. Weak passenger confidence in travel and declining economic conditions among many countries will impact the total number of passengers travelling, in both the domestic and international segments. Passenger numbers are estimated to reach pre-pandemic levels only after a couple of years (p. 1).

These passenger numbers will only increase when customers feel more confident in the travel process and the COVID19 vaccination plan.

To reestablish a relationship with customers and to potentially boost earnings, airlines and airports should consider incorporating digital opportunities ("The Global

Airport and Airline Industry: Post-Pandemic (COVID19) Growth Opportunity Analysis, 2020 and Beyond" 2020). These opportunities can help the airlines build social presence with their customer base while ensuring the smooth transition of air travel. This re-establishment can also be enhanced by increased levels of collaboration between customer relationship management systems (CRMs), hardware providers, and other business entities to facilitate communication and provide increased levels of customer service opportunities. These pandemic communication opportunities can be more cost effective for airlines and airports by enabling them to utilize existing resources and refining processes.

11.3 Domestic Travel, Communication, and COVID-19

Domestic travel in countries in Latin America as well as the United States, China, and India, is projected to increase as travel bans are internally lifted in those counties ("The Global Airport and Airline Industry: Post-Pandemic (COVID19) Growth Opportunity Analysis", 2020). However, international travel is contingent upon other factors like vaccination access and international economic conditions. The pandemic continues to make an impact on the airlines in the United States. As a comparison to the economic loss experienced during the September 11 time period, the airlines are facing a grim outlook. The six largest airlines in the United States have experienced economic losses nearly twice as big as inflation-adjusted losses those same airlines (including their mergers) had after 9/11 (McCartney 2020). These airlines were not profitable until six years after the September 11th tragedy.

Airlines are communicating with their passengers through verbal and non-verbal messaging strategies. These strategies are focused on passenger spacing and cleanliness (McCartney 2020). Some airlines like Delta do not enable passengers to book the middle seats. As a result, Delta Airlines had the lowest passenger load numbers (41%) in July, August and September (Fig. 11.1).

In comparison, American Airlines did not block the middle seats and they experienced a larger passenger load at almost 59% of its capacity during that same time period. This number represented the highest load of the largest airlines.

11.4 Social Media, Pandemic Communication, and the Airline Industry

On Twitter, the airlines produced various social media messages to keep customers informed during the pandemic. The most effective pandemic communication messages received a high level of engagement during the pandemic. Engagement techniques such as comments, retweets, and favorites indicate consumer interest in the content of the message. Comments are usually the highest form of engagement,

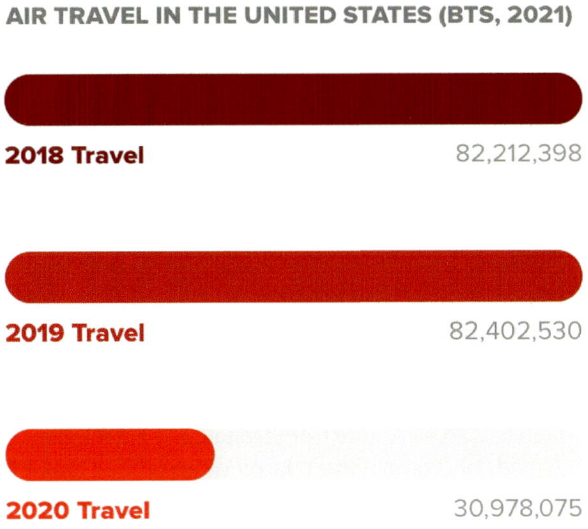

Fig. 11.1 Air travel in the United States. *Source* (U.S. Department of Transportation—Bureau of Transportation Statistics 2021)

retweets are the second highest form of engagement, and favorites represent the lowest form of engagement. If a Twitter message receives multiple engagements, it is more likely to be seen by an increased number of Twitter users.

11.4.1 Top Pandemic Communication Messages from American Airlines (@AmericanAir)

October 25, 2020—You're never fully dressed without a mask. Plus, it helps slow the spread of #COVID19. With a short video featuring the following text: "Fact: No. 8—Wearing a surgical mask on a plane may reduce infection risk to less than 1% (Harvard T.H Chan School of Public Health)" (With a link to more information on the harvard.edu website).

Pandemic Communication Engagement—This tweet received 10 comments, 30 retweets, and 127 favorites. November 8, 2020—The more the merrier when it comes to research about the safety of flying during #COVID19. Check out this recent study.

"Quote Tweet—Department of Defense Flag of United States (@DeptofDefense)—Oct 16, 2020.

Thanks to a DOD study with industry—the largest of its kind—we know the risk of Coronavirus during air travel can be reduced—refreshing the cabin air every two minutes, HEPA air filters, and top to bottom air flow all protect passengers." https://ustranscom.mil/cmd/docs/TRANSCOM%20Report%20Final.pdf.

Pandemic Communication Engagement—This tweet received 26 comments, 20 retweets, and 114 favorites.

October 17, 2020—What is it like flying during the pandemic? Hear from Capt. Peter Loparnos about his experience with COVID-19 and the protocols we've put in place to keep our customers and team safe. Watch the full documentary on @CNBC today at 3 pm ET. #YouAreWhyWeFly http://bit.ly/AACNBCFlying (with a picture of Captain Peter Loparnos).

Pandemic Communication Engagement—This tweet received 25 comments, 28 retweets, and 126 favorites.

11.4.2 Top Pandemic Communication Messages from Delta Airlines (@Delta)

March 6, 2020—A note from Delta CEO Ed Bastian. Learn more about our response to Covid-19 (coronavirus): http://dl.aero/6018Tju44 (with a picture of a letter from the Delta Airlines CEO community and a link to the letter on the Delta Airlines website).

Pandemic Communication Engagement—This tweet received 259 comments, 336 retweets, and 1200 favorites.

March 21, 2020—Delta is waiving all change fees for travel impacted by coronavirus. This applies to all domestic and international travel departing in March, April or May 2020, as well as all tickets purchased in March 2020.

Full details at http://delta.com/coronavirus (with an image titled "Simplified Waivers for Flight Changes" and a link to the Delta Airlines website) Pandemic Communication Engagement—This tweet received 142 comments, 121 retweets, and 421 favorites.

March 26, 2020—Due to coronavirus, flight schedules are rapidly evolving. Downloading the Fly Delta app lets you manage flights and get real-time notifications about any changes, sent right to your mobile device. (With a picture of the Fly Delta App and a link to the Delta Airlines website).

Pandemic Communication Engagement—This tweet received 42 comments, 32 retweets, and 169 favorites.

11.4.3 Top Pandemic Communication Messages from United Airlines (@United)

February 27, 2020—We have issued a travel waiver for Northern Italy due to the impact of Coronavirus. If you're traveling through April 30, you can change your flight at no cost on our app, by direct messaging us here or calling 1-800-864-8331. http://uafly.co/2vmnkSk (with an "important notice" gif featuring the United Airlines logo).

Pandemic Communication Engagement—This tweet received 57 comments, 45 retweets, and 106 favorites.

November 22, 2020—This year, United Cargo has already helped fly over 190 million pounds of supplies to the front lines of the COVID-19 pandemic. Now our COVID Vaccine Readiness Task Force is prepared to transport vaccines as soon as they are available. (With a video featuring an employee loading a cargo plane with "When the vaccine is ready, we are ready" as dynamic text).

Pandemic Communication Engagement—This tweet received 57 comments, 20 quote tweets, and 357 favorites.

February 28, 2020—Coronavirus schedule updates: We are suspending some service to Tokyo Narita, Osaka, Singapore and Seoul. China and Hong Kong flight suspensions are extended through 4/30. We will stay in close contact with the CDC as we continue to evaluate our schedule. http://uafly.co/2wW60Ec (with an "important notice" gif featuring the United Airlines logo).

Pandemic Communication Engagement—This tweet received 164 comments, 441 retweets, and 329 favorites.

11.4.4 Top Pandemic Communication Messages from Southwest Airlines (@SouthwestAir)

October 2, 2020—For nearly two years, Kent S. commuted from San Jose to Burbank every week. He worked as the Head of Cinematography on an animated feature, and when the pandemic brought his weekly commute to a halt, he made a video to look back on his journey, 173 flights in total. (with a 2:55 min aviation video) Pandemic Communication Engagement—This tweet received 16 comments, 26 retweets, and 156 favorites.

September 29, 2020—But, it's not just in regard to air travel that research shows masks are an effective way to slow the spread of the coronavirus. Experts continue to agree that mask use is one of the most powerful tools we have as a society to support public health.

https://cdc.gov/coronavirus/2019-ncov/prevent-getting-sick/diy-cloth-face-coverings.html (4/5).

Pandemic Communication Engagement—This tweet received 12 comments, 10 retweets, and 98 favorites.

July 25, 2020—Learn about the creative ways in which our Employees have supported communities and individuals impacted by the COVID-19 pandemic. #SouthwestHeartStrong (with a picture of Southwest Airlines employees gathering food for the community and a link to an article on the Southwest Airlines website) Pandemic Communication Engagement—This tweet received 4 comments, 6 retweets, and 80 favorites.

These pandemic communication messages provide a glimpse into the more effective types of messages posted by the airlines from March 2020 to January 2021. The most effective types of messages included a picture, gif, video to engage viewers. Some of the messaging also included an industry hashtag (i.e. #SouthwestHeart-Strong) while other messages contained external links to airline websites. The earliest pandemic communication message was posted by United Airlines regarding travel changes. The most engaging pandemic communication message was posted by Delta Airlines on March 6, 2020 and featured a letter from CEO Ed Bastian.

11.5 Social Media, Pandemic Communication, and the Hotel Industry

The hotel industry in the United States has been greatly affected by COVID19. Nationwide, the total hotel occupancy was 48% in the third week of October 2020. According to STR, a data company focused on the hotel industry (McCartney 2020), these hoteliers are dependent on airline travel to transport business and leisure travelers to their cities. As cities continue to temporarily close tourist attractions, Broadway shows, and other large scale events, vacationers have less reasons to visit the large cities like the Big Apple.

Hotel chains like Hyatt Hotels that rely on the business traveler are adopting new formats to engage current and new customers. Over two-thirds of Hyatt's bookings are from business travel and the hotel chain is considering adopting unique business formats that will attract businesspeople who are willing to travel. This strategy also focuses on engaging the travel and their co-workers who may join a virtual meeting hosted by the business traveler (Ecker 2020). These business formats will enable Hyatt properties to offer meetings in various spaces and formats through virtual and live experiences. These experiences have the potential to become part of the long-term business traveler engagement strategy.

11.5.1 Pandemic Communication Messages from U.S. Hotels

Hilton Hotels and Marriott Hotels did not post any pandemic communication messages focused on the COVID19 pandemic. The hotel chains were potentially more focused on the development of pandemic processes and did not maintain a focus on communication with customers. Hyatt Hotels (@Hyatt) was the only hotel chain out of the largest U.S. hotel chains to post a focused pandemic communication message.

April 29, 2020—Guided by our purpose of care and established operational excellence, we are proud to announce Hyatt's Global Care and Cleanliness Commitment, introducing steps that will help ensure safety and wellbeing during the COVID-19 pandemic and beyond. http://hyatt.com/info/global-care-and-cleanliness-commitment (With a picture of the Hyatt Hotel with a heart outline in downtown St. Louis).

Pandemic Communication Engagement—This tweet received 31 comments, 4 quote tweets, and 77 favorites.

The travel and tourism industry should consider adopting pandemic communication messages that address the travelers' fears while providing potential solutions. When customers realize or experience the severity of the pandemic, some may become fearful of traveling during and after the COVID19. These travel fears can also have long term impacts on the travel which can result in travel avoidance long after the travel ban is over (Zheng et al. 2021). Travelers need to be reassured that their safety is a priority after the pandemic ends. This reassurance can be given in the form of clearing procedures, redesign of tourism experiences, and contactless customer experiences (Sigala 2020).

11.6 Best Practices for Pandemic Communication Messages

The travel and tourism industry can craft communication messages that can help travelers become more prepared and knowledgeable about travel during and after a pandemic. These types of pandemic communication messages may include the following:

- Indoor mask requirement.
- The corporation or company is adhering to strict cleanliness guidelines utilizing specific cleaning products.
- Promoting hand sanitizer and hand washing stations.
- Providing the current percentage-based capacity of a hotel property or airplane.
- Providing meal delivery options and a realistic preview of the restaurant or aviation menu.

Additional types of messages include:

- Strict measures regarding the control of social distancing will be enforced (Zheng et al. 2021).
- Providing social distancing information in real-time (i.e. number of tourists at different scenic spots and the number of individuals on an aircraft) (Zheng et al. 2021).
- Providing problem-focused or emotion-focused information focused on how to practice COVID19 prevention strategies when traveling (Zheng et al. 2021).

These types of pandemic communication messages can positively impact tourists' perceived efficacy in COVID-19 prevention (Zheng et al. 2021).

11.7 Conclusion

From examples of effective social media messages to a cohesive list of best practices for pandemic communication in the travel industry, this chapter offered strategies for enhancing traveler confidence in the travel industry. COVID19 presented various challenges for many industries in the United States, but it also presented a wealth of opportunities for industries like the travel industry. The pandemic helped the industry become more innovative, adopt a focus on cleanliness, and increase their focus on customer-centric messaging. Airlines addressed the COVID19 crisis directly through social media and marketing outreach campaigns, while their hotel counterparts adopted a pandemic communication strategy more generally focused.

References

A4A reports states with air service and travel hardest hit by COVID-19 pandemic (2020). Airline Industry Information (M2)

Alward K (2020) COVID-19 Coronavirus and United States air force precautions. Mobility Forum: the J Air Mobility Command's Mag 29(2):24–25

Dial J (2020) How hotels can protect themselves from COVID-19 lawsuits. Hotel Management (21582122), 235(9), 20. Retrieved from http://www.hotelmanagement.net

Ecker D (2020) Why Hyatt is more vulnerable to covid than most: the pandemic is hitting high-end hotels hardest. Crain's Chicago Bus 43(33):1

Felkai PP (2021) How to travel after the COVID-19 pandemic?. Int J Travel Med Global Health 9(1):1–3. Retrieved from http://www.ijtmgh.com

IATA Medical Advisory Group (2020) Restarting aviation following COVID-19: medical evidence for various strategies being discussed as at 27 April 2020. Retrieved from https://www.iata.org/contentassets/f1163430bba94512a583eb6d6b24aa56/covid-medical-evidence-for-strategies-200423.pdf

McCartney S (2020) How coronavirus ravaged travel in 2020; the numbers tell the story of the damage done to airlines, hotels and the rest of the industry—and how far the climb back will be. Wall Street Journal (Online)

Research and Markets (2020) Airline brands & the pandemic 2020: benchmarking COVID-19 policies & communications—researchAndMarkets.com. Business Wire

Ross T (2021) U.K. still hasn't signed any deals for covid quarantine hotels. Bloomberg.Com. Retrieved from http://www.bloomberg.com/

Sider A (2020) Covid-19 Resurgence threatens travel rebound. Wall Street J Online Edition

Sigala M (2020) Tourism and COVID-19: impacts and implications for advancing and resetting industry and research. J Bus Res 117:312–321. https://doi.org/10.1016/j.jbusres.2020.06.015

The global airport and airline industry: post-pandemic (COVID19) growth opportunity analysis, 2020 and beyond (2020). M2PressWIRE

U.S. Department of Transportation—Bureau of Transportation Statistics (2021) Passengers: all carriers—all airports. Retrieved from http://www.transtats.bts.gov

World Travel and Tourism Council (2020) Recovery scenarios 2020 and economic impact from COVID19: America's recovery scenarios updates. Retrieved from https://wttc.org/Research/Economic-Impact/Recovery-Scenarios/moduleId/1902/

Zheng D, Luo Q, Richie BW (2021) Afraid to travel after COVID-19? Self-protection, coping and resilience against pandemic 'travel fear'. Tourism Manag 83. https://doi.org/10.1016/j.tourman.2020.104261

Part IV
Rhetoric, Prophylactics, and Public Resiliency

Chapter 12
Health Campaign or War Campaign? Donald Trump's Metaphoric Narrative on COVID-19

Esmaeil Esfandiary

Abstract Perhaps the first question one can ask about the Trump administration's messaging strategy on COVID-19 is the narrative the president constructed around the virus. How did he portray the virus? How did he label it? What were the implications of his narrative? This chapter looks into the first few weeks of the president's daily press conferences (during COVID) to analyze his narrative regarding COVID-19. More specifically, metaphoric criticism will be used to examine the meaning and implications of recurring metaphors in Trump's narrative: His repeated metaphors that the virus is an "invisible" or "hidden enemy" and that we are in a "war" with this "unseen enemy." The chapter ends with a discussion of how such metaphors turned a public health campaign into a war campaign giving him the power and legitimacy of a wartime president; and how a totally different picture of the whole issue could have been portrayed.

Keywords COVID-19 · Pandemic · Metaphoric criticism · White House press conferences · Trump's messaging

12.1 Introduction

By mid-March 2020, it had become increasingly clear that the novel CoronaVirus (COVID-19) was hitting the US on a massive scale. At that point, the Trump administration felt the urge to mobilize a task force to coordinate and organize government's response to the outbreak. The central communication channel of the newly-shaped government task force was the president's daily press conferences—which started on March 16th and continued for months. In these press conferences, President Trump (along with other government officials, epidemiologists, and relevant healthcare practitioners) would talk about the latest measures taken by the government in its campaign against the virus. In these press conferences, the president spoke every single day in front of the press; and major networks and outlets gave him

E. Esfandiary (✉)
Tuskagee University, Tuskegee, AL, USA
e-mail: eesfandiary@tuskegee.edu

D. M. Berube (ed.), *Pandemic Communication and Resilience*, Risk, Systems and Decisions, https://doi.org/10.1007/978-3-030-77344-1_12

live coverage. Their impact was amplified by public panic and thirst for information at the beginning of the pandemic—the time when the entire country went into a sudden lockdown and tens of millions of Americans lost their jobs. Given their impact and reach, these press conferences became the main communication channel for the administration in its public health campaign against the virus.

One of the first questions one can ask about the administration's messaging strategy in these daily press conferences is the very narrative the president constructed around the virus. How did he portray the virus? How did he label it? What were the implications of his narrative? This chapter looks into the first few weeks of the president's daily press conferences to study his narrative regarding COVID-19. In his press conferences, Trump repeatedly called the virus an "invisible" or "hidden enemy" and stated that we are in a "war" with this "unseen enemy." This chapter will use metaphoric analysis to examine the meanings and implications of such prominent metaphors in Trump's narrative on COVID-19. It will be argued that his metaphors turned a public health campaign into a war campaign—giving him the power and legitimacy of a wartime president. Finally, it will be argued that this picture was not the "natural," or the only possible way, to portray the pandemic; that alternative narratives could have portrayed a totally different picture of the whole issue.

12.2 Strategic Narratives and the Framing of Critical Public Issues

What is called a "strategic narrative" and how do "strategic narratives" work? In *Strategic Narratives: Communication Power and the New World Order*, Miskimmon et al. (2013) provide us with an understanding of strategic narratives and their role in projecting power in today's politics. "Strategic narratives are a means for political actors to construct a shared meaning of the past, present, and future ... to shape the behavior of domestic and international actors" (p. 2). But this is not just true about narratives regarding one's own identity: "actors work to frame their own character and that of others, by selecting and highlighting some facets of their history or actions in order to promote a particular interpretation and evaluation of their character" (p. 5).

Strategic narratives are constructed upon collective memories adopted from history and culture. Miskimmon et al. refer to "[h]istory, analogies, metaphors, symbols, and images" as the resources used to construct these narratives (p. 7). Similarly, Monroe Price (2014) argues that "A state is, in part, a collection of stories connected to power. Remembered traditions, obligations and laws—all stories in themselves—shape internal and external perceptions of a state and the range of its efficacy" (p. 41).

Strategic narratives have a twofold power effect. They are "an instrument of power in the traditional Weberian or *behavioral* sense of A getting B to do what B otherwise would not" (Miskimmon et al. 2013, p. 17); but they also constitute "... identity of [the] actors and the meaning of the system" (ibid). For example, a carefully formulated narrative might imply that America has long been taken advantage of by other countries due to its wealth, leadership role, and generosity. This narrative can help persuade allied states to spend more and participate more actively in collective security and military ventures. But such a narrative also defines the "real world" for pertinent actors and ascribe identities and roles on them. By this narrative, for instance, the identity and leadership role of the US would be seen in a totally different way: That the US should not be a "generous" leader anymore; that it should put its own interests first (if it doesn't want to be taken advantage of).

It is, thus, true that facts exist on the ground, but strategic narratives connect these facts with ideologies and histories in order to give them meaning—to construct a coherent *reality*. In other words, a combination of both hard facts and discursive practices enable narratives to strategically construct the "real world." In that sense, alternative narratives can help construct alternative political realities as well. According to Miskimmon et al., "if they [states] are unhappy about the prevailing way in which borders, human rights, collective action, the basis of rule, or any other foundation of political order is done, then these can be contested. And this involves forming an alternative narrative ... that puts these core aspects in a new light" (2013, p. 16). In a similar vein, Price argues "[w]hether it is the clash of civilizations, the loss of "values," or the need to protect jobs or economies, competition for national narratives of legitimacy—for good and for ill—will persist" (2014, p. 60).

Promoting one's own legitimacy is probably the core of how strategic narratives project and amplify one's power. At the end of the day, the desired outcome of a strategic narrative for a leader or a politician is to amplify their power and status. This chapter argues that Trump's metaphoric narrative on COVID-19 handed him more legitimacy and power in his role as the president of the United States. This will be discussed in more detail throughout the conclusion section. Before that, the meanings and functions of metaphors will be addressed in next section.

12.3 Metaphors and Metaphoric Criticism

According to Merriam-Webster dictionary, Metaphor is a "*figure of speech* in which a word or phrase literally denoting one kind of object or idea is used in place of another to suggest a likeness or *analogy* between them" (Merriam-Webster Dictionary). Lakoff and Johnson explain that: "*The essence of metaphor is understanding and experiencing one kind of thing in terms of another*" (1980, p. 5). George Lakoff (2003) uses the example of "International Community" to explain how metaphors work. The metaphor "International Community" portrays the collection of countries, or nation-states, as a community. The *community* metaphor has important implications and meanings. In communities, there are leaders, there are hierarchies (with higher and lower members), there are good and bad members, the members are expected to behave in certain ways and follow certain rules when it comes to the decision-making process and the community's shared future. Most importantly, this metaphor implies that nation-states are like persons who constitute members of this community.

According to Lakoff (2003),

> The Nation as Person metaphor is pervasive, powerful, and part of an elaborate metaphor system. It is part of an International Community metaphor, in which there are friendly nations, hostile nations, rogue states, and so on... In the International Community, peopled by Nation-Persons, there are Nation-adults and Nation-children, with maturity metaphorically understood as industrialization. The children are the "developing" nations of the Third World, in the process of industrializing, who need to be taught how to develop properly and to be disciplined. (para. 4).

Metaphors are important because they can shape both our understanding of social realities *and* our future course of action. "Metaphors have entailments through which they highlight and make coherent certain aspects of our experience" (Lakoff and Johnson 1980, p. 156). Metaphors are used widely—though often unconsciously—in almost all areas of human knowledge and understanding. The fact that they are used unconsciously and uncritically makes them even more powerful (and potentially dangerous). This is particularly true because "[i]n allowing us to focus on one aspect of a concept ... a metaphorical concept can keep us from focusing on other aspects of the concept that are inconsistent with that metaphor" (p. 10).

Metaphors are pervasive in our culture, language and communication; "[metaphor] is central to our understanding of ourselves, our culture, and the world at large" (Lakoff and Johnson 1980, p. 214). Not only do metaphors pervade our language and linguistic choices, but they also frame our mindset and thought process. This is particularly true about metaphors that are so common in our culture that are not even noticed. "To study metaphor is to be confronted with hidden aspects of one's own mind and one's own culture... that one has a worldview, that one's imagination is constrained, and that metaphor plays an enormous role in shaping one's everyday understanding of everyday events" (p. 214). Such analysis is the aim of metaphoric criticism.

In metaphoric criticism, metaphors are considered as the "means by which arguments are expressed. Moreover, metaphors may provide insight into a speaker's motives or an audience's social reality" (ed. Burgchardt 2005, p. 305). In other words, because it unconsciously frames thoughts and mindsets, metaphor by itself is an important means of persuasion. This chapter analyzes the metaphors used by Donald Trump based on this definition.

In order to analyze Trump's metaphoric narrative on COVID-19, the metaphors he used were retrieved first. The transcripts of the first round of his daily press conferences (between March 13th and April 18th, 2020) were pulled from the official WhiteHouse.com. They were then studied and analyzed for any metaphor he used in reference to COVID-19. The time span for this search belongs to the early phase of the pandemic where public panic and information thirst about the pandemic was at its height. In the next section, Trump's metaphors will be analyzed.

12.4 Trump's Metaphoric Narrative: Our War on the Invisible Enemy

The major, overarching metaphor in President Trump's narrative on COVID-19 is his portrayal of the virus as an "invisible enemy" who is in "war" with us. From his very first press conferences, Trump repeatedly refers to the virus as an "hidden" or "invisible" enemy "waging war" on us. Here are a few examples of such remarks. On March 17th, he says "We have to fight that invisible enemy ..." (Trump 2020a, para. 10); On March 18th, "we're going to defeat the invisible enemy ..." (Trump 2020b, para. 20). On March 20th, he states "Americans ... are rallying together to defeat the unseen enemy striking our nation" (Trump 2020c, para. 25); On March 22nd, "It's now attacking—the enemy is attacking 144 countries ..." (Trump 2020d, para. 52). On March 27th, he declares "we got hit by the invisible enemy ..." (Trump 2020e, para. 44) while on April 3rd he states "we are in a war. The invisible enemy—remember" (Trump 2020g, para. 28). The metaphor of virus as the enemy permeates all of Trump's press conferences. In the 33 press conferences analyzed in this study, just the terms invisible or hidden enemy are repeated 31 times.

Other military metaphors are extensively used in his press conferences as well. In reference to the virus and the pandemic, metaphors such as war, battle, fight, frontline, attack, strike, defeat, soldier etc. shape the central theme of his narrative: "we will win this battle, we will defeat this enemy ... we're attacking the enemy on all fronts" (Trump 2020h, April 6, para. 8); "Our nation is engaged in a historic battle against the invisible enemy. To win this fight, we have undertaken the greatest national mobilization since World War Two" (Trump 2020i, April 16, para. 1); "... nurses [are] on the frontlines of the battle against the virus" (Trump 2020b, March 18, para. 16); "It's like military people going into battle, going into war ..." (Trump 2020f, March 31, para. 31); "healthcare workers who are the soldiers of this war ..." (April 6, para. 71).

In the 33 press conferences analyzed here, Trump uses the terms battle, frontline, fight and war (or warrior) more than 130 times! He uses other military metaphors extensively as well: attack, strike, defeat, front, soldier and so on. Considering his constant use of military metaphors it seems that Trump emphatically wants everyone to see COVID-19 as an enemy and the health campaign against it as a military campaign against this enemy. Why should Trump want to portray the pandemic in this way? From the perspective of "strategic narratives," how does this metaphoric portrayal shape the identity of involved parties? And how does that serve Trump's political interests? Next section will address these questions.

12.5 Conclusion: Military Campaign and War-Time President

As we saw in the previous section, President Trump's narrative draws heavily on military metaphors to portray his administration's health campaign on COVID-19 as a military campaign against an invisible enemy. He constantly uses such metaphors to create the image of a military campaign (rather than a health campaign). What are the meanings and implications of such an image? Why does Trump insist on framing the pandemic, and the virus, in military terms? How does this framing serve him politically? As discussed in the literature (on strategic narratives and metaphors), his emphatic framing should not be assumed to be natural or coincidental; other possibilities to frame the pandemic should not be ignored either. As George Lakoff puts it, "[t]hat is what [metaphorical] framing is about. Framing is about getting language that fits your worldview. It is not just [haphazard] language" (2004, p. 4).

From its very first days, the outbreak was used as an opportunity for publicity and promotion by the administration. The very introduction of *daily* presidential press conferences—going on for hours every single day, broadcast by major networks and watched by millions of Americans—was a testament to such a use. Some observers even called these press conferences as Trump's new nightly show or a replacement for his campaign rallies (Kruse 2020; Dale 2020). In this context, significance of his framing becomes even more important. Going back to the literature, one should ask, how would the image of a military campaign serve Trump?

If US government's unprecedented health campaign against the virus is seen as a military campaign, then all elements of a military campaign applies to the situation as well: In a military battle, first and foremost, the president assumes the role of an undisputed commander-in-chief: he orders and everyone else follows. A general climate of public mobilization and an expectation of national unity, patriotism, and determination hands the president more authoritarian powers and enables him to construct national consensus around his own ideas. Dissent, criticism and difference is less and less tolerated—at times labelled as unpatriotic. Altogether, circumstances of a military battle would naturally give additional authority and power to the president, and more legitimacy and leeway to him to use it the way he pleases.

On the other hand, the role of officials, healthcare workers, press, pundits and others on the "frontlines" is reduced to foot soldiers—who are supposed to selflessly support the nation's battle against the enemy. All of a sudden, the normal civilian systems become more of a mobilized structure under the president's chain of command. Also, in military battles and wars, there is not much open space for debate, critique or disagreement. In the midst of a battle, commanders' mistakes are usually not subject to open discussion—national interest often dictates sweeping them under the rug. Generally, optimism, trust and even sacrifice for the leadership become the new normal. Put together, these circumstances boost the leadership's status and power while it reduces their normal threshold of accountability and liability.

This seems to be the reason why Donald Trump wanted to create an atmosphere of war. Still, one might ask, what is "unnatural" about this set of metaphors and the narrative they construct? What is the problem with seeing our campaign against the pandemic as a military battle? Of course, some similarities exist between the two, and such metaphors might initially seem like a natural fit for the situation—people are losing their lives and a concerted, national effort to "fight" the virus is necessary. Still, substantial differences exist as well. After all, a public health campaign is not a military campaign. A quick look at leadership styles in other western countries that were more successful in controlling the pandemic illustrates that, as a matter of fact, less authoritarian styles and narratives were more effective in containing the virus.

For instance, in countries such as Germany and New Zealand, where political leadership was much more effective in containing COVID-19, leadership became more accountable—not less—from the outset. Those leaders practiced listening more than speaking. They listened to the experts and let experienced scientists and healthcare practitioners take the lead. They showed more respect to scientists and more empathy to their peoples (Friedman 2020; Kretchmer 2020). In this sense, their leadership styles turned less authoritarian and more participative and compassionate; they gave up some of their authority to share it with qualified experts. They stood alongside scientists and healthcare professionals and exercised humility and empathy. Therefore, a more authoritarian style of leadership is not necessarily the only "natural" response to an outbreak. In this sense, Trump's metaphoric narrative portraying the pandemic as a military battle was not a necessary or natural messaging or leadership strategy. Instead, it was a deliberate, strategic effort to portray the situation in a way that would hand him more power and make the system more authoritarian. Persuasive military metaphors helped him make such a portrayal look more natural and legitimate. Again, as Lakoff (2004) puts it, "[t]hat is what [metaphorical] framing is about. Framing is about getting language that fits your worldview" (p. 4).

References

Burgchardt CR (ed) (2005) Readings in rhetorical criticism, Strata, State College

Dale D (2020) Trump uses daily coronavirus briefings to replace campaign rallies. CNN 24 March. Available from: https://www.cnn.com/2020/03/23/politics/trump-coronavirus-briefings-rallies/index.html. Accessed 3 March 2021

Friedman U (2020) New Zealand's prime minister may be the most effective leader on the planet. The Atlantic 19 April. Available from: https://www.theatlantic.com/politics/archive/2020/04/jacinda-ardern-new-zealand-leadership-coronavirus/610237/. Accessed 3 March 2021

Kretchmer, H (2020) 3 leadership lessons from the age of coronavirus, World Economic Forum 19 August. Available from: https://www.weforum.org/agenda/2020/08/coronavirus-leadership-women-leaders-jacinda-ardern/. Accessed 3 March 2021

Kruse M (2020) Trump turns a crisis into his new nightly TV show. Politico 25 March. Available from: https://www.politico.com/news/magazine/2020/03/25/trump-coronavirus-white-house-briefing-room-press-conference-147571. Accessed 3 March 2021

Lakoff G (2003) 'Metaphor and war, again', AlterNet. Available from: http://www.alternet.org/story/15414/metaphor_and_war%2C_again. Accessed 1 Sept 2015

Lakoff G (2004) Don't think of an elephant. pp 3–34

Lakoff G, Johnson M (1980) Metaphors we live by. University of Chicago Press, London

Miskimmon A, O'Loughlin B, Roselle L (2013) Strategic narratives: communication power and the new world order. Routledge, New York
Price M (2014) Free expression, globalism and the new strategic communication. Cambridge University Press, New York
Trump DJ (2020a) Remarks by President Trump, Vice President Pence, and members of the coronavirus task force in press briefing, 17 March 2020, White House. Available from: https://trumpwhitehouse.archives.gov/briefings-statements/remarks-president-trump-vice-president-pence-members-coronavirus-task-force-press-briefing-4/. Accessed 3 March 2021
Trump DJ (2020b) Remarks by President Trump, Vice President Pence, and members of the coronavirus task force in press briefing, 18 March 2020, White House. Available from: https://trumpwhitehouse.archives.gov/briefings-statements/remarks-president-trump-vice-president-pence-members-coronavirus-task-force-press-briefing-5/. Accessed 3 March 2021
Trump DJ (2020c) Remarks by President Trump, Vice President Pence, and members of the coronavirus task force in press briefing, 20 March 2020, White House. Available from: https://trumpwhitehouse.archives.gov/briefings-statements/remarks-president-trump-vice-president-pence-members-c-oronavirus-task-force-press-briefing/. Accessed 3 March 2021
Trump DJ (2020d) Remarks by President Trump, Vice President Pence, and members of the coronavirus task force in press briefing, 22 March 2020, White House. Available from: https://trumpwhitehouse.archives.gov/briefings-statements/remarks-president-trump-vice-president-pence-members-coronavirus-task-force-press-briefing-8/. Accessed 3 March 2021
Trump DJ (2020e) Remarks by President Trump, Vice President Pence, and members of the coronavirus task force in press briefing, 27 March 2020, White House. Available from: https://trumpwhitehouse.archives.gov/briefings-statements/remarks-president-trump-vice-president-pence-members-coronavirus-task-force-press-briefing-13/. Accessed 3 March 2021
Trump DJ (2020f) Remarks by President Trump, Vice President Pence, and members of the coronavirus task force in press briefing, 31 March 2020, White House. Available from: https://trumpwhitehouse.archives.gov/briefings-statements/remarks-president-trump-vice-president-pence-members-coronavirus-task-force-press-briefing-15/. Accessed 3 March 2021
Trump DJ (2020g) Remarks by President Trump, Vice President Pence, and members of the coronavirus task force in press briefing, 3 April 2020, White House. Available from: https://trumpwhitehouse.archives.gov/briefings-statements/remarks-president-trump-vice-president-pence-members-coronavirus-task-force-press-briefing-18/. Accessed 3 March 2021
Trump DJ (2020h) Remarks by President Trump, Vice President Pence, and members of the coronavirus task force in press briefing, 6 April 2020, White House. Available from: https://trumpwhitehouse.archives.gov/briefings-statements/remarks-president-trump-vice-president-pence-members-coronavirus-task-force-press-briefing-21/ Accessed 3 March 2021
Trump DJ (2020i) Remarks by President Trump, Vice President Pence, and members of the coronavirus task force in press briefing, 16 April 2020, White House. Available from: https://trumpwhitehouse.archives.gov/briefings-statements/remarks-president-trump-vice-president-pence-members-coronavirus-task-force-press-briefing-27/. Accessed 3 March 2021

Chapter 13
How Does My Mask Look? Nonverbal Communication Through Decorative Mask-Wearing

M. Eilene Wollslager

Abstract The CDC (How to select, wear, clean your mask, 2020) recommends wearing masks to slow the spread of COVID-19. Many states and municipalities require mask-wearing. An ABC News/Ipsos Poll Jackson et al. (Americans slow to change out-of-home behavior, 2020) reported that 87% of people in the U.S. wore a mask in public in the previous week. This study explores the role of decorative masks in mask-wearing compliance among adults. It also examines how decorative masks are used as nonverbal artifacts to communicate an aspect of the wearer's individuality. A survey of 235 individuals across the U.S. found that most individuals (72%) believe their mask communicates something nonverbally to others. Messages such as "showing concern for others" (86%) and "a desire to slow the spread of COVID-19 (84%) were supported as possible messages mask-wearing communicates. Sex did have a significant effect on the choice of mask selection. A significant percentage of women and those identifying as intersex chose fashion masks more frequently than medical masks when compared to men.

Keywords Mask-wearing · COVID-19 · Coronavirus · Nonverbal artifacts · Ask-wearing compliance

13.1 Introduction

As 2020 began, few people would expect that mask-wearing would be a part of daily life. The COVID-19 pandemic changed many social practices—one of the most controversial has been mask-wearing to stop the virus's spread. While experts vacillated on the benefits of mask-wearing in slowing the spread of COVID-19 (Berezow 2020, March 31) ultimately, the CDC recommended mask-wearing (2020, August 27). Several studies demonstrate the efficacy of mask-wearing in reducing the transmission of COVID-19 (Zhang et al. 2020; Zeng et al. 2020). In response, many states and municipalities mandated mask-wearing. As of December 2020, 38 states had some mask mandate (Hubbard 2020, December 2).

M. E. Wollslager (✉)
Regis University, Denver, USA

D. M. Berube (ed.), *Pandemic Communication and Resilience*, Risk, Systems and Decisions, https://doi.org/10.1007/978-3-030-77344-1_13

Surveys of the public indicate that in the U.S., mask-wearing is commonplace in response to the pandemic. An ABC News/Ipsos Poll (Jackson et al. 2020, July 24) reported 87% of U.S. citizens wore a mask in public in the previous week. This data comes from self-reports, which may or may not be accurate. An observational study found the actual level of mask-wearing was much lower (41%), but that number increased with mandatory entry requirements into public establishments. While there may be a discrepancy in the actual number of people in the U.S. wearing masks, the fact remains that individuals are using masks at some level.

A large ($N = 8317$) international study (Clark et al. 2020) looked at compliance with COVID-19 precautions. Researchers found that compliance was slightly higher for women and that the primary determinant for compliance was a trust that the efforts would prevent the spread of the virus. Trust in the government and keeping oneself safe were not predictors. While this study provided insight into the motivations surrounding all precautions (not just mask-wearing), it did not explore what individuals felt they were communicating with wearing masks.

Rothman (2020) identifies two main concentrations of mask-wearing research. The first, and most prevalent, involves the efficacy of mask-wearing in preventing the spread of COVID-19. Less common is social behaviour surrounding mask-wearing.

Communication research regarding mask-wearing is sparse. Existing communication research deals primarily with crisis communication (Malecki et al. 2020, June 16; Wu et al. 2020; Christensen and Lægreid 2020), social media (Furini et al. 2020; Obiała et al. 2020; Cato et al. 2021) and communication challenges resulting from mask-wearing (Marler and Dittoa 2020; Hampton et al. 2020). For deaf and hard of hearing, mask-wearing provides an additional barrier to communication (Atcherson et al. 2020).

While there is evidence to suggest that mask-wearing can hinder communication, what if mask-wearing could also communicate nonverbally? This study attempts to fill a gap in the discussion of nonverbal communication and mask-wearing—primarily how does the choice of mask communicate nonverbal of the wearer's self-image? It posits that mask-wearing has become more than a health practice; it also communicates nonverbal aspects of the wearer's self-image. This study will also explore whether fashion masks improve compliance and positivity about mask-wearing.

13.2 Nonverbal Communication and Clothing

Many scholars from a range of disciplines offer definitions of what constitutes nonverbal communication. Like all definitions of communication, the results are flawed as it is impossible to define communication with itself. Rodriguez (2015) defines non-verbal communication as "The process between two parties to share any messages by a means other than words, that is, expressing information through any non-linguistic code, that is, using the diversity of signs, symbols or other stimuli of that magnitude lacking syntax, the result of which is a decoding and an immediate psychological response from the receiver" (7).

What constitutes clothing is also debated. Argyle (1972) uses the term "appearance" which includes clothing, hair, skin, physique and body condition. Argyle notes that much time and money is spent on manipulating appearance and that people send messages about themselves by "wearing the appropriate costume" (248).

13.3 Fashion and Mask-Wearing

Before 2020, most face masks in the U.S. were worn by healthcare workers, athletes, robbers and Trick-Or-Treaters. Once the COVID-19 pandemic hit full-force in March 2020, the public learned about PPE's (Personal Protective Equipment). Mask-wearing mandates followed to halt the spread of the novel virus. Before long, fashion masks became common in the U.S.

Fashion masks, for the purpose of this study, are cloth masks that have a design or color. This contrasts with the plain, disposable paper medical masks or the N95 masks. The design or color may be any size or color. It may include add-ons such as rhinestones or embroidery. Cloth neck gaiters, like fashion masks, come in a variety of colors and patterns.

Fashion masks existed before the advent of the COVID-19 pandemic. "People have used fashion face masks throughout history, to send a symbolic message to those around them" (Meyer 2020, April 29).

Globally fashion masks were available, particularly in high pollution areas, for decades. They are relatively new to the U.S. where public mask-wearing was not common prior to the pandemic; however, in areas where there are fewer cultural barriers to mask-wearing, such as Asia, the fashion mask is not novel.

Fashion masks are now big business. In the U.S. in 2020, the fashion mask market was valued at $865 million, almost double what it was in 2019. The market is expected to grow 22.7% annually with the U.S. market expected to reach $2.3 billion by 2027 (Grand View Research 2020, May).

While fashion masks primarily started as homemade colored fabric masks in the U.S., marketers quickly jumped on the trend to mix fashion and public health. Designers began making increasingly creative designs. In Japan, consumers purchase jewel-embellished masks for $9,600. Louis Vuitton sells signature face shields for $961. Brides can buy fashion masks to coordinate with their gowns, and consumers can purchase bikinis with matching masks for the beach (Moore 2020, November 25).

13.4 Fashion as Nonverbal Communication

Lemon (2007) argued that fashion is a form of nonverbal communication. Ludmila (2020) said, "The aesthetic and artistic approach of fashion is complemented by the communicative aspect of the phenomenon" and is a continuous form of communication. It would seem that fashion masks would fit into this category of nonverbal communication.

Dolan (Parker 2020, May 14) said that even though we are in public with our faces covered, our nonverbal communication continues through the choice of masks and how they are worn. This study concludes that masks are nonverbal artifacts that offer insights into aspects of the individual's personality.

13.5 Research Questions

This review of nonverbal communication and mask-wearing leads to the following research questions.

RQ1 How much do individuals use fashion masks as a communication artifact? Individuals use other artifacts such as jewelry and clothing to communicate nonverbally, so as an extension, masks might also be a communication artifact.

RQ2 Is there a relationship between compliance and fashion mask-wearing? There has been research exploring whether decorative bandages reduce discomfort in children with mixed results (Johnston et al. 1993; Schiff et al. 2001, August). If compliance increases for those who chose fashion masks, it might be something to consider.

RQ3 Does sex play a role in the selection of medical or fashion masks? One might think that sex might affect mask choice. Women are stereotypically more interested in fashion than are men. However, this may or may not be true for mask-wearing if individuals consider it more of a medical artifact rather than a personal one.

RQ4 What messages do individuals believe mask-wearing communicates? If individuals feel that their mask expressed something about them nonverbally, then what exactly does it communicate?

RQ5 Which group identifies more with nonverbal communication through mask-wearing—those wearing disposable medical versus fashion masks? Since medical masks traditionally serve a more utilitarian purpose than fashion masks, would disposable mask-wearers feel that their mask choice communicates less about themselves than those choosing decorative masks?

13.6 Methodology

There were 235 individuals responding to the survey invitation from SurveyMonkey. This sample was purchased from the company. Two individuals opted out after reading the informed consent, leaving $N = 233$.

There were slightly more women (53%, n = 124) than men (47%, n = 109). The largest group of participants were ages 35–54 (43%. n = 98) followed by the 18–34 age group (32%, n = 75) and the 55+ age group (25%, n = 58). Most were college graduates (64%, n = 148), or had high school degrees (28%, n = 64) with the remaining 8% (n = 14) who either did not respond or did not have a high school degree. Those earning less than $50,000 (42%, n = 98) comprised the largest group closely followed by those earning $50,000–74,999 (41%, n = 96) of participants. Individuals earning $75,000–124,999 were next (21%. N = 49) leaving earners making $125.000+ making up 6% (n = 20). There were 10% (n = 22) that did not share their income.

SurveyMonkey collected responses from across the U.S., with representation in all regions. Given the sample size, the margin of error is 6.55% at a confidence level of 95%.

13.7 Procedure

Participants were given a 20-question survey during a two-day period in January 2021. The survey (see Appendix) included an informed consent followed by six questions regarding mask-wearing habits. Two of these six questions were 5-point Likert scales concerning the frequency and the likelihood of wearing a fashion mask. The other four questions concerned the type of masks worn, the masks' designs, reasons for wearing a paper mask, and the number of masks owned. Upon review Q4 had two categories that were the same (plan cloth mask and single-colored cloth mask), so the data from these two categories was merged.

The next six questions concerned mask-wearing and communication. There were five Likert scale questions (5-point) regarding how much the individual's mask choice communicates something about them. These questions involved how much the mask communicated something about the person, how their masks make them feel, the likelihood of continued mask-wearing, and how much they felt their mask communicated concern for others and how much it communicated a desire to stop the spread of the virus. One open-ended question allowed respondents to explain what they felt their mask communicated about them.

The final demographic questions covered age, sex and education. SurveyMonkey included five demographic questions for their purposes, including a different categorization of age and income, a binary sex question (did not include intersex or a prefer not to answer category), region of the U.S. and a question on the device-type used for the survey. The aggregate data included these results.

For the study's purposes, fashion masks were operationalized as single-colored or masks bearing an embellishment or design. Paper masks are disposable or the N-95 or KN-95. Question selections provided opportunities to specify the type of mask or indicate that they did not wear colored or designer masks.

13.8 Results

Data from the online questionnaire were transferred from the SurveyMonkey website to SPSS. Next, the data was cleaned and analyzed. The data were screened for missing responses, outliers, normality, linearity, and homoscedasticity using methods and rules as described by Tabachnick and Fidell (2001). Missing data were replaced with non-missing variable means (Green and Salkind 2008). Variables were transformed as necessary, including reverse coding for independent t-test sample questions for comparison analysis. Descriptive statistics were used to analyze information about participant demographics and media use habits and results from sample groups in the study.

All respondents indicated they wore a mask. The vast majority (92%, n = 212) reported wearing a mask always or usually. About 7% (n = 16) wore a mask sometimes, with 1% (n = 3) reporting they wore a mask rarely and none said they never wore a mask (M = 1.40, SD = 0.72).

When considering mask choices (more than one type could be selected), participants wore plain (single-colored) cloth masks the most frequently (34%, n = 79), fashion masks (27%, n = 61), disposable paper masks (26%, n = 60), N95 or KN95 masks (6%, n = 14) and masks with a company logo (3%, n = 8). Very few respondents wore pull-up gaiters (2%, n = 4) or selected other (2%, n = 5).

When asked about the likelihood of wearing a single-color or mask with a design, most people were very likely (42%, n = 97) or likely (23%, n = 53) to choose a cloth mask with a single color or design. Twenty-four percent (n = 55) answered they were neither likely nor unlikely to wear a single-colored or designer mask, with 6% (n = 15) and 5% (n = 11) indicating they were unlikely or very unlikely, respectively, to wear this type of mask (M = 2.10, SD = 1.17).

The types of cloth masks chosen by participants (n = 231) were primarily single-colored ones (56%, n = 128). When reviewing the comments by those indicating "other" types of designs all but one of the 15 responses could be included in the provided categories. Only one person said that there was not particular one design that they wore. The Artistic designs were the most common (25%, n = 57) type of fashion mask worn, followed by sports teams (9%, n = 20), company logos (5%, n = 10), sequined/bling (3%, n = 6) and game logos (2%, n = 4).

Respondents have several masks from which to choose, with 24% (n = 56) having more than 10 masks. The next most common amount was 3–4 masks (21%, n = 48), 5–6 masks (18%, n = 42), 7–8 masks (10%, n = 23), varied amount—disposable masks (10%, n = 23), 1–2 (9%, n = 22), 9–10 (7%, n = 16), 1 person preferred not to answer.

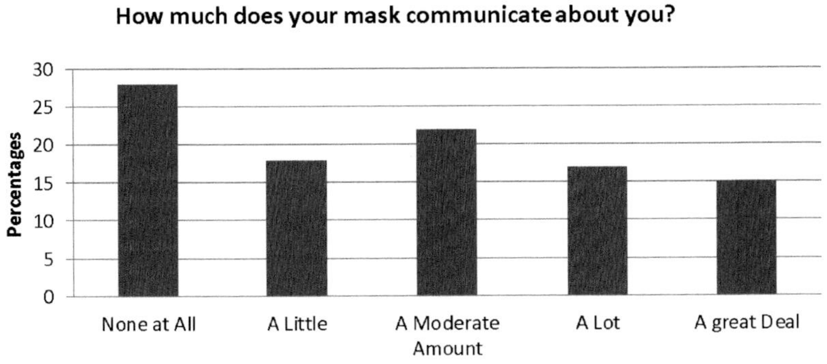

Fig. 13.1 Frequencies analysis for masks as communication (n = 231)

Regarding *RQ1* on the use of masks as a communication artifact, 28% thought their mask communicated nothing at all. The remaining 72% ($M = 1.6$, $SD = 0.92$). Acknowledged that their masks communicated something about themselves (see Fig. 13.1).

Two subsequent survey questions supported this data. When asked if they felt their masks showed courtesy to others, 86% agreed ($M = 1$, $SD = 0.92$). Participants also felt wearing their mask demonstrated a desire to stop COVID-19 (84%, $M = 1.5$, $SD = 0.92$).

In consideration of *RQ2*, a Pearson product-moment correlation was computed to assess the relationship between mask-wearing compliance and choosing fashion masks. There was an insignificant positive correlation ($r = 0.04$, $n = 231$, $p = 0.57$).

RQ3 explored whether there would be a statistical significance between men ($n = 107$) and women ($n = 120$) or intersex ($n = 4$) in selecting fashion masks. A significant difference exists in the scores for men ($M = 2.3$, $SD = 1.10$), women ($M = 2.26$, $SD = 1.16$) and intersex ($M = 3.5$, $SD = 1.00$); $t(230) = 27.41$, $p < 0.001$. Women and intersex were more likely to wear fashion masks than men.

RQ4 considered the open-ended responses about what masks communicated; 13% ($n = 31$) said their masks communicated their interest in "safety" either for themselves or others. Another 13% ($n = 30$) said their mask communicated nothing. The subsequent most frequent response was "colors," 5% ($n = 12$) referencing a favorite or matching color. One individual wrote, "It (mask) conveys that I am being cautious and it may also inadvertently (and correctly or incorrectly) be perceived to convey my political leanings. The color (solid black, usually) may also convey a sort of professional or no-nonsense air." Another respondent said their mask communicated, "That I like to be fashionable, not boring, and want to match my mask to my outfit." One other participant said their mask communicated. "That I'm a cowboy. But that's already clear from how I dress, so the mask is medical masks really unimportant in that regard." Another expressed that the mask reflected their interest. "I have a Stitch mask because I love Lilo and Stitch. I have a KSU mask Because I go to school there."

Some participants expressed that the mask did not communicate anything but then provided something that the mask communicated. For example, one individual said, "Nothing, just that I know how to properly wear my mask and not have my nose sticking out. I actually really care about not getting COVID-19 again." Two individuals expressed how much they disliked wearing masks. One person said, "They are stupid, pointless and I don't wear them unless I have to."

RQ5 explored the relationship between the choice of mask and the awareness of a mask as nonverbal communication. Individuals who wore paper and N95 medical masks were combined, as were all types of cloth and fashion masks. There was a significant relationship between medical masks ($M = 1.66$, $SD = 0.48$) and fashion masks ($M = 2.91$, $SD = 1.61$) and awareness of nonverbal communication $t(230) = 35.36$, $p < 0.001$.

13.9 Discussion

One of the limitations of this study was the small sample size ($N = 233$). A larger national sample would provide increased reliability in determining U.S. mask-wearing trends and motivation. Self-reports of compliance should be further compared with field studies as self-reports are not always accurate. This study did not explore the motivations behind mask-selection beyond their use as a nonverbal communication artifact. There are undoubtedly other factors that influence compliance beyond what the mask communicates to others.

Future studies may want to explore how other nonverbal communication is affected by mask-wearing. Do eye contact, touch or other nonverbal communication increase or decrease because masks obscure the mouth during interactions? Another line of inquiry could explore how mask-wearing encourages or discourages verbal interaction.

Much like the studies on children and decorative bandages (Johnston et al. 1993; Schiff et al. 2001, August), this study did not find a relationship between fashion masks and compliance. Fashion masks may not improve compliance, but if the vast majority of the public wears masks in compliance with mandates, as this survey suggests, promoting masks as fashion may not be needed to increase mask-wearing.

The majority of participants expressed that their masks communicated something about their personalities or preferences. Individuals seem to recognize that their selection of masks contributes to their nonverbal communication with others. Safety was not the only concern when choosing masks. As one respondent said, their mask communicates that "I take my health seriously and my style."

Women and those identifying as intersex seemed to be more concerned with fashion than safety when compared to men. This study indicated that a significant number of women chose fashion masks than did men.

With the challenges presented by the COVID-19 pandemic, it seems that individuals have "made the best of it" in their choice of masks. Whether communicating that they care for themselves or others or whatever communication obstacles mask-wearing creates, it also provides opportunities for self-expression and showing concern for others.

Appendix

Survey questions.

1. Statement of Consent
   I have read this form (or have had it read to me). I have been encouraged to ask questions. I have received answers to my questions. I give my consent to participate in this study. I understand the risks and discomforts associated with the above study and understand that I may quit the study at any time without penalty.

Yes
No

2. How often do you wear a mask because of COVID-19?

Always	Usually	Sometimes	Rarely	Never

3. If you wear a disposable mask, why is this your preferred choice (select all that apply)?

Convenient	Don't like cloth masks
Safer	Cooler to wear
Work provides	Fits better
Other (please specify)	

4. When you wear a mask in public, what type(s) of mask do you wear most often?

Disposable paper mask	Cloth mask with work/company logo
Plain cloth mask	N95 or KN95 mask
Solid colored cloth mask (all red, black, blue, etc.)	Solid colored gaiter—pull up covering
Cloth mask with design or embellishment	Gaiter with design or embellishment
Other (please specify)	

5. How likely are you to wear a colored cloth mask or a mask with a design?

Very likely	Likely	Neither likely nor unlikely	Unlikely	Very unlikely

6. What is the design of the mask you wear most often?

Plain color	Artistic design
Sports team	Company logo
Sequined/bling	Game logo
Other (please specify)	

7. How many masks do you own?

1–2
3–4
5–6
7–8
9–10
More than 10
Varies-using disposable masks
None
Prefer not to answer

8. How much does your mask communicate something about you?

A great deal	A lot	A moderate amount	A little	None at all

9. What do you feel your masks communicate about you?
10. How does your mask choice make you feel about wearing your mask?

Completely happy to wear it	Somewhat happy to wear it	Neutral about wearing it	Somewhat unhappy to wear it	Completely unhappy to wear it

11. How likely are you to continue wearing your mask?

Very likely	Likely	Neither likely nor unlikely	Unlikely	Very unlikely

12. How much do you agree or disagree that wearing your mask demonstrates courtesy to others?

Strongly agree	Agree	Neither agree nor disagree	Disagree	Strongly disagree

13. How much do you agree or disagree that wearing your mask shows your desire to stop the spread of COVID-19?

Strongly agree	Agree	Neither agree nor disagree	Disagree	Strongly disagree

14. What is your sex?

Female
Male
Inter sex
Prefer not to answer

15. What is your age?

18–24
25–34
35–44
45–54
55–64
65–74
75 or older
Prefer not to answer

16. What is the highest level of education you have completed?

Did not attend school
Graduated from middle school
Graduated from high school
Graduated from college
Some graduate school
Completed graduate school
Prefer not to answer

Survey Monkey added five of their questions to the survey. Some were repetitive with different categories.

A. Age (<18, 18–29, 30–44, 45–60, >60)
B. Sex (male and female)
C. Income $0–$200,000+
D. Region (New England, Middle Atlantic, East North Central, West North Central, South Atlantic, East South Central, West South Central, Mountain, Pacific)
E. Device type used (IOS Phone/Tablet, Android Phone/Tablet, Other Phone/Tablet, Windows Desktop/Laptop, MacOS Desktop/Laptop, Other).

References

Argyle M (1972) Non-verbal communication in human social interaction. In: Hinde RA (eds) Non-verbal communication. Cambridge University Press, pp 242–348

Atcherson SR, Finley ER, McDowell BR, Watson C (2020) In the search for transparent face coverings during the COVID-19 pandemic. Audiol Today 32(6):20–27

Berezow A (2020, Mar 31) The great face mask debate: to wear or not to wear? American council on science and health. Retrieved 28 Dec, 2020 from https://www.acsh.org/news/2020/03/31/great-face-mask-debate-wear-or-not-wear-14675.

Cato S, Iida T, Ishida K, Ito A, Katsumata H, McElwain KM, Shoji M (2021) The bright and dark sides of social media usage during the COVID-19 pandemic: survey evidence from Japan. Int J Disaster Risk Reduct 54. https://doi.org/10.1016/j.ijdrr.2020.102034

CDC (2020, Aug 27) How to select, wear, clean your mask. CDC.gov. https://www.cdc.gov/coronavirus/2019-ncov/prevent-getting-sick/about-face-coverings.html

Christensen T, Lægreid P (2020) The coronavirus crisis—crisis communication, meaning-making, and reputation management. Int Public Manag J 23(5):713–729. https://doi.org/10.1080/10967494.2020.1812455

Clark C, Davila A, Regis M, Kraus S (2020) Predictors of COVID-19 voluntary compliance behaviors: an international investigation. Global Trans 2:76–82. https://doi.org/10.1016/j.glt.2020.06.003

Furini M, Mirri S, Montangero M, Prandi C (2020) Untangling between fake-news and truth in social media to understand the Covid-19 Coronavirus. In: 2020 IEEE symposium on computers and communications (ISCC), computers and communications (ISCC), pp 1–6. https://doi-org.dml.regis.edu/10.1109/ISCC50000.2020.9219663

Grand View Research (2020, May) Fashion face mask market size, share & trends analysis report by product (anti-pollution, non anti-pollution), by distribution channel (offline, online), by region, and segment forecasts, 2020—2027. Retrieved from https://www.grandviewresearch.com/industry-analysis/fashion-face-mask-market/segmentation

Green SB, Salkind NJ (2018) Using SPSS for windows and Macintosh: analyzing and understanding data, 8th edn. Pearson, New York

Hampton T, Crunkhorn R, Lowe N, Bhat J, Hogg E, Afifi W, Sharma S (2020) The negative impact of wearing personal protective equipment on communication during coronavirus disease 2019. J Laryngol Otol 134(7):577–581. https://doi.org/10.1017/S0022215120001437

Hubbard K (2020, Dec 2) These states have COVID-19 mask mandates. U.S. News & World Report. Retrieved 28 Dec, 2020 from https://www.usnews.com/news/best-states/articles/these-are-the-states-with-mask-mandates#del

Jackson C, Newall M, Jinhee Y (2020, July 24) Americans slow to change out-of-home behavior, even as coronavirus concerns intensify, ABC/Ipsos Poll. Retrieved 10 Sept, 2020 from https://www.ipsos.com/en-us/news-polls/abc-news-coronavirus-poll

Johnston CC, Stevens B, Arbess G (1993, June 8) The effect of the sight of blood and use of decorative adhesive bandages on pain intensity ratings by preschool children. J Pediatr Nurs 3:147–151 PMID: 8340879

Lemon J (1990) Fashion and style as non-verbal communication. South African J Commun Theory Res 16(2):19–26. https://doi.org/10.1080/02500169008537718

Malecki KM, Keating JA, Safdar N (2020, June 16) Crisis communication and public perception of COVID-19 risk in the era of social media. Clin Infect Dis. https://doi.org/10.1093/cid/ciaa758.

Marler H, Ditton A (2021, Apr 29) "I'm smiling back at you": exploring the impact of mask-wearing on communication in healthcare. Int J Lang Commun Disord 56(1):205–214. https://doi.org/10.1111/1460-6984.12578

Meyer M (2020, Apr 29) Fashion face masks—a short history before COVID 19. WTVOX. Retrieved 27 Dec, 2020 from https://wtvox.com/fashion/short-history-of-fashion-face-masks

Moore C (2020, Nov 25) Japanese face masks being sold $9.6G feature pearls, crystals, more. Fox News. Retrieved 27 Dec, 2020 from https://www.foxnews.com/lifestyle/japanese-face-masks-9-6g-pearls-crystals-diamonds

Obiała J, Obiała K, Mańczak M, Owoc J, Olszewski R (2020) COVID-19 misinformation: accuracy of articles about coronavirus prevention mostly shared on social media. Health Policy Technol. https://doi.org/10.1016/j.hlpt.2020.10.007

Parker P (2020, May 14) Extended noon news interview for Thursday 5/14. WISR. https://www.wisr680.com/extended-noon-news-interview-for-thursday-5-14/?doing_wp_cron=1611619742.7583990097045898437500

Rodriguez J (2015) Approach to the anthropomorphic factor: the body as a stimulus in non-verbal communication (Graduate Thesis). Politecnico Grancolombiano University Institution, Bogota, Colombia

Rothman SB (2020) COVID-19 community mask wearing: lessons from foot-binding and infibulation. Eubios J Asian Int Bioethics 30(7):354. https://www.eubios.info/EJAIB92020.pdf

Schiff WB, Holtz KD, Peterson N, Rakusan T (2001, Aug) Effect of an intervention to reduce procedural pain and distress for children with HIV infection. J Pediatr Psychol 26(7):417–427. https://doi.org/10.1093/jpepsy/26.7.417

Tabachnick BG, Fidell LS (2018) Using multivariate statistics, 7th edn. Pearson, Boston

Wu AW, Connors C, Everly GS (2020) COVID-19: peer support and crisis communication strategies to promote institutional resilience. Ann Intern Med 172:822–823. https://doi.org/10.7326/M20-1236

Zeng N, Li Z, Ng S, Chen D, Zhou H (2020) Epidemiology reveals mask-wearing by the public is crucial for COVID-19 control. Med Microecol 4. https://doi.org/10.1016/j.medmic.2020.100015
Zhang K, Vilches TN, Tariq M, Galvani AP, Moghadas SM (2020) The impact of mask-wearing and shelter-in-place on COVID-19 outbreaks in the United States. Int J Infect Dis 101:334–341. https://doi.org/10.1016/j.ijid.2020.10.002

Chapter 14
Masks Don't Work but You Should Get One: Circulation of the Science of Masking During the Covid-19 Pandemic

Ekaterina Bogomoletc, Jean Goodwin, and Andrew R. Binder

Abstract The Covid-19 pandemic quickly increased public demand for scientific information, thus providing a stress test that can reveal whether, and how, the science communication environment is able to meet those demands. The topic of mask use by the general public emerged as particularly fraught. Informed by the intermediate agenda setting theory, framing theory, and research on flows of information, our chapter examines how scientific information about masks was disseminated and interpreted by the mainstream media and Twitter users during the early months of the Covid-19 pandemic. The study revealed that neither news media nor Twitter were securely in the lead when it comes to disseminating scientific articles. The analysis also demonstrated that both mainstream media and Twitter users cited the same scholarly articles in support of opposite positions regarding masks, and that media publications were more likely to communicate the uncertainty of the science than Twitter posts.

Keywords Framing · Information flows · Intermedia agenda setting · Facemasks · Covid-19 · Science communication

14.1 Introduction

The emergence of the Covid-19 pandemic in February, 2020 quickly increased public demand for scientific information, thus providing a stress test that can reveal whether, and how, the science communication environment (Kahan 2017) is able to meet those demands. The topic of mask use by the general public emerged as particularly fraught. Americans received conflicting and changing official guidance regarding mask efficacy, until in April the CDC (US Centers for Disease Control) finally recommended that everyone wear a mask, including even a homemade face covering, in public places. Such adjustments were not easy, leaving people with diverging views on mask use even before the issue became caught up in partisan divides. While it might

---

E. Bogomoletc (✉) · J. Goodwin · A. R. Binder
North Carolina State University, Raleigh, NC, USA
e-mail: ebogomo@ncsu.edu

D. M. Berube (ed.), *Pandemic Communication and Resilience*, Risk, Systems and Decisions, https://doi.org/10.1007/978-3-030-77344-1_14

be tempting to blame lay publics for not reacting quickly enough to official guidelines, or leaders for not promoting the correct behavior, the deliberations about mask usage also underscored the growing complexity of the science communication environment, where individuals can seek information directly from published research, from traditional and new media, and from each other.

In the current chapter, we look at the ways scientific information about masks was disseminated and interpreted during the early months of the Covid-19 pandemic. First, we provide the overall context of the debate and discuss how inconsistent messaging from health officials together with the new media environment created the conditions nurturing uncertainty regarding mask behavior. Second, we discuss scholarship and theoretical frameworks such as framing theory, agenda setting theory, and research on information flows that informed our study. We then present the results of our study, tracing the dissemination and interpretation of six widely shared research articles on the efficacy of cloth masks. We examine both the trends in media and social media coverage of mask research during the pandemic. Ultimately, we address the broader question of information flows through the science communication environment during the pandemic.

14.2 The Context of the Debate

14.2.1 Ambiguous Science

Public and official discussion about use of masks[1] emerged soon after the first reports in January, 2020, of what we now call the Covid-19 pandemic. Although the use of masks to limit the spread of disease in community settings had become customary in many Asian countries (MacIntyre and Chughtai 2015), there had been little research into the efficacy of the practice. As will be detailed below, these studies failed to provide definitive evidence of the efficacy of homemade masks in non-medical settings. For example, some provided data on different materials without considering fit or the exigencies of daily life, while others tested masks in actual hospital settings, but comparing them with medical masks, not with going unmasked. While none of these articles rejected mask use by the general public, none of them gave the practice a ringing endorsement. The uncertainty only got worse due to the inconsistent messaging of public health officials.

[1] Unless further specified, we use the term "mask" to refer to cloth face coverings used by the general public, the recommendation eventually adopted by public health officials and thus the technology of interest for our study.

14.2.2 Inconsistent Messaging

Throughout the first stages of the pandemic, public health officials changed their position regarding masks multiple times, aggravating public uncertainty regarding the issue (see Fig. 14.1).

Early official recommendations urged mask use only when there were clear signs of infection; the general public was told specifically *not* to mask. This stance began receiving pushback in March, with an influential *New York Times* editorial by Zeynip Tufekci (2020), an online campaign urging #Masks4All, and a movement among "makers" to answer a Million Mask Challenge to produce homemade masks for hospitals. But public health officials were facing multiple challenges. They needed to avoid provoking the hoarding that would exacerbate the shortage of personal protection equipment (PPE) for health care workers. At the same time, they needed to address the issue of spread of the virus as well as the growing public anxiety and public request for recommendations on personal protection. And they had to make recommendations with limited information about how Covid-19 was transmitted and about the efficacy of masks, especially homemade masks in community settings. The attempts to find a balance between these challenges would result in CDC finally taking a clear stand regarding the masks by focusing on homemade "cloth covers" at the very beginning of April. In addition, CDC published links to DIY (Do-It-Yourself) tutorials on how to make face covers as well as a video tutorial with Surgeon General. The video went viral and received over four and one-half million views by December 2020.

At the international level, WHO went through a similar process, finally recommending mask wearing on June 5, 2020. Caught between the need to prevent mask hoarding, provide safety guidance for lay public and find the best science-driven solution, the organizations produced inconsistent messaging which only fueled the debate. Moreover, as noted by Tufekci (2020), "the message became counterproductive... because it seemed as though authorities were shaping the message around managing the scarcity rather than confronting the reality of the situation" (para. 2). One might speculate that such lack of guidance could force lay publics to seek guidance elsewhere. However, when seeking information beyond official sources, people faced the challenge of navigating the so-called "infodemic".

14.2.3 Infodemic

In addition to navigating the spread of the virus itself, the Covid-19 crisis brought up the issue of "infodemic": a situation characterized by "too much information including false or misleading information in digital and physical environments during a disease outbreak" (WHO 2020). Some scholars and public health experts blamed social media for allowing misinformation to spread and called them "an engine of

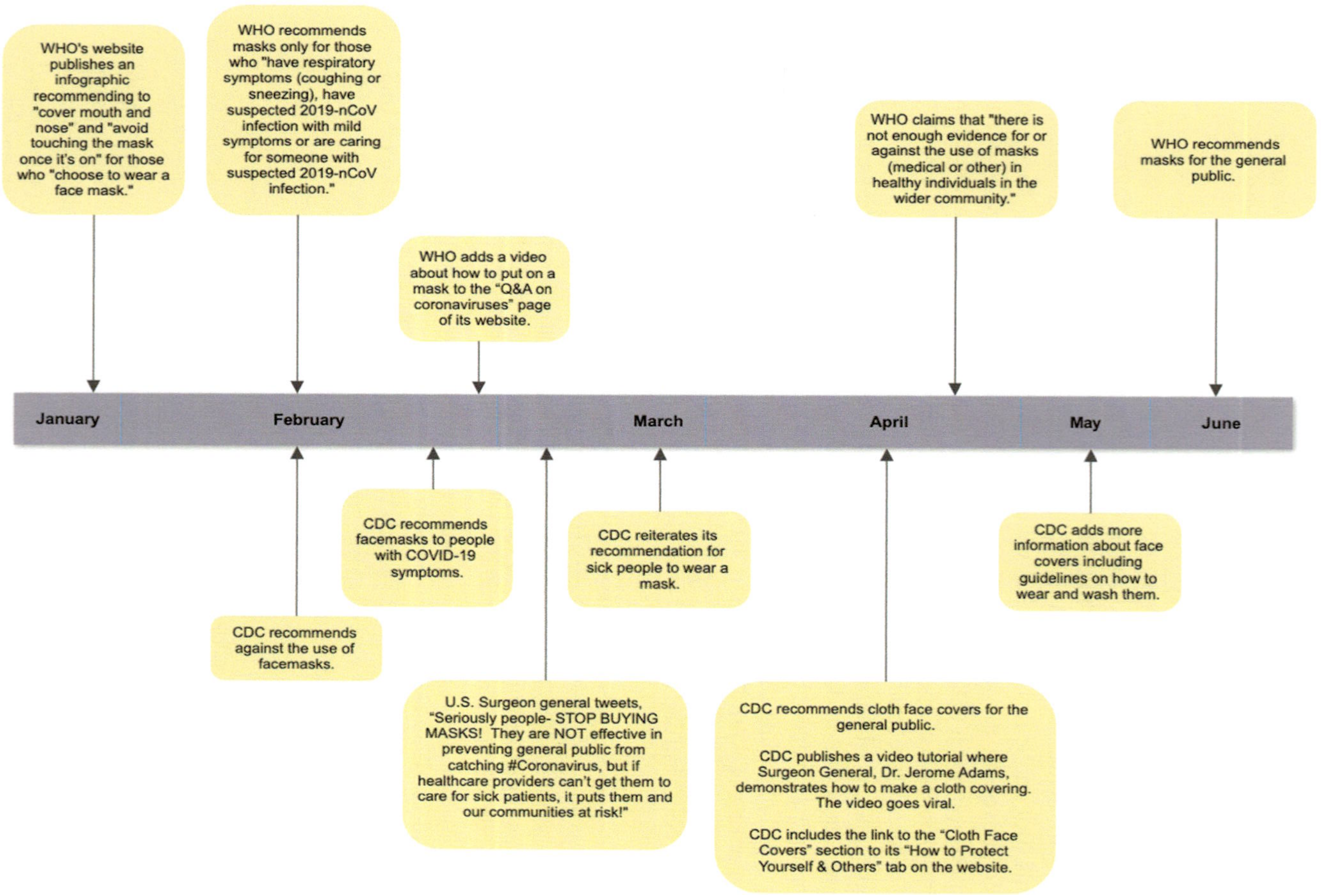

Fig. 14.1 The timeline of the changes in the positions of CDC and WHO (World Health Organization) on masks (based on the internet archive wayback machine)

untruths" (Buchanan 2020, p. 894). They suggested that while the issue of misinformation existed during previous pandemics, social media aggravated this problem for the Covid-19 pandemic by increasing the speed of information distribution and amplifying purveyors of false information, such as anti-vaccination views (Buchanan 2020; Zarocostas 2020). As a result, WHO had to put extra efforts into addressing social media rumors and promoting evidence-based information online (Zarocostas 2020).

However, the issue of infodemic is not limited by the affordances of social media platforms. Among the factors contributing to the infodemic, scholars also named low health literacy (Alvarez-Risco et al. 2020; Chong et al. 2020), increased free time (Alvarez-Risco et al. 2020), and a shift in the audiences of scientific information (The Lancet Infectious Diseases 2020). The latter refers to the situation in which scientists faced the challenge of "reaching experts and non-experts alike in an emotionally charged global environment" (The Lancet Infectious Diseases 2020, p. 875). In other words, while in the pre-pandemic world, scientists would usually target the expert community which is familiar both with scientific and publishing processes, the Covid-19 crisis brought scholarly articles to the attention of lay publics who might not have training or experience in dealing with scientific research.

Taken together, these factors have put media and lay publics in the position where they needed to search for sources in addition to official advisories, interpret original research articles, and adapt their views and behavior to the quickly changing science communication environment. Informed by research on flows of information (Bucchi 2017), by intermediate agenda setting theory (McCombs 2005), and by framing theory (Cacciatore et al. 2016), we conducted a study that examines trends in the use of research articles on masks by the media and by the general public. More specifically, the chapter tracks differences in interpretation of the six widely cited articles on facemask efficacy during the public debate on masks.

14.3 Literature Review

14.3.1 Information Flows

While we are still developing an understanding of information flows within the emerging online science communication environment, the general shape of the changes have become clear. According to what Bucchi (1996) has termed the "canonical account" of science communication, forty years ago scientific knowledge was available only to those who had physical access to a research library and the knowledge of how to use complex indexes to locate relevant articles. Most people therefore received topical science information through the "narrow communications channel" provided by print and television journalism (Mazur 1981, p. 109). These gatekeepers transmitted only a tiny fraction of scientists' output (Suleski and Ibaraki 2010), selected to fulfill journalistic norms such as conflict and novelty (Dunwoody 2014).

In particular, in what became a "compelling issue within the domain of science news" (Jensen 2008, p. 349), science journalists often stripped studies of their qualifications and limitations (Nelkin 1987; Singer 1990). The simplifying required for science news can easily cross over into oversimplifying and sensationalizing, potentially contributing to public misconceptions about science (see, e.g., Gustafson and Rice 2020; Jensen et al. 2013; Retzbach and Maier 2015).

The internet continues to disrupt this traditional system. Gatekeepers have experienced swift declines, with science news coverage shrinking and science journalism units being downsized or eliminated, in what Scheufele (2013) has characterized as a "crumbling science-public infrastructure" (see also Schäfer 2017). Close to 70% of Americans now say that they turn to online sources to find information about a specific science topic (NSB 2020). There they find and use (Su et al. 2015) a much broader array of information sources than previously available: online editions of traditional media, new online-only publications, advocates and advertisers of all stripes, and of course each other, through the affordances of social media (Brossard and Scheufele 2013; Schäfer 2017; NASEM 2016). Scientists encouraged by the open science movement (Grand et al. 2012), by the desire to gain public attention to their work (Bucchi 1996), and by long-standing calls for increased science communication (Yeo and Brossard 2017) are also now making their work directly accessible to anyone with an internet connection. As Brossard concludes, "a simple Google search can give anyone access to virtually unlimited information about a specific scientific topic" (Brossard 2013).

In evaluating these rapid and large-scale changes, researchers have frequently noted a dilemma (e.g., Bubela et al. 2009; Bucchi 2017; Southwell 2017; Trench 2008). Does direct access to information reduce the impact of media biases, improve public understanding of science and lead to better decision-making? Or does the decline of traditional gatekeeping enable a proliferation of mis- and disinformation from unqualified sources and drive polarization? The Covid-19 pandemic has raised these questions in a particularly pressing way (Rosenberg et al. 2020). In the early months of the pandemic, Americans were putting together information from multiple sources, including the government (88%), television (74%), social media (74%), newspapers (70%) and websites (68%) (Ali et al. 2020). Analyses of online sources, however, suggest that the quality of information they found online was often low, whether on websites (Cuan-Baltazar et al. 2020), YouTube (Szmuda et al. 2020), or Twitter (Al-Rakhami and Al-Amri 2020). Prevention measures, including the use of face masks, emerged as a popular topic of discussion on social media (Xue et al. 2020). But one review found that less than half of websites included correct information on the use of masks, with even public health websites varying substantially from WHO guidance (Hernández-García and Giménez-Júlvez 2020).

Not much is known about either the supply or demand side of science info online: what is available, from what sources, with what perspectives, and why and how differently situated individuals access it (Xenos 2017). In this study, we target *good* information–published research articles, examining how it flowed through and was interpreted by gatekeepers and among social media communicators. In particular, we explore whether communicators retained the limitations and qualifications of the studies, since the failure to communicate scientific uncertainty seems especially significant during a pandemic when the demand for evidence-based guidance dramatically increases while the science itself remains uncertain and inconclusive.

14.3.2 Intermedia Agenda Setting

Agenda setting theory was initially introduced to explain the way media affect public perceptions of the salience of certain issues. The theory suggests that media direct the public to what to think *about* (McCombs 2005): the more media cover an issue, the higher importance the audience assigns to this issue. The theory was later developed in several directions, one of which is concerned with *intermedia agenda setting* effects (McCombs 2005). Studies in this area investigate whether traditional media affect each other's agenda (Denham 2014; Vliegenthart and Walgrave 2008); whether online media affect each other's agenda (Vargo and Guo 2017); how traditional media interact with online media in terms of setting the agenda (Vonbun et al. 2016); how online media interact with social media, and in particular whether there is such a thing as reverse agenda setting, i.e., whether social media users dictate to the media what to communicate about (Groshek and Groshek 2013).

It is worth noting that a large part of this research is concerned with topics outside of health communication. However, several findings offered by existing scholarship on intermedia agenda setting may be relevant to understanding the circulation of science during the pandemic. First, research on political news demonstrated that publications in mainstream media both affect the topics discussed by social media users and impact the way these topics are being discussed (Kim et al. 2016). At the same time, Rogstad (2016) demonstrated that while a large portion of Twitter posts include references to mainstream media, Twitter users also cover some issues that do not receive as much attention on media, e.g., environmental topics. This might indicate the presence of an alternative agenda offered by social media platforms.

Second, research provides conflicting findings when it comes to "reverse agenda setting", i.e., the ability of social media to shape the agenda of mainstream media. A study by Valenzuela et al. (2017) suggested that Twitter might inspire news coverage of disasters by TV journalists. At the same time, Groshek and Groshek (2013) suggested that "the potential for SNSs [social networking sites] to directly shape media agendas does exist but only sporadically and on certain topics" (p. 24). More specifically, the researchers demonstrated the differences between Facebook and Twitter in their ability to predict traditional media agenda when it comes to the topics of culture and politics.

Finally, some researchers have examined the possibility of mutual influence between social and mainstream media (Conway et al. 2015; Wang and Guo 2015). For example, Conway et al. (2015) demonstrated that during elections, traditional media and Twitter "appear to have a symbiotic relationship that varies in intensity and duration depending on the issues being analyzed" (p. 374). The scholars highlighted "legitimacy" as one of the important characteristics of traditional media that allow them to keep their position in the agenda setting process (Conway et al. 2015, p. 375). Similarly, in their study on media and public discussion of genetically modified mosquitoes, Wang and Guo (2018) suggested that Twitter and online media informed each other's agenda. Interestingly, the channels switched the roles depending on the stage of the debate with Twitter leading the discussion at the beginning and mainstream media taking the lead only later, "when the public became aware of the issue." (Wang & Guo 2018, p. 947). Moreover, the study demonstrated that while the channels affect each other's agenda in terms of volume of the discussion, they were independent when it comes to interpretation of the issue.

14.3.3 Mass Media Framing Theory

Framing theory, according to mass media and communication scholars, suggests that a speaker's or writer's choices in language and presentation about an issue influences the way an audience interprets the issue (Entman 1993; Scheufele 1999). As noted by Nisbet and Mooney (2007), frames "allow citizens to rapidly identify why an issue matters, who might be responsible, and what should be done" (p. 56). Cacciatore et al. (2016) discuss two dominant ways to approach understanding and definition of framing: a sociological definition (emphasis framing) and an equivalence-based definition. *Emphasis framing* refers to "manipulating the content of a communication" (Cacciatore et al. 2016, p. 8). From this perspective, framing "involves emphasizing one set of considerations over another" (Cacciatore et al. 2016, p. 10), i.e., focusing on a specific side of a discussed phenomenon. For example, when discussing face masks, one might focus on mask efficacy, individual freedom, or even the fashion side of mask wearing thus highlighting different aspects of the same issue. *Equivalence-based framing* refers to "manipulating the presentation of logically equivalent information" (Cacciatore et al. 2016, p. 8). Studying framing from this perspective, researchers often examine the effects of gain vs. loss frames (e.g., Kahneman and Tversky 1979). Gain frames refer to highlighting benefits of a certain action while with the loss frames, drawbacks of the lack of action would be highlighted. For example, for a gain frame, a journalist might suggest that wearing a mask might help someone to stay healthy. For a loss frame, the same information might be presented as a warning that the lack of a mask might lead to getting sick.

Interestingly, research has reached conflicting results when it comes to equivalence framing of preventive health measures. Updegraff et al. (2011) demonstrated a higher efficiency of gain frames over lack of messages for promoting the use of hand sanitizers during the H1N1 pandemic in 2009. A study by Nan et al. (2012), on the other hand, found that loss frames were more efficient in promoting H1N1 vaccine among people with lower levels of perceived efficacy of the vaccines. At the same time, Guidry et al. (2018) did not find a significant impact of either gain or loss frame messages on people's intention to get vaccinated against Zika. However, the study demonstrated the positive effect of gain frames on "intermediate" constructs that might potentially lead to a higher likelihood of vaccination (e.g., perceived benefits of the vaccine). Finally, Adonis et al. (2016) tested the effects of gain vs. loss framing when communicating about cervical cancer screening via emails and found no significant differences in the preventive behavior among the groups receiving differently framed messages, including messages with a "neutral" frame. Overall, one might conclude that based on the existing scholarship, the effects of framing on preventive health behavior are context-dependent and oftentimes, mediated by various additional factors.

It is worth noting that the practice of framing scientific findings has received conflicting reactions in the academic community, with some scholars suggesting that it raised ethical concerns related to manipulating one's audience (Nisbet 2009). According to this perspective, framing might compromise "objectivity" of science communicated to lay publics. However, researchers also argue that framing is inevitable; all messages are framed in some way, including those communicating scientific information (Sprain 2018). Indeed, when discussing scientific issues like mask efficacy, both journalists and social media users are forced to present information from academic articles in a brief, condensed manner. They have to select and emphasize certain aspects of the studies that they are citing, and especially when the study is ambiguous, they have to accentuate either gains or losses of the recommended behavior for a broader audience. In other words, they must engage in framing of scientific findings.

The aspect of framing related to selecting information from scholarly articles and providing one's audience with interpretation and implications of the studies informs our chapter. In the debate over mask wearing, this selection process could take the form of either type of framing: emphasis or equivalence. In the category of emphasis framing, journalists and social media users might select certain pieces of information but not others, e.g., "this study found no evidence that masks prevent exposure." In contrast, an equivalence-framing interpretation could lead to statements like, "according to this study, if you do not wear a mask you are much more likely to contract coronavirus." The specific type of framing is not so important to our research; rather we highlight these different categories as to illustrate the different kinds of framing that might appear in media and social media reports on scientific studies of masks and infectious disease.

14.3.4 Summary

Informed by the scholarship discussed above, the current exploratory study aims to answer the following questions:

RQ1: What were the trends and relationships in the volume of discussion of the scholarly articles on mask efficacy (1) in online media and (2) on Twitter?

RQ2: How is the science of masking presented online?

RQ2.1 How did the interpretation of the articles as pro- or anti-mask by online media differ from the interpretation of the articles by Twitter users?

RQ2.2: To what extent do traditional and social media communicators retain qualifications and limitations when communicating studies on mask efficacy?

14.4 Methods

14.4.1 Sampling

To answer our research questions, we collected a corpus of discourse about mask use from Altmetric, a tool that provides information about the number of mentions of a scholarly article across multiple media, including public policy documents, blogs, mainstream media, social media, and other platforms (Altmetric 2020). More specifically, we looked at the circulation of the science of masking in online media. In order to do this, we selected six articles on mask efficacy published before the pandemic. The process of selecting the articles included two steps: (1) a comprehensive search for the articles on mask efficacy through the library and Google search using such key words as "mask", "mask efficacy", "cloth covering" as well as search through the references of the studies that were found first and (2) selection of the top popular articles based on their Altmetric scores as of May 2020. The Altmetric scores are based on a number of factors and reflect "the amount of attention that [a study] has received" (Altmetric 2020). Based on our search, six articles quickly emerged as some of the most circulated:

- Article 1: van der Sande et al. (2008): Tests on volunteers of a variety of mask types revealed that all provided some protection, with wide variations among individuals and types.
- Article 2: Rengasamy et al. (2010): Tests of the penetration of particles through readily available materials showed them to provide only marginal protection in contrast with N95 masks.
- Article 3: MacIntyre et al. (2015): Health care workers in Vietnam experienced increased influenza-like illness when wearing cloth masks in comparison to their normal regime, and lab tests revealed cloth masks to be much more porous than medical masks.
- Article 4: MacIntyre and Chughtai (2015): A review article noted the paucity of studies and called for more research.
- Article 5: Dato et al. (2006): Physical testing showed an 8 layer t-shirt mask provided a measurable level of protection, although less than an N95 mask.
- Article 6: Davies et al. (2013): A variety of tests performed on masks that might be made at home showed they provided limited protection, and led the authors to conclude that they should be used only as a last resort.

After selecting the research articles, we collected mentions of the studies online. For this project, we focused on mainstream media publications and Twitter posts appearing in English. The mainstream media database on Altmetric is composed of media mentions collected from more than two thousand media outlets (Altmetric 2020). The publications are collected based on the presence of a link to the study of interest. The database also includes reprints of media publications that may or may not contain the link. So, the search mechanism is based on the link, scholarly identifiers (such as DOI) and a text mining algorithm. Twitter data includes tweets containing a link to the study. For this project, we relied on the Altmetric Explorer tool and a script for data collection that uses Twitter API.

14.4.2 Time Frame

We collected the mentions that appeared from January 1, 2020 to April 14, 2020. The time frame was chosen for two reasons. First, we wanted to understand better how social understanding of masks shifted during the period of rising awareness and information flows in the first few months of the Covid-19 pandemic. Second, we were concerned about the politicization of the mask debate at the later stages of the pandemic and the extent to which those characteristics would introduce confounding variables that were outside of the scope of our study. Ultimately, the time frame is based on (1) the dynamic of the volume of the public discussion which we initially assessed based on the number of tweets mentioning masks and (2) on the need to include part of the discussion after the change in the mask position of CDC (10 days after health officials recommended wearing the masks).

14.4.3 Measures and Analysis

Each of our research questions required its own approach in terms of measures and analysis. To answer Research Question 1, we started with the data from Altmetric that provided two different measures of volume: the number of mentions within media publications and the number of mentions with hyperlinks to the articles on Twitter. In Table 14.1, we report the descriptive statistics for each of research articles on both these metrics. We also report the number of days, out of the 105 days in our study period, in which there were *any* mentions of the articles as raw numbers and percentages.

These descriptive data already tell us that the six articles received different amounts of attention in media reports and Twitter posts. For example, Research Article #6 received them most media coverage (322 mentions over 105 days), but Research Article #3 received the most Twitter mentions (18,593). In contrast, Research Article #5 received fewer media and Twitter mentions, but received fairly consistent attention–at least one mention per day for 104 days–across the study time period. The descriptive data demonstrate just how variable mentions by media outlets and by social media users can be.

Table 14.1 Descriptive statistics for overall media mentions and Twitter mentions

Research article	Media mentions overall			Twitter mentions overall		
	Mentions	% of Total	Unique days	Mentions	% of Total	Unique days
Article #1	117	15.92	40	3,312	11.39	56
Article #2	73	9.93	15	399	1.37	69
Article #3	181	24.63	36	18,593	63.92	76
Article #4	20	2.72	10	238	0.82	73
Article #5	22	2.99	10	2,523	8.67	104
Article #6	322	43.81	25	4,024	13.83	69
All Articles	735	100.00	61	29,089	100.00	104

With our aim of understanding the relationship between a media mention and a Twitter mention for each article, we needed to look at the data in a longitudinal format. To prepare these descriptive data for an analysis over time, we transformed the data in two ways. First, we transformed the unit of analysis from the six different articles into each unique day from January 1 to April 14, 2020. Based on these 105 days, we reorganized the data to reflect the different points in time when each article was receiving attention. With these data reorganized, we can report the average daily mentions in both media and Twitter across the 105 days (see Table 14.2).

Second, we encountered a high level of variability and skewness in these variables, and these abnormalities raised several questions. Could we consider media mentions to be equivalent, on a ratio-level measurement scale, to Twitter mentions? The answer here seems to be "no." The number of media publications likely to mention one of these research articles, out of all media publications, would seem intuitively to be substantially less than the number of individual Twitter users likely to mention these articles. We decided the best way to investigate the relationship between media mentions and Twitter mentions within each of the six research articles was to standardize each of the variables, which normalizes them in a way that they can be more

Table 14.2 Descriptive statistics for daily mentions in media and in Twitter mentions

Research article	Daily media mentions			Twitter mentions		
	Mean	Median	SD	Mean	Median	SD
Article #1	1.11	0.00	2.70	31.54	1.00	59.56
Article #2	0.70	0.00	2.50	3.80	1.00	6.54
Article #3	1.72	0.00	5.84	177.08	2.00	918.08
Article #4	0.19	0.00	0.87	2.27	1.00	6.61
Article #5	0.21	0.00	0.86	24.03	2.00	75.66
Article #6	3.07	0.00	8.90	38.32	1.00	106.55
All Articles	7.00	1.00	16.45	277.04	19.00	970.54

comparable. Importantly, we also considered each article to be somewhat independent in its likelihood of garnering attention from both media outlets and Twitter users. Therefore, we completed the standardization process within the measurements for each article, subtracting the mean value for each article's mention variable (Table 14.2) from the mean and dividing by the standard deviation. The resulting variables will allow us to plot and to understand the relationships between the variables over time with much more clarity than if we plotted the raw numbers. In other words, we were able to analyze more clearly the relationship between media and Twitter mentions *within* the individual articles while also drawing inferences *across* the six articles in terms of the patterns of those media/Twitter relationships.

Our second research question necessitated a different approach to the data. In this section of our inquiry, we were more interested in understanding the content of the mentions–how the articles were being used to support or oppose the use of masks in the Covid-19 pandemic. In this section, each mention was treated as the unit of analysis. Since we were interested in the content of the mentions, we did not analyze duplicates (i.e., reprints for media and retweets for Twitter mentions). For media publications ($n = 234$), we operationalized mention as the paragraph mentioning the article along with one paragraph before and one paragraph after the paragraph mentioning the article. Including two extra paragraphs provided us with the context necessary for coding the communicator's interpretation of the results of the studies. For Twitter content ($n = 4{,}775$), we use the entire text of a tweet.

The publications and tweets were coded based on their interpretation of the scholarly articles, i.e., whether they are used as evidence supporting a specific position regarding mask efficacy or usage (anti-mask, pro-mask, the science is unclear/conflicted about mask efficacy, or the mention does not provide enough information to interpret the position, e.g. because it only contains a link to the study). The publications and tweets were also coded for the presence of qualifying language, i.e., whether the interpretation of the articles was presented with any level of uncertainty or limitation regarding the results, scientific recommendations, or mask efficacy (no qualifying language, qualifying language). Media mentions were also coded for the communicator's overall position regarding mask efficacy based on the headline and the first two paragraphs (positions as above). We did not code the tweets for the overall position because we assumed that it would oftentimes overlap with the interpretation of the studies due to the character limit imposed by the social network.

To secure reliability, the entire dataset was coded separately by the authors in segments, with disagreements periodically resolved through discussion until interpretive convergence was achieved (Saldaña 2016). Descriptive statistics were run using Excel to answer the research questions.

14.5 Results

14.5.1 Volume of the Discussion

Our first research question focused on a closer look at the trends and relationships between the attention paid to research articles about mask efficacy in two different science communication environments: online news websites and the social media platform Twitter. As outlined in the Methods section, we assume for our analysis that mentions of these six articles in either environment arise independently. The assumption allows us to examine the relative relationship–based on standardized variables–between news and social media mentions for each article in turn.

The trends and relationships for Article #1 (human testing of a variety of mask types) are depicted in Fig. 14.2. The first mention of this research article occurred on day 31 within our data set, January 31, 2020, in online news media, followed by several more mentions in that news environment. From day 61 (March 1st) through day 91 (March 31st), the article was mentioned off-and-on in media publications while its Twitter mentions increased dramatically. The mentions of this article peaked on day 94 (April 3rd, when an announcement from CDC was expected) in media publications before diminishing in both environments by mid-April. The relationship between the two trends indicates that the amount of attention on Twitter outpaced media attention throughout most of the 105 days of our study period.

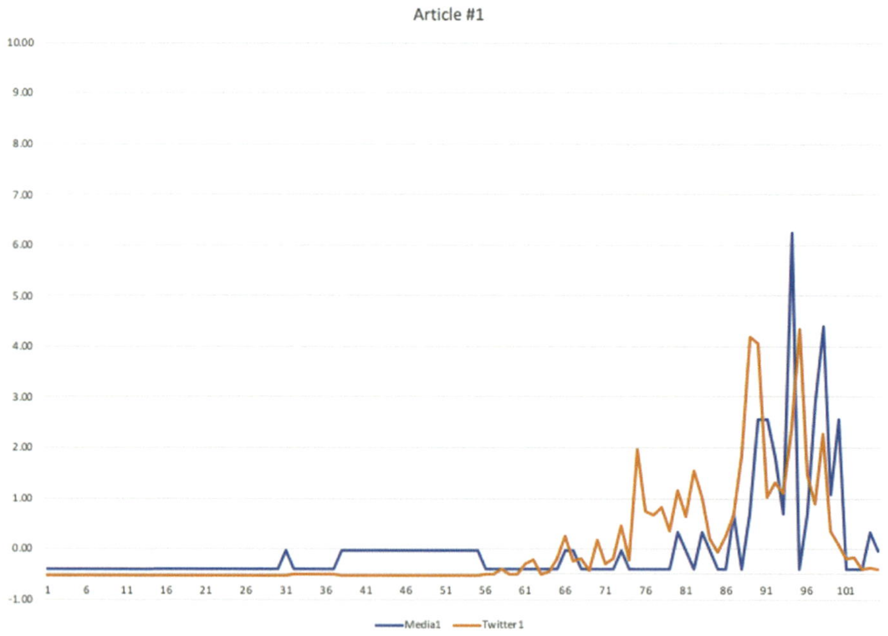

Fig. 14.2 The trends and relationships for Article #1

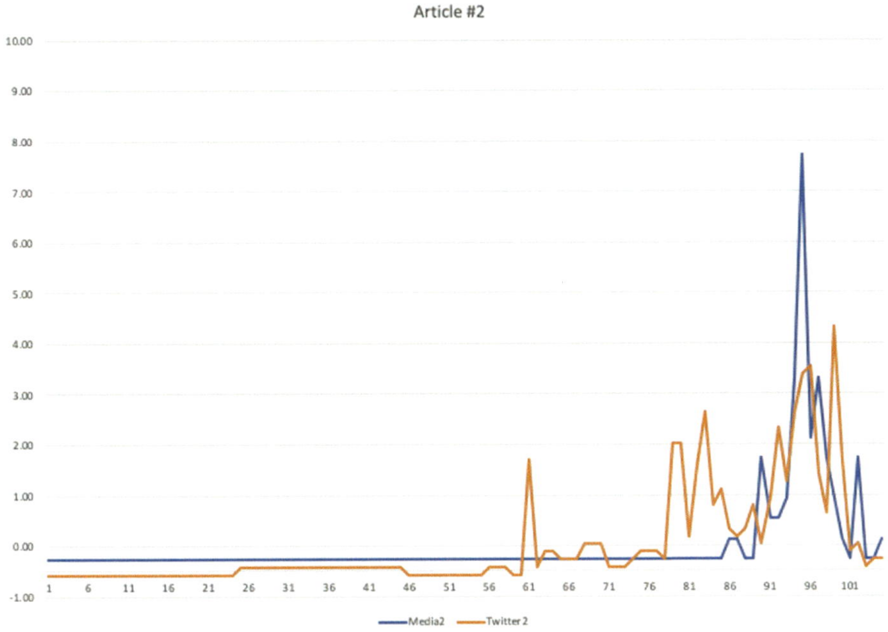

Fig. 14.3 The trends and relationships for Article #2

The trends and relationships for Article #2 (physical testing of potential mask materials) are shown in Fig. 14.3. This study of mask efficacy first received attention on Twitter in late January and continued to gain attention on the social media platform, with several discrete peaks, until day 87 (March 27th). At that point, attention shifted from Twitter to media publications, where it suddenly increased and peaked on day 95 (April 4th), followed by the biggest peak on Twitter on day 99 (April 8th). The data patterns for this research article indicate a larger amount of attention from Twitter preceding the peak in media mentions.

The trends and relationships for Article #3 (experimental study of effects of differing mask use by healthcare workers in Vietnam; Fig. 14.4) were quite different from the first two, featuring longer periods of low attention punctuated by two main peaks. This research article gained its greatest share of online news attention at two different points in time. The first peak was by day 29 (January 29th), followed by a very quiet period, until a resurgence in mentions peaking on day 94 (April 3rd). This research article only received a large amount of attention on Twitter in a more limited time period, between days 82 and 93 (March 22nd and April 2nd). For this research article on masks, online media showed the greater attention first, with virtually no complementary attention on Twitter, followed by a peak in both information environments by early April.

Article #4 (a review) received little average attention per day across the 105 day period. There were two notable peaks on Twitter, the first spanning days 30 and 32 (January 30th and February 1st) and the second, larger peak isolated to day 61

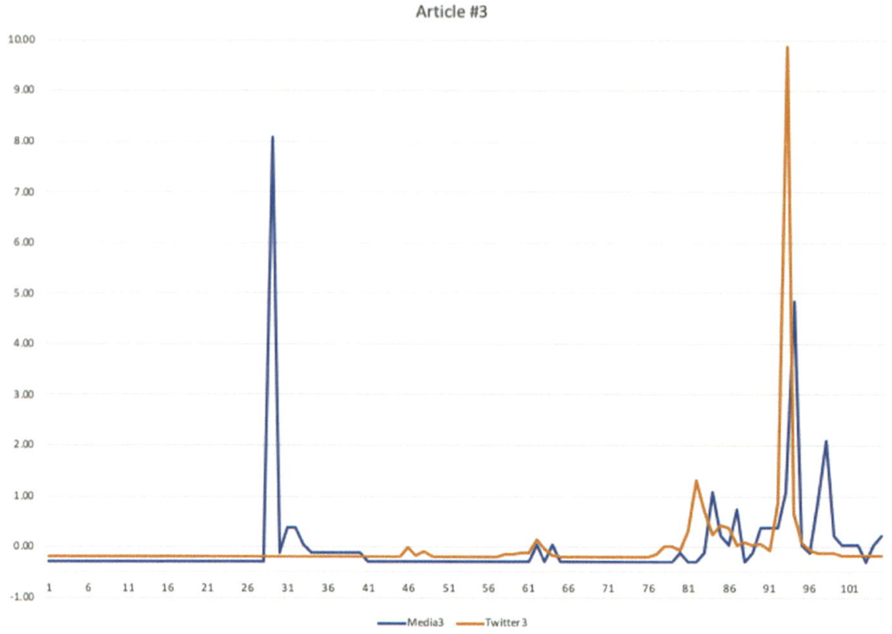

Fig. 14.4 The trends and relationships for Article #3

(March 1st). There was a small peak in attention in online news on day 56 (February 25th)–immediately preceding the bigger peak of attention on Twitter. The larger peak in online news occurred on day 98 (April 7th) with no corresponding attention on Twitter (Fig. 14.5).

The trends for Article #5 (physical testing of a t-shirt mask) are shown in Fig. 14.6. The Twitter attention was relatively limited between days 79 and 81 (March 19th and March 21st). Online news attention was a little more spread out, with two peaks: day 88 (March 28th) and days 102 and 103 (April 11th and 12th). For this research article, Twitter attention occurred and peaked very late in the study period followed by the steadier attention from online news media.

Finally, Article #6 (a multi-methods study with human and physical testing of different masks), like the previous article, received most attention very late in the study period–in fact, this article received no attention at all until day 77 (March 17th). Twitter attention grew quickly, peaking on day 88 (March 28th). Media mentions followed closely behind, peaking on day 91 (March 31st). The parallel growth and decline in attention in both information environments is notable. Attention was first gained on Twitter and very nearly matched in its intensity and decline in the following days in online news media (Fig. 14.7).

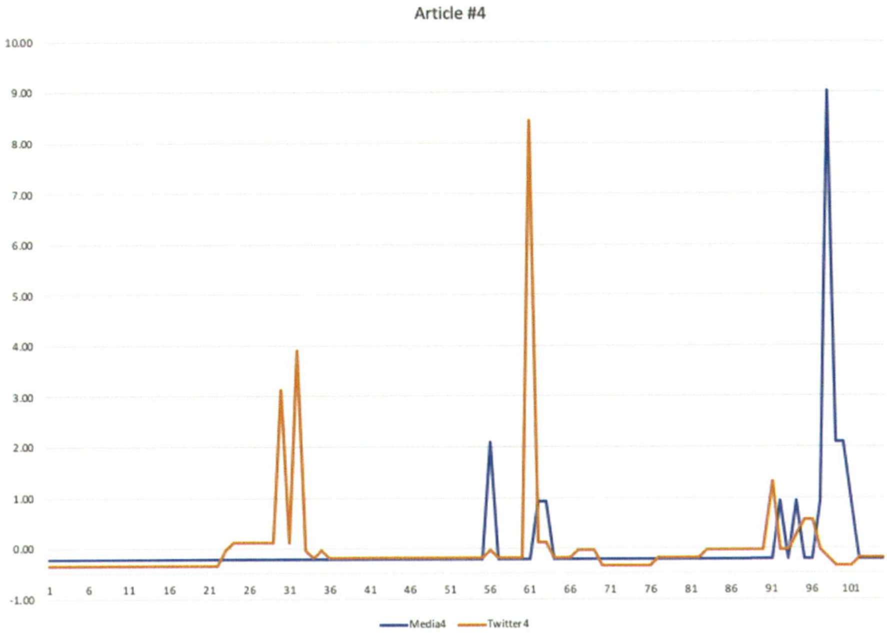

Fig. 14.5 The trends and relationships for Article #4

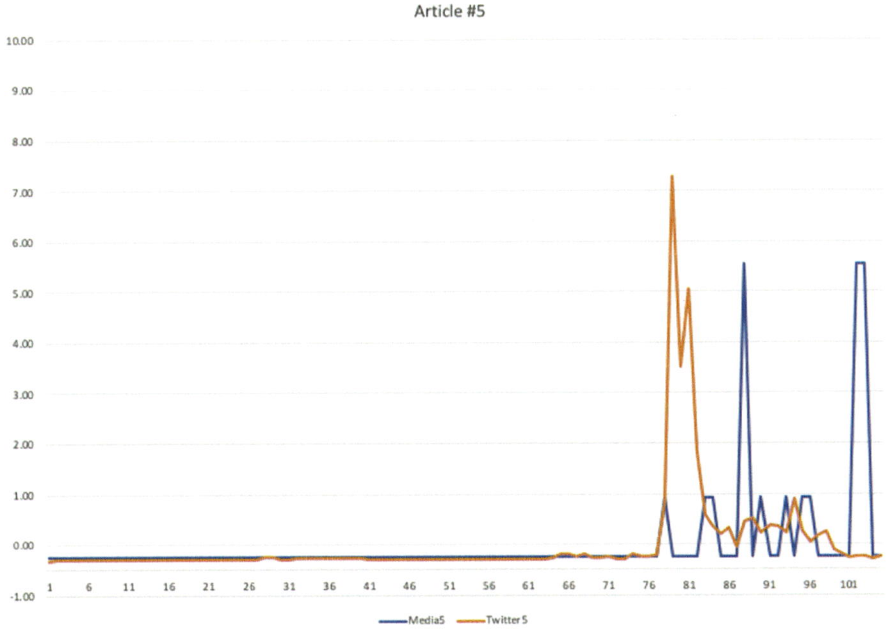

Fig. 14.6 The trends and relationships for Article #5

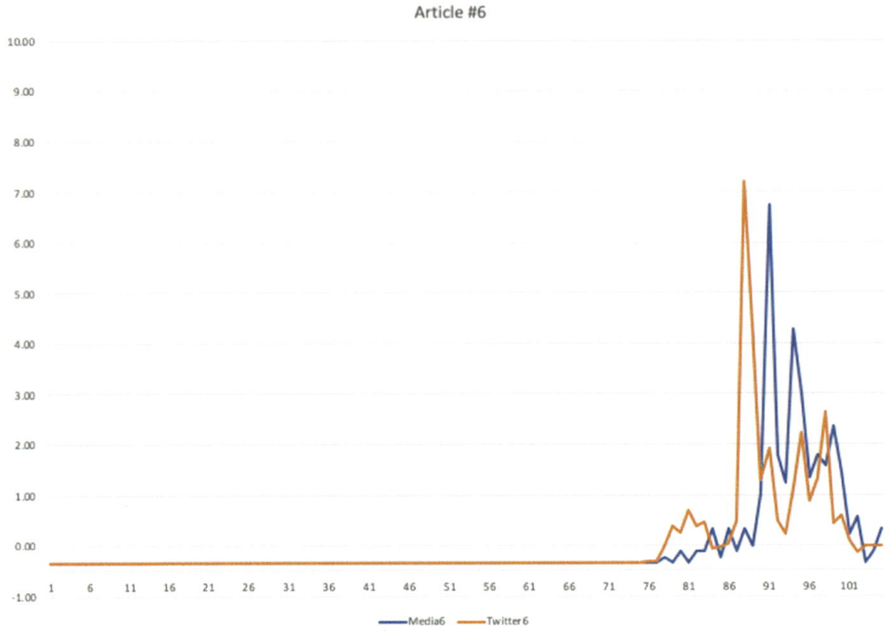

Fig. 14.7 The trends and relationships for Article #6

14.5.2 Content of the Mentions

For our second research question we looked at the differences in interpretation of the research articles by media and Twitter communicators. As mentioned in the methods section, we looked both at the interpretation of the results (a position regarding mask efficacy/usage assigned to the study) and at the presence of uncertainty in such interpretations.

Interpretation of the articles. Overall, the majority of the scholarly articles of interest appeared in pro-mask media publications (62%), followed by publications without a determinable stand regarding masks (28%), and publications suggesting that science does not provide a clear answer (8%). Interestingly, only 2% of the media publications expressed a clear anti-mask stand (see Table 14.3). When it comes to the interpretation of the research articles themselves, 73% of the media publications interpreted the target articles as providing evidence supporting the pro-mask position, 13% of the publications used them as evidence supporting the anti-mask position, 12% of the publications did not communicate a clear stand when citing the articles, and 1% used them to demonstrate that science is not clear and/or is conflicted about mask efficacy. When it comes to interpretation of the studies themselves, 73% of the citations on media used the scholarly articles as pro-mask evidence, 13% interpreted them as evidence against masks, 1% suggested that science was conflicted, and 12% provided no determinable stand (see Table 14.4).

Table 14.3 Overall position on facemasks of media publications

	Pro-mask	Anti-mask	Unclear/Conflicted science	No determinable stand	Total
Article #1	23	0	3	8	34
	(68%)	(0%)	(9%)	(24%)	(100%)
Article #2	12	0	1	4	17
	(71%)	(0%)	(6%)	(24%)	(100%)
Article #3	15	1	4	20	40
	(38%)	(3%)	(10%)	(50%)	(100%)
Article #4	3	1	1	0	5
	(60%)	(20%)	(20%)	(0%)	(100%)
Article #5	5	0	0	1	6
	(83%)	(0%)	(0%)	(17%)	(100%)
Article #6	87	3	9	33	132
	(66%)	(2%)	(7%)	(25%)	(100%)
Total	145	5	18	66	234
	(62%)	(2%)	(8%)	(28%)	(100%)

Table 14.4 Interpretation of the studies by the media

	Pro-mask	Anti-mask	Unclear/Conflicted science	No determinable stand	Total
Article #1	32	0	0	2	34
	(94%)	(0%)	(0%)	(6%)	(100%)
Article #2	13	2	0	2	17
	(76%)	(12%)	(0%)	(12%)	(100%)
Article #3	4	22	1	13	40
	(10%)	(55%)	(3%)	(33%)	(100%)
Article #4	2	2	1	0	5
	(40%)	(40%)	(20%)	(0%)	(100%)
Article #5	5	0	0	1	6
	(83%)	(0%)	(0%)	(17%)	(100%)
Article #6	115	5	1	11	132
	(87%)	(4%)	(1%)	(8%)	(100%)
Total	171	31	3	29	234
	(73%)	(13%)	(1%)	(12%)	(100%)

Table 14.5 Interpretation of the studies by Twitter users

	Pro-mask	Anti-mask	Unclear/Conflicted Science	No determinable stand	Total
Article #1	599	7	5	409	1020
	(59%)	(1%)	(0%)	(40%)	(100%)
Article #2	42	28	1	109	180
	(23%)	(16%)	(1%)	(61%)	(100%)
Article #3	109	677	19	1027	1832
	(6%)	(37%)	(1%)	(56%)	(100%)
Article #4	7	3	3	18	31
	(23%)	(10%)	(10%)	(58%)	(100%)
Article #5	424	2	15	474	915
	(46%)	(0%)	(2%)	(52%)	(100%)
Article #6	406	22	6	363	797
	(51%)	(3%)	(1%)	(46%)	(100%)
Total	1587	739	49	2400	4775
	(33%)	(15%)	(1%)	(50%)	(100%)

As explained in the methods section, we only determined the interpretation of the articles, not the overall position, when working with the Twitter data. On Twitter, most of the citations were cited in tweets without a determinable stand regarding masks (50%). Roughly one-third of the tweets (33%) cited the studies as those supporting a pro-mask position, followed by anti-mask interpretation (15%). Only 1% of the tweets suggested that science was conflicted regarding mask efficacy (see Table 14.5).

As demonstrated in Table 14.4, Article #1 (human testing) was mostly cited by the media as evidence for a pro-mask position (N = 32). It was also cited twice without indicating a clear position that the article was supposed to support (N = 2). While the majority of tweets also cited the article as pro-mask evidence (N = 599), a number of citations did not specify the position that the article supported (N = 409), some tweets used the article to demonstrate that science was conflicted about mask efficacy (N = 5), and some even interpreted it as anti-mask (N = 7).

Similar to the first article, media mostly cited Article #2 (physical testing) when supporting a pro-mask position (N = 13). However, the article was also used as anti-mask evidence (N = 2) and in citations without a clear position (N = 2). Interestingly, Twitter users mostly cited the same article without providing a clear interpretation (N = 109). Moreover, the article got a number of pro- (N = 42) and anti-mask (N = 28) citations. One tweet also cited the article as evidence suggesting that research is conflicted about mask efficacy (N = 1).

More than a half of media publications citing Article #3 (experimental study showing that healthcare workers wearing cloth masks exhibited more influenza-like symptoms than the control group using normal practices) presented it as evidence against mask efficacy (N = 22). Some media also cited the article without clearly specifying the position that it supports (N = 13), as evidence for mask usage (N = 4) and as evidence that research was conflicted about mask efficacy (N = 1). Just like in the case of Article #2, Twitter users mostly cited Article #3 without specifying a clear position that the article supports (N = 1027), followed by citations used as anti-mask evidence (N = 677), pro-mask evidence (N = 109), and claims that science was conflicted about mask efficacy (N = 19).

Article #4 (literature review) presents an interesting case where the article was used as evidence for (N = 2) and against (N = 2) masks in the same number of publications. The article was also cited as evidence of conflicted science (N = 1). Most of the citations on Twitter did not provide enough information to decide the author's interpretation of the article (N = 18). The article was also presented as evidence for mask efficacy (N = 7), against mask efficacy (N = 3), and as evidence demonstrating that science is conflicted (N = 3).

Media mostly cited Article #5 (t-shirt mask) as evidence supporting mask efficacy (N = 5). The article was also cited without expressing a specific position (N = 1). On Twitter, the article got almost equal number of mentions without expressing a specific position (N = 474) and pro-mask mentions (N = 424). The article was also used to demonstrate that science is conflicted (N = 15), and as anti-mask evidence (N = 2).

Finally, Article #6 (multi-methods testing) received most media citations as pro-mask evidence (N = 115). It was also cited without a clear position (N = 11), as anti-mask evidence (N = 5), and as evidence of conflicted science (N = 1). Twitter users mostly cited the article either as pro-mask evidence (N = 406) or without expressing a specific position (N = 363). The article was also mentioned as anti-mask evidence (N = 22) and as evidence of conflicted science (N = 6).

Expression of uncertainty. Not only did Twitter users and mainstream media demonstrate differences in interpretations of the articles but they also presented those interpretations differently (see Table 14.6).

Most of the media publications (N = 31) cited Article #1 with qualifying language thus suggesting some uncertainty regarding the claims about mask efficacy. Only one publication cited the study without qualifying language (N = 1). While the article was frequently cited by Twitter users along with qualifying language (N = 458), a large proportion of tweets also referenced the study without expressing uncertainty regarding mask efficacy (N = 153).

Similarly, all the media mentions of the Article #2 and Article #4 included qualifying language. On Twitter, Article #2 was cited both within the tweets expressing uncertainty (N = 57) and tweets without qualifying language (N = 14). Article #4 mostly received mentions along with qualifying language on Twitter (N = 12) but also was cited without language indicating uncertainty (N = 1).

Most of the media mentions of Article #3 included qualifying language (N = 25) with only two mentions not expressing uncertainty (N = 2). On Twitter, the same article received a number of mentions both with (N = 596) and without (N = 209) qualifying language. The same was true for Article #5 which was mostly cited along with qualifying language by the media (N = 4). One media citation of the article was without expression of uncertainty (N = 1). Twitter mentions of this article include both tweets with (N = 176) and without qualifying language (N = 265).

Table 14.6 The presence of qualifying language on mainstream media and Twitter

	Media			Twitter		
	Present	Not present	Total	Present	Not present	Total
Article #1	31	1	32	458	153	611
	(97%)	(3%)	(100%)	(75%)	(25%)	(100%)
Article #2	15	0	15	57	14	71
	(100%)	(0%)	(100%)	(80%)	(20%)	(100%)
Article #3	25	2	27	596	209	805
	(93%)	(7%)	(100%)	(74%)	(26%)	(100%)
Article #4	5	0	5	12	1	13
	(100%)	(0%)	(100%)	(92%)	(8%)	(100%)
Article #5	4	1	5	176	265	441
	(80%)	(20%)	(100%)	(40%)	(60%)	(100%)
Article #6	116	5	121	349	85	434
	(96%)	(4%)	(100%)	(80%)	(20%)	(100%)
Total	196	9	205	1648	727	2375
	(96%)	(4%)	(100%)	(69%)	(31%)	(100%)

Table 14.7 The presence of qualifying language in media mentions based on the position assigned to a study

	Pro-mask mentions			Anti-mask mentions		
	Present	Not present	Total	Present	Not present	Total
Article #1	31	1	32	0	0	0
	(97%)	(3%)	(100%)	(0%)	(0%)	(0%)
Article #2	13	0	13	2	0	2
	(100%)	(0%)	(100%)	(100%)	(0%)	(100%)
Article #3	2	2	4	22	0	22
	(50%)	(50%)	(100%)	(100%)	(0%)	(100%)
Article #4	2	0	2	2	0	2
	(100%)	(0%)	(100%)	(100%)	(0%)	(100%)
Article #5	4	1	5	0	0	0
	(80%)	(20%)	(100%)	(0%)	(0%)	(0%)
Article #6	110	5	115	5	0	5
	(96%)	(4%)	(100%)	(100%)	(0%)	(100%)
Total	162	9	171	31	0	31
	(95%)	(5%)	(100%)	(100%)	(0%)	(100%)

Finally, media mostly cited Article #6 along with qualifying language (N = 116). Only five mentions of the article in mainstream media did not express uncertainty (N = 5). Twitter users, on the other hand, cited the article with (N = 349) and without (N = 85) expressing the uncertainty.

As demonstrated in Table 14.6, out of the media publications that did communicate some position (pro-mask, anti-mask or unclear/conflicted science), 96% used qualifying language in the paragraphs citing the research articles indicating some level of uncertainty regarding mask (in)efficacy. Interestingly, there were no media publications that would cite the research articles as evidence against masks without using qualifying language (see Table 14.7). In other words, whenever journalists suggested that the research articles of interest demonstrated mask inefficacy, they would communicate some level of uncertainty regarding this position. While most of the media publications citing research articles as those supporting a pro-mask position also used qualifying language, some of them cited the research articles without communicating uncertainty. This includes Article 1 (N = 1), Article 3 (N = 2), Article 5 (N = 1), and Article 6 (N = 5) were also cited as evidence supporting a pro-mask position without being accompanied by language communicating uncertainty.

While Twitter users were also more likely to use qualifying language when communicating one of the three position-based interpretations, the percentage of tweets citing the research articles without qualifying language was higher than the percentage of media publications (31% and 4% respectively). When cited as pro-mask evidence on Twitter, five out of six articles would have citations used without qualifying language (see Table 14.8). Moreover, Twitter users would not hesitate to

Table 14.8 The presence of qualifying language in Twitter mentions based on the position assigned to a study

	Pro-mask mentions			Anti-mask mentions		
	Present	Not present	Total	Present	Not present	Total
Article #1	446	153	599	7	0	7
	(74%)	(26%)	(100%)	(100%)	(0%)	(100%)
Article #2	38	4	42	18	10	28
	(90%)	(10%)	(100%)	(64%)	(36%)	(100%)
Article #3	101	8	109	476	201	677
	(93%)	(7%)	(100%)	(70%)	(30%)	(100%)
Article #4	7	0	11	2	1	3
	(100%)	(0%)	(100%)	(67%)	(33%)	(100%)
Article #5	160	264	424	1	1	2
	(38%)	(62%)	(100%)	(50%)	(50%)	(100%)
Article #6	327	79	406	16	6	22
	(81%)	(19%)	(100%)	(73%)	(27%)	(100%)
Total	1079	508	1587	520	219	739
	(68%)	(32%)	(100%)	(70%)	(30%)	(100%)

cite articles as evidence against masks without expressing uncertainty. In the case of Twitter, 30% of the anti-mask citations were published without being accompanied by qualifying language, as opposed to 0% in media publications.

14.6 Discussion

The current chapter explored the circulation of scientific knowledge in the mask debate during the Covid-19 pandemic. More specifically, we looked at the longitudinal trends of citing six popular articles to see how the studies gained attention of mainstream media and Twitter users. We also examined trends in interpreting the studies from the perspective of the assigned position and (un)certainty expressed along with this position by mainstream media and Twitter users.

14.6.1 Independent Agendas on Mainstream Media and Twitter

The articles of interest circulated on mainstream and social media in several different ways both in terms of where the articles first appeared (mainstream vs. social media), and the dynamic of distribution (steep peaks vs. relatively low yet constant attention). While some articles were first mentioned on Twitter and then got picked up by the mainstream media, others received mainstream media attention first and only after that, began to be noticed by Twitter users. In other words, our results show that neither news media nor Twitter are securely in the lead when it comes to disseminating scientific studies. Although this does not directly demonstrate that either mainstream media or Twitter users shape each other's agenda, it might indicate the mutual influence revealed in previous research (Conway et al. 2015; Wang and Guo 2015) or at least the existence of separate agendas that remain relatively independent. These complexities raise questions about the implications of possible interdependence of mainstream and social media agendas when it comes to scientific information during a pandemic. Who decides what studies get cited and amplified? Why are some peaks in one type of media followed by increased attention in another media type, while others are not? Should journalists monitor and address scientific information disseminated by social media users, or should they aim to preserve their gatekeeping role, emphasizing only the studies they assess to be valuable? "Treatment" of the online infodemic will require a deeper understanding of the interactions of specific media within the broader science communication environment.

Another interesting finding is related to the dynamic of the dissemination of the research articles. Our study showed that while some articles on mask efficacy were receiving smaller yet constant attention on mainstream and social media throughout the three-month period, others demonstrated sharp increases in public interest that might or might not be followed by a sharp decline. From the perspective of agenda setting theory, it raises questions about the most productive attention model in terms of promoting scientific knowledge during infodemics accompanying health crises. If the media do dictate what to think about (McCombs 2005), what would be more beneficial for promoting health interventions such as masks? Would it be better if relevant scientific articles remain on the agenda for a longer period of time but with lower levels of mainstream and social media attention, or if they receive higher levels of attention but within a short period of time? More importantly, is this a real dichotomy? Is there a way to make sure that research is disseminated and discussed both with high volume and while the issue (such as mask use) remains relevant?

14.6.2 Various Ways to Interpret the Same Articles

When it comes to interpretation of the studies, several interesting trends emerge across the articles. First, we found that the same research article could be cited as supporting both pro- and anti-mask positions, both on Twitter and by mainstream media. This result highlights the flexibility of interpretation of scientific studies. Needless to say, this raises significant concerns about information flows and the spread of scientific (mis)information online during a pandemic. Where Al-Rakhami and Al-Amri (2020) suggest that communicators should "contribute to sharing credible content from reputable sources on the web" (p. 155,969), our results reveal that credible content alone is insufficient. Even the best sources–scientific studies–can be variably interpreted. Both traditional and social media communicators must not only share, but also frame the content they are sharing. Since such framing may affect audience's understanding of the science and their willingness to take preventive measures to protect their own as well as public health (Nan et al. 2012; Nisbet 2009; Sprain 2018; Updegraff et al. 2011), communicators who want to sustain the science communication environment need to take care in conveying accurate interpretations of the sources they are disseminating. This raises a number of questions related to scientists' and science communicators' responsibilities when it comes to framing studies. What should a researcher do when their audience grows during a health crisis, and a study originally directed at their scientific peers is now circulating among lay publics actively seeking information? Since framing of their findings will inevitably happen (Sprain, 2018), would it be beneficial for society to have scientists frame their own research, e.g., through carefully crafted plain language abstracts? In this case, for example, two of the research teams which had authored target articles had to step up and publish additional guidance to emphasize that despite the qualified language of their conclusions, their studies in fact supported widespread use of cloth masks during the pandemic (Davies et al. 2020; MacIntyre et al. 2020).

Second, our results show that Twitter communicators frequently mention a study without making explicit any conclusion to be drawn from it, often simply posting a link. This paucity of reasoning is especially in contrast to the greater degree of explicitness of media communicators. On the one hand, this might be explained by technological affordances of the social network: the limited number of characters per tweet does not make room for elaboration. It is also likely that audiences can at times infer a communicator's position on the issue from other evidence in the tweet thread, not included in our data. Even so, however, it appears that Twitter is not promoting among its audience the kind of explicitness and reasoning required for sound decision-making in a pandemic. One might speculate that such lack of elaboration might further exacerbate the issue of infodemic as Twitter users are contributing to overwhelming amounts of information regarding health interventions without ensuring proper understanding of what that information means. We would, however, warn against interpreting this result as an attempt to demonize social media. More research is needed to determine the level of sophistication in argumentation that Twitter users rely on when dealing with studies, and the effects of various levels of sophistication on public perception of science and one's willingness to enact health interventions.

14.6.3 Twitter Users Are More Confident in Their Interpretation of Research

Finally, our results show that Twitter communicators are less likely to include qualifications when interpreting studies compared to media communicators. This difference is especially salient in the case of anti-mask interpretations; where the media would always use qualifying language, about one-third of the tweets provided their anti-mask interpretation of the studies with no indicators of uncertainty. The finding brings up the old concerns related to oversimplification of science (Gustafson and Rice 2020; Jensen 2008; Retzbach and Maier 2015). Moreover, the issue gets a new twist in the social media environment. While the original concern focused on journalists' over-simplifying scientific findings or the lay public, now we deal with simplification done by the lay publics themselves. Given the number of people seeking information about the pandemic on social media (Ali et al. 2020), this again might be a problematic practice contributing to infodemics. At the same time, our study demonstrated that the vast majority of media publications did use some sort of qualifying language when presenting the results of the studies. This gives hope for a productive discussion between scientists and the lay public when mediated by the traditional gatekeepers. Taken together, these trends raise questions about the need for the scientific community to consider both journalists and a broader audience when communicating their findings.

14.6.4 Limitations

While we believe that our chapter opens up a promising discussion on the topic of circulation of scientific information during a pandemic, the study has several limitations. First, the study only focuses on media articles and tweets published from January 2020 to mid-April 2020. In other words, we did not study the issue after it was politicized, which likely changed patterns in media use and interpretation. Second, we did not examine the transmission of mis- or disinformation. All of our target studies were legitimate, peer-reviewed research published before the pandemic. In that sense, our study represents a "worst case scenario," showing how even the best information can receive variable attention and support diverging viewpoints. The analysis of "the dark side" of the mask debate would have likely revealed additional, perhaps divergent, patterns in public use and interpretation of scientific information.

14.6.5 Future Possibilities

In addition to the results on our research questions, our data points the way to additional lines for investigation. The specific techniques communicators use to frame the studies beyond merely tilting towards a pro- or anti-mask position deserves additional attention. Even in the brief format afforded by Twitter, communicators could emphasize particular aspects of the study they mentioned. For example, an intriguing example of framing scientific results came up in the discussions of different kinds of protection provided by the masks. While some users focused on mask efficacy in terms of *protection of the wearer*, others emphasized the *source-control efficacy* of masks (i.e., protection of others from the wearer); yet others suggested that "*protection goes both ways.*" (A1_t104). The same trend could be observed in media publications. For example, when interpreting Article #6, one media publication emphasized the source-control potential of the masks, "*According to public health experts, fabric or cloth masks are not intended to protect the wearer from getting infected, but prevents them from spreading the virus*" (A6_m19). The same article was used as evidence against masks by a publication emphasizing protection of the wearer, "A 2013 U.K. study that looked at masks made from cotton T-shirts found that the homemade masks were not effective protection in a flu pandemic" (A6_m23).

Sometimes, the processes of emphasizing certain aspects of the research findings would take an extreme form of omitting the part of the conclusions that did not fit the speaker's position. An interesting example of this practice also came from Article #6 which originally provided the following conclusion: "Our findings suggest that a homemade mask should only be considered as a last resort to prevent droplet transmission from infected individuals, but it would be better than no protection." (Davies et al. 2013, p. 413). While some media publications and tweets provided the full quote, others included only half–either the "last resort" or the "better than nothing" phrase. In other words, the ambiguity of the conclusion allowed both mainstream media and Twitter users to select what part of the sentence they would focus on, thus framing the study as evidence for or against the masks.

Another intriguing example of different ways of framing the same information occurs when interpreting specific numbers drawn from the studies. For example, when discussing Article #3, a number of Twitter users cited the reported 97% penetration rate as evidence of the ineffectiveness of masks: "*Mandy, Penetration of cloth masks by particles is almost 97%. This is instilling a false sense of security which is dangerous... Cloth is unacceptable.*" (A3_t1319). But another user asserted the opposite perspective, suggesting that "*3% reduction is 3% reduction. this is a game of small incremental improvements that add up*" (A3_t1730). Like an optical illusion, the same result can be seen both as 3% protection and as 97% penetration, depending on the position. Interestingly, while the media would oftentimes explicitly take a side when interpreting the studies, some Twitter users could simply post the numbers thus encouraging others to decide if the offered protection is enough.

These observations open up a promising discussion on the ways media and lay publics interpret scholarly articles. More research is needed to identify the reasons behind opposite ways to interpret the same results. We would encourage exploring factors that relate to scientific pieces (e.g., ambiguous conclusions), science communicators (e.g., science literacy, personal characteristics) and the context (e.g., stage of the pandemic).

14.6.6 New Environment Comes with New Challenges

In this exploratory study, we have examined online circulation of ambiguous research on mask efficacy in the early months of the pandemic. We have documented a complex science communication environment, with the quickly changing scientific evidence, growing number of actors having direct access to scientific research, more channels for communicating science, and higher speed of information dissemination. While it might be easy to place blame on lay publics or health officials for failing the mask response at the first stages of the pandemic, our chapter demonstrates that the issue is more complex and requires a comprehensive approach to ensure successful public response to future health crises. In the situation of a pandemic, health guidelines may and most likely, *will* change thus confusing both journalists and lay publics. In fact, mask-related guidelines are changing even as we are writing this chapter with CDC now recommending two layers of masks (CDC 2021). With the growing movement for open science, lay publics will be likely getting more information directly from scholarly journals, and the higher levels of internet penetration will only increase the amounts of scientific information coming from media and fellow internet users. We suggest that within this context, the findings of our study call for reconsideration of science communication practices that would allow for a productive public discussion and timely implementation of health interventions during a pandemic.

Acknowledgements We would like to thank Altmetric (Digital Science), for providing us with access to the Altmetric Explorer tool. The tool allowed us to collect media mentions and Twitter data.

References

Adonis L, Paramanund J, Basu D, Luiz J (2017) Framing preventive care messaging and cervical cancer screening in a health-insured population in south africa: Implications for population-based communication? J Health Psychol 22(11):1365–1375. https://doi.org/10.1177/1359105316628735

Altmetric (2020) Sources of attention. Retrieved from: https://www.altmetric.com/about-our-data/our-sources/

Al-Rakhami MS, Al-Amri AM (2020) Lies kill, facts save: detecting COVID-19 misinformation in Twitter. IEEE Access 8:155961–155970

Ali SH, Foreman J, Tozan Y, Capasso A, Jones AM, DiClemente RJ (2020) Trends and predictors of COVID-19 information sources and their relationship with knowledge and beliefs related to the pandemic: nationwide cross-sectional study. JMIR Public Health Surveill 6(4):e21071

Alvarez-Risco A, Mejia CR, Delgado-Zegarra J, Del-Aguila-Arcentales S, Arce-Esquivel AA, Valladares-Garrido MJ, Rosas Del Portal M, Villegas LF, Curioso WH, Sekar MC, Yáñez JA (2020) The peru approach against the COVID-19 infodemic: Insights and strategies. Am J Trop Med Hyg 103(2):583–586. https://doi.org/10.4269/ajtmh.20-0536

Bucchi M (1996) When scientists turn to the public: alternative routes in science communication. Public Underst Sci 5:375–394

Bubela T, Nisbet MC, Borchelt R, Brunger F, Critchley C, Einsiedel E, Geller G, Gupta A, Hampel J, Hyde-Lay R, Jandciu EW, Ashley Jones S, Kolopack P, Lane S, Lougheed T, Nerlich B, Ogbogu U, O'Riordan K, Ouellette C, Spear M, Strauss S, Thavaratnam T, Willemse L, Caulfield T (2009) Science communication reconsidered. Nat Biotechnol 27(6):514–518

Bucchi M (2017) Microblogging as an extension of science reporting. Public Underst Sci 26(8):953–968

Brossard D (2013) New media landscapes and the science information consumer. Proc Natl Acad Sci 110(Suppl. 3):1–6

Brossard D, Scheufele DA (2013) Science, new media, and the public. Science 33(6115):40–41

Buchanan M (2020) Managing the infodemic. Nat Phys 16(9):894–894. https://doi.org/10.1038/s41567-020-01039-5

Conway BA, Kenski K, Wang D (2015) The rise of twitter in the political campaign: searching for intermedia Agenda-setting effects in the presidential primary. J Comput-Mediat Commun 20(4):363–380. https://doi.org/10.1111/jcc4.12124

Cacciatore MA, Scheufele DA, Iyengar S (2016) The end of framing as we know it... and the future of media effects. Mass Commun Soc 19(1):7–23. https://doi.org/10.1080/15205436.2015.1068811

Chong YY, Cheng HY, Chan HYL, Chien WT, Wong SYS (2020) COVID-19 pandemic, infodemic and the role of eHealth literacy. Int J Nurs Stud 108:103644–103644. https://doi.org/10.1016/j.ijnurstu.2020.103644

Cuan-Baltazar JY, Muñoz-Perez MJ, Robledo-Vega C, Pérez-Zepeda MF, Soto-Vega E (2020) Misinformation of COVID-19 on the internet: infodemiology study. JMIR Public Health Surveill 6(2):e18444

CDC (2021) Maximizing fit for cloth and medical procedure masks to improve performance and reduce SARS-CoV-2 transmission and exposure, 2021. Retrieved from: https://www.cdc.gov/mmwr/volumes/70/wr/mm7007e1.htm?s_cid=mm7007e1_w

Dato VM, Hostler D, Hahn ME (2006) Simple respiratory mask. Emerg Infect Dis 12(6):1033–1034. https://doi.org/10.3201/eid1206.051468
Davies A, Thompson K, Giri K, Kafatos G, Walker J, Bennett A (2013) Testing the efficacy of homemade masks: would they protect in an influenza pandemic? Disaster Med Public Health Prep 7(4):413–418. https://doi.org/10.1017/dmp.2013.43
Denham BE (2014) Intermedia attribute agenda setting in the new york times: the case of animal abuse in U.S. horse racing. Journalism Mass Commun Q 91(1):17–37. https://doi.org/10.1177/1077699013514415
Dunwoody S (2014) Science journalism. In: Bucchi M, Trench B (eds) Routledge handbook of public communication of science and technology. 27–39
Davies A, Thompson K, Giri K, Kafatos (2020) Frequently asked questions—homemade facemasks study. Retrieved from: https://www.researchgate.net/publication/340253593_Frequently_asked_questions_-_homemade_facemasks_studypdf
Diseases TLI (2020) The COVID-19 infodemic. Lancet Infect Dis 20(8):875–875. https://doi.org/10.1016/S1473-3099(20)30565-X
Entman RM (1993) Framing: toward clarification of a fractured paradigm. J Commun 43(4):51–58. https://doi.org/10.1111/j.1460-2466.1993.tb01304.x
Grand A, Wilkinson C, Bultitude K, Winfield AF (2012) Open science: a new "trust technology"? Sci Commun 34(5):679–689
Groshek J, Groshek MC (2013) Agenda trending: reciprocity and the predictive capacity of social networking sites in intermedia agenda setting across topics over time. Media Commun (Lisboa) 1(1):15. https://doi.org/10.12924/mac2013.01010015
Guidry JPD, Carlyle KE, LaRose JG, Perrin P, Ryan M, Messner M, Adams J (2018) Framing and visual type: effect on future zika vaccine uptake intent. J Public Health Res 7(1):1162–1162. https://doi.org/10.4081/jphr.2018.1162
Gustafson A, Rice RE (2020) A review of the effects of uncertainty in public science communication. Public Underst Sci (bristol, England) 29(6):614–633. https://doi.org/10.1177/0963662520942122
Hernández-García I, Giménez-Júlvez T (2020) Assessment of health information about COVID-19 prevention on the internet: infodemiological study. JMIR Public Health Surveill 6(2):e18717
Jensen JD (2008) Scientific uncertainty in news coverage of cancer research: effects of hedging on scientists' and journalists' credibility. Hum Commun Res 34(3):347–369. https://doi.org/10.1111/j.1468-2958.2008.00324.x
Jensen JD, Krakow M, John KK, Liu M (2013) Against conventional wisdom: when the public, the media, and medical practice collide. BMC Med Inform Decis Making 13 Suppl 3(S3):S4–S4. https://doi.org/10.1186/1472-6947-13-S3-S4
Kahneman D, Tversky A (1979) Prospect theory: an analysis of decision under risk. Econometrica 47(2):263–292. https://doi.org/10.2307/1914185
Kim Y, Gonzenbach WJ, Vargo CJ, Kim Y (2016) First and second levels of intermedia agenda setting: political advertising, newspapers, and Twitter during the 2012 U.S. presidential election. Int J Commun (Online). 4550
Kahan DM (2017) On the sources of ordinary science knowledge and extraordinary science ignorance. In: Kahan D, Scheufele DA, Jamieson KH (eds) The Oxford handbook of the science of science communication. Oxford University Press, Oxford, pp 35–49
Mazur A (1981) Media coverage and public opinion on scientific controversies. J Commun 31(2):106–115
McCombs M (2005) A look at agenda-setting: past, present and future. Journalism Stud (london, England) 6(4):543–557. https://doi.org/10.1080/14616700500250438
MacIntyre CR, Chughtai AA (2015) Facemasks for the prevention of infection in healthcare and community settings. BMJ: British Med J 350(apr09 1):h694–h694. https://doi.org/10.1136/bmj.h694

MacIntyre CR, Seale H, Dung TC, Hien NT, Nga PT, Chughtai AA, Rahman B, Dwyer DE, Wang Q (2015) A cluster randomised trial of cloth masks compared with medical masks in healthcare workers. BMJ Open 5(4):e006577–e006577. https://doi.org/10.1136/bmjopen-2014-006577

MacIntyre CR, Tham CD, Seale H, Chughtai AA (2020) COVID-19, shortages of masks and the use of cloth masks as a last resort. Retrieved from: https://bit.ly/3p9g3v8

Nelkin D (1987) Selling science: how the press covers science and technology. W.H. Freeman, New York

Nisbet MC, Mooney C (2007) Science and society. Framing science. Science (am Assoc Advan Sci) 316(5821):56–56. https://doi.org/10.1126/science.1142030

Nisbet MC (2009) The ethics of framing science. In: Nerlich B, Elliott R, Larson B (eds) Communicating biological sciences: ethical and metaphorical dimensions. Taylor and Francis, New York, NY

Nan X, Xie B, Madden K (2012) Acceptability of the H1N1 vaccine among older adults: the interplay of message framing and perceived vaccine safety and efficacy. Health Commun 27(6):559–568. https://doi.org/10.1080/10410236.2011.617243

National Academies of Sciences, Engineering, and Medicine (2016) Communicating science effectively: a research agenda. The National Academies Press, Washington, DC. https://doi.org/10.17226/23674

National Science Board, National Science Foundation (2020) Science and technology: public attitudes, knowledge, and interest. Science and engineering indicators 2020. NSB-2020-7. Alexandria, VA. Available at https://ncses.nsf.gov/pubs/nsb20207/

Pulido CM, Villarejo-Carballido B, Redondo-Sama G, Gómez A (2020) COVID-19 infodemic: more retweets for science-based information on coronavirus than for false information. Int Sociol 35(4):377–392. https://doi.org/10.1177/0268580920914755

Rengasamy S, Eimer B, Shaffer RE (2010) Simple respiratory protection—evaluation of the filtration performance of cloth masks and common fabric materials against 20–1000 nm size particles. Ann Occup Hyg 54(7):789–798. https://doi.org/10.1093/annhyg/meq044

Retzbach A, Maier M (2015) Communicating scientific uncertainty: media effects on public engagement with science. Commun Res 42(3):429–456. https://doi.org/10.1177/0093650214534967

Rogstad I (2016) Is Twitter just rehashing? Intermedia agenda setting between Twitter and mainstream media. J Inform Tech Polit 13(2):142–158. https://doi.org/10.1080/19331681.2016.1160263

Rosenberg H, Syed S, Rezaie S (2020) The Twitter pandemic: the critical role of Twitter in the dissemination of medical information and misinformation during the COVID-19 pandemic. Can J Emerg Med 22(4):418–421

Singer E (1990) A question of accuracy: how journalists and scientists report research on hazards. J Commun 40:102–116

Scheufele D (1999) Framing as a theory of media effects. J Commun 49(1):103–122. https://doi.org/10.1111/j.1460-2466.1999.tb02784.x

Suleski J, Ibaraki M (2010) Scientists are talking, but mostly to each other: a quantitative analysis of research represented in mass media. Public Underst Sci 19(1):115–125

Scheufele DA (2013) Communicating science in social settings. Proc Natl Acad Sci 110(Supplement 3):14040–14047

Su LYF, Akin H, Brossard D, Scheufele DA, Xenos MA (2015) Science news consumption patterns and their implications for public understanding of science. Journalism Mass Commun Q 92(3):597–616

Southwell B (2017) Promoting popular understanding of science and health through social networks. In: Jamieson KH, Kahan D, Scheufele DA (eds) The Oxford handbook of the science of science communication. Oxford University Press, Oxford, pp 223–231

Schäfer MS (2017) How changing media structures are affecting science news coverage. In: Kahan D, Scheufele DA, Jamieson KH (eds) The Oxford handbook of the science of science communication. Oxford University Press, Oxford, pp 51–57

Sprain L (2018) Framing science for democratic engagement. In: Priest S, Goodwin J, Dahlstrom MF (eds) Ethics and practice in science communication. University of Chicago Press, Chicago, IL

Szmuda T, Syed MT, Singh A, Ali S, Özdemir C, Słoniewski P (2020) YouTube as a source of patient information for coronavirus disease (COVID-19): a content-quality and audience engagement analysis. Rev Med Virol 30(5):e2132

Trench, B. (2008). Internet: turning science communication inside-out? In: Bucchi M, Trench B (eds), Routledge handbook of public communication of science and technology. Routledge, pp 27–39

Tufekci (2020) Why telling people they don't need masks backfired. Accessed 21 Mar 2021. https://www.nytimes.com/2020/03/17/opinion/coronavirus-face-masks.html

Updegraff JA, Emanuel AS, Gallagher KM, Steinman CT (2011) Framing flu prevention-an experimental field test of signs promoting hand hygiene during the 2009–2010 H1N1 pandemic. Health Psychol 30(3):295–299. https://doi.org/10.1037/a0023125

van der Sande M, Teunis P, Sabel R (2008) Professional and home-made face masks reduce exposure to respiratory infections among the general population. PLoS ONE 3(7):e2618–e2618. https://doi.org/10.1371/journal.pone.0002618

Vliegenthart R, Walgrave S (2008) The contingency of intermedia agenda setting: a longitudinal study in Belgium. Journalism Mass Commun Q 85(4):860–877. https://doi.org/10.1177/107769900808500409

Vonbun R, Königslöw KK, Schoenbach K (2016) Intermedia agenda-setting in a multimedia news environment. Journalism (london, England) 17(8):1054–1073. https://doi.org/10.1177/1464884915595475

Vargo CJ, Guo L (2017) Networks, big data, and intermedia agenda setting: an analysis of traditional, partisan, and emerging online US news. Journalism Mass Commun Q 94(4):1031–1055. https://doi.org/10.1177/1077699016679976.

Valenzuela S, Puente S, Flores PM (2017) Comparing disaster news on twitter and television: an intermedia agenda setting perspective. J Broadcast Electron Media 61(4):615–637. https://doi.org/10.1080/08838151.2017.1344673

Wang W, Guo L (2018) Framing genetically modified mosquitoes in the online news and twitter: intermedia frame setting in the issue-attention cycle. Public Understand Sci (bristol, England) 27(8):937–951. https://doi.org/10.1177/0963662518799564

WHO (2020). Infodemic. Accessed 1 Mar 2021. https://www.who.int/health-topics/infodemic#tab=tab_1

Xenos MA (2017) Citizens making sense of science issues: supply and demand factors for science news and information in the digital age. In: Kahan D, Scheufele DA, Jamieson KH (eds) Oxford handbook on the science of science communication. Oxford University Press, Oxford, pp 283–289

Xue J, Chen J, Chen C, Zheng C, Li S, Zhu, T. (2020). Public discourse and sentiment during the COVID 19 pandemic: using latent dirichlet allocation for topic modeling on Twitter. PloS one, 15(9):e0239441

Yeo SK, Brossard D (2017) The (changing) nature of scientist–media interactions: a cross-national analysis. In: Kahan D, Scheufele DA, Jamieson KH (eds) The Oxford handbook of the science of science communication. Oxford University Press, Oxford, pp 261–272

Zarocostas J (2020) How to fight an infodemic. Lancet (british Edition) 395(10225):676–676. https://doi.org/10.1016/S0140-6736(20)30461-X

Part V
Resilient Women and Underrepresented Populations

Chapter 15
Pandemics, Perception, and Risk

Elizabeth Sperry and Gina Lane

Abstract This chapter argues that intersectional attitudes rhetorically influence perceptions of risk in disease outbreaks in the United States. Our examination of media reporting during the 2016 Zika epidemic reveals how public reaction to those outbreaks was manipulated through reification of socially constructed narratives of race, gender and class. Using intersectionality as a theoretical lens offers a means of understanding how public perception of the risks of contracting Zika was downplayed by a variety of media frames that reinforced white, male hegemonic views of pregnancy and motherhood. The concerns of white, privileged women were both ridiculed and oversimplified, while the most likely victims of Zika--non-white, lower-income women--were "symbolically annihilated" and subject to erasure. Although the risk of contracting Zika has diminished significantly since 2016, the paper concludes that the framing of risk during Zika can begin to explain the reluctance of a significant proportion of the population to take seriously the risk of COVID-19.

Keywords Zika · Gender · Hegemony · Intersectionality · Pregnancy · Motherhood · Erasure

15.1 Risk Misperception in Pandemic Contexts

Epidemiological risk is often analyzed and addressed in terms of concrete factors such as age, population density, pre-existing medical vulnerabilities, the extent to which a pathogen is contagious, the degree to which a government will enforce containment measures, and the like. These factors are important, but so too is one that has received far less attention: perception. The course of a pandemic is sensitive to perceptions of agents' vulnerability to contagion based on their gender, race, and class. Perceptions of risk surrounding intersectional marginalization and privilege often misdirect attention from risk mitigation strategies that would more effectively

E. Sperry (✉) · G. Lane
William Jewell College, Liberty, USA
e-mail: sperrye@william.jewell.edu

D. M. Berube (ed.), *Pandemic Communication and Resilience*, Risk, Systems and Decisions, https://doi.org/10.1007/978-3-030-77344-1_15

minimize contagion. Thus, misperceptions of risk based on social location perversely increase actual risks in pandemic contexts for all members of a society.

Since 2014, three disease outbreaks in the United States have received vastly different public responses. Despite a death count exceeding 500,000, reaction to the ongoing COVID-19 pandemic has been highly politicized, with half the country supporting mask mandates, social distancing, and quarantines; and the other half arguing that COVID warnings are overly restrictive and harmful to the economy (Brenan 2020). In comparison, the Ebola scare in 2014, which affected a handful of patients and led to two deaths, created a widespread panic that was entirely out of proportion to its risk. Despite official assurances that quarantine was unnecessary, over 70% of the public supported mandatory quarantines for those returning to the United States from their work with Ebola patients in West Africa (Dann 2014). One study found a strong correlation between media coverage of Ebola and internet searches by the general public, leading the authors to conclude that the Ebola scare contained "significant evidence of (media) contagion" (Towers et al. 2015).

When the Zika virus outbreak began in 2015, news writers drew the inevitable parallels to Ebola (see, for example, Kanen and Haberkorn 2016). However, the public did not exhibit the same level of panic, despite dramatic stories of Zika babies born with microcephaly, and other causes for medical concern. The Zika epidemic was far more significant than the tiny number of Ebola cases in the United States, with the CDC reporting 5168 cases of Zika in the U.S. and nearly 36,000 cases in Puerto Rico (CDC "Case Counts"). Although most of the cases in the United States were acquired through travel, with local mosquito-borne transmission limited to less than three hundred cases in Florida and Texas, the Zika virus put at risk approximately 3400 infants born in the United States and its territories (Smoots et al.), with 5–10% of those infants diagnosed with Zika-associated birth defects (CDC "Data and Statistics").

This study analyzes a public health crisis, using an interdisciplinary approach, and seeking to understand barriers to accurate perceptions of community health and risk. Although the Zika virus did not become widespread within the United States, the lack of a mass public reaction so soon after a nearly panicked reaction to a handful of Ebola cases a few years earlier raised questions about why the response varied so greatly. Keller (2016) believed that the public did not anxiously respond to Zika because of the media's exaggerated Ebola coverage, but we posit a different explanation. As professors of philosophy and rhetoric, we align ourselves with C. Crenshaw's view that we should be "increasingly sensitive to interdisciplinary feminist scholarship that values pluralized differences among women and seeks to understand how these differences are ideologically valued or devalued in the texts we examine" (1997, 220). Our analysis examines the response to epidemics in the United States and finds further support for the view that "the way a society responds to problems of disease will reveal its deepest cultural, social, and moral values" (Brandt 1988, 415).

We use feminist and critical race theory analyses of oppression, intersectionality, and the social construction of forms of ignorance to argue that in pandemic contexts, oppression and privilege interact with perceived risk to produce an increase in the actual risk of contagion for all community members. Enhanced awareness of the role played by these non-concrete perceptual factors can motivate community health authorities to address the causes of misdirected risk perception. Our application of this theoretical material to media coverage of the 2016 Zika outbreak will illustrate how privilege blinds us to otherwise foreseeable disease risks. Mistaken perceptions of disease vulnerability resulting from oppression and privilege increase the actual vulnerability of all members of a society. In other words, a less sexist, racist, and classist society is one in which pandemics are more effectively contained, for the betterment of all.

Our case begins with an explanation of the philosophical concepts of oppression, privilege, intersectionality, and the epistemology of ignorance. Next we draw on Metzl (2019) to show that oppression and privilege do have known, concrete effects on health outcomes; and that perceptions of oppression and privilege also have concrete effects on health outcomes. We turn then to media accounts of Zika, using rhetorical criticism to investigate the perceptions of social standing that undergird our account. We show how the concepts that emerge from our analysis—Trivialization, Erasure, and Othering; and Risk Othering, Risk Denial, Disowned Risk, and Risk Reification—result from careful examination of media messages and philosophical analysis. These concepts combine to provide a picture of risk misperception in epidemic contexts, a picture that is concretely instantiated via rhetorical criticism and deepened via philosophical analysis.

15.2 Theoretical Background

Perhaps the most powerful philosophical analysis of oppression is found in Marilyn Frye's classic essay of the same name. Oppression, as she explains, literally "presses" (Frye 1983, 2). It constricts and limits. Oppressed persons experience barriers to opportunity, security, self-development, agency, and other socially-supported goods. Some social barriers are frustrating, but nevertheless navigable. They are like a fence, perhaps: one can climb over the fence, or find a gate. Even though one is not given straightforward access, one can navigate the barrier if one is resourceful and diligent. Oppression, Frye explains, is *not* like this; it is not a single, simple barrier. Instead, it is an interlocking system of barriers, more like a birdcage. If one looks very narrowly at one isolated bar in the cage, one may wonder why the bird does not simply fly around it. Unfortunately, however, there are intersecting bars in the spaces where one might have tried to move around the barrier. Thus, oppression functions as "a network of systematically related barriers ... which, by their relations to each other, are as confining as the solid walls of a dungeon" (Frye 1983, 5).

As Frye argues, privileged people may suffer, simply because human life includes suffering. But a rich white man who breaks his leg while skiing, and experiences pain and a long recuperation, is not suffering because of oppression, because of intersecting barriers to his development and flourishing; rather he is suffering because bad things happen soon or later to all of us. Oppression makes bad things happen (or prevents good things from happening) disproportionately for socially devalued people, and makes them happen via socially generated and supported practices. Privilege, therefore, is the absence of interlocking barriers to socially-managed goods; meanwhile, privilege disproportionately funnels these goods to its bearers. White people, men, and the financially well-off are granted educational, economic, and professional developmental opportunities that are structurally denied to BIPOC persons, women, and the poor.

Thus, oppression is already, in a sense, intersectional. Oppression confronts its targets with intersecting socially-produced barriers. But what happens when a person is simultaneously marginalized in multiple ways? One might be Black *and* female, or gender-queer *and* poor *and* disabled. As Kimberle Crenshaw showed in her groundbreaking 1989 essay, "Demarginalizing the Intersection of Race and Sex," being marginalized in more than one way means one is oppressed more intensively. This is intersectionality of a different sort: the intersection in one person of multiple forms of marginalization[1] may change both the intensity and the nature of one's oppression. A Black woman will experience both sexism and racism, and the sexism will make the racism worse; meanwhile, the racism will make the sexism worse. In addition, the intersectionally oppressed person regularly encounters distinctive barriers, forms of marginalization that are not shared with those whose oppression is monistic. For instance, Black women have been the targets of sexual assault "not as women generally, but as Black women specifically: their femaleness made them sexually vulnerable to racist domination, while their Blackness effectively denied them any protection" (Crenshaw 1989, 158–59). Accordingly, those who are oppressed along a single vector may be uninformed about the experiences of the intersectionally oppressed.[2] In order to move toward greater justice and equality, we will need to address our ignorance about the effects and manifestations of oppression, including intersectional oppressions.

Consequently, we draw on work in the epistemology of ignorance, a concept first introduced[3] by Charles Mills in *The Racial Contract* (1997).[4] Many feminist philosophers and critical race theorists have expanded and applied Mills' concept.[5] Historically, western philosophy has denied that knowledge is socially produced and mediated; traditional epistemology insists that we know because reality impresses itself and its concepts upon us. Gradually, however, philosophers have begun acknowledging the reality that we know because we are taught by human others what to learn and how to learn it. This realization, however, suggests the further insight that others may mis-teach us. If knowledge is socially produced and mediated, so too can

[1] We use "marginalization" throughout this essay as a synonym for oppression.

[2] For instance, white feminism has persistently overlooked the problems and experiences of BIPOC women.

[3] As Mills explains, Carole Pateman's *The Sexual Contract* (1988) is a precursor to his project. Nonetheless, his is the first fully explicit presentation of the epistemology of ignorance.

[4] See page 18.

[5] See, for instance, the essays by Harding, Heldke, Ortega, Tuana, all in the 2006 special issue of *Hypatia* addressing feminist epistemologies of ignorance; and work by Jose Medina, Miranda Fricker, Linda Martin Alcoff, Lorraine Code, and Lucius Outlaw.

be ignorance. Such ignorance, as Mills argues, "cover[s] both false belief and the absence of true belief" (Mills 2007, 16).

Ignorance is often taught in ways that uphold pre-existing power relations. Certain forms of ignorance are "systematically produced and sustained to misrepresent reality in ways that not coincidentally sustain patterns of racial [and class and gender] privilege" (Townley 2011, x). People are less likely to fight for social justice if they have been systematically misinformed about the ways in which a more equitable society would produce greater flourishing for almost everyone, or if they have been systematically misinformed about the deeply painful and unfair struggles experienced by oppressed persons. If I do not understand the brutal dehumanization experienced by enslaved Africans, indigenous persons, and persons of color in the Jim Crow American South, then I may wrongly hold BIPOC individuals personally responsible if they struggle with generational poverty or suboptimal health. If I do not understand the pervasiveness and trauma of sexual assault, I will not understand why many women do not go out alone after dark, or are wary of unmonitored encounters with men. Socially produced knowledge and ignorance are the result of agreement about "what counts as moral and factual knowledge of the world" (Mills 1997, p. 17). In privileged social circles, it is acceptable to be ignorant about the experiences of BIPOC persons and women, because such facts are not perceived as actual knowledge of the world.

The problem of ignorance is exacerbated because it is a problem we don't know we have. I don't know what I don't know about the struggles and successes of marginalized people. My socially-mediated perceptions of marginalized others feel to me like facts, particularly if I am a privileged person accustomed to having my worldview affirmed. Socially-mediated ignorance protects my privilege and reinforces the marginalized status of others. My ignorance of others' struggles functions as implicit proof that there are no such struggles. I will see oppressed persons' oppression as the result of their own character failings, while remaining oblivious to the birdcage that produced their oppression. Furthermore, I will not see the ways in which my own perceptions reinforce the birdcage's wires. My socially-mediated perceptions will reassure me that those perceptions are all I need to know about the experiences and natures of women, people of color, and the poor. Correspondingly, I will remain ignorant of my own privilege, and I will see the socially-facilitated goods I enjoy as a testament to my own intrinsic merit. As Mills explains, ignorance produced and maintained by power "produce[s] the ironic outcome that [privileged people] will in general be unable to understand the world that they themselves have made" (1997, 18).

Socially-mediated misperceptions of oppressed persons, of intersectionally oppressed persons, and of privileged persons together play a large role in perpetuating injustice and inequality. As we will argue, they also play a large role in facilitating the spread of infection during epidemics. Analyses drawn from the epistemology of ignorance can show how injustice is perpetuated by beliefs. Such analyses can also "redirect your vision [and] help you see what has been there all along" (Mills 1997, 2). Investigating our socially-produced ignorance is more than an abstract endeavor, since "the lack of appropriate concepts can hinder learning, interfere with memory, block inferences, obstruct explanation, and perpetuate problems" (Mills 1997, 7). Below we work to unblock inferences and explanations, in order to facilitate improved responses to disease outbreaks.

15.3 Oppression and Medical Outcomes

We argue that misperceptions of marginalization and privilege produce subpar medical outcomes during epidemics, thanks to ignorance of and inattention to the actual risks of various behaviors and governmental responses. It is important to recall, then, that marginalization and misperceptions of oppression produce deficient medical outcomes in non-epidemic contexts too. The 2019 book *Dying of Whiteness* made this case through extensive statistical analysis and intensive interviews with people in Missouri, Tennessee, and Kansas. As Jonathan Metzl writes, "I found support for a set of political positions that directly harmed [voters'] own health and well-being or the health and well-being of their own families" (Metzl 2019, 2). He concludes that this support for medically dangerous policies resulted from "[d]ogma that... aligned with beliefs about a racial hierarchy that overtly and implicitly aimed to keep white Americans hovering above Mexicans, welfare queens, and other nonwhite others" (Metzl, 3). Misperceptions about marginalized people, Metzl found, caused those who were not marginalized[6] to vote for policies that would penalize, or at least fail to benefit, the marginalized; and these voters did so despite the fact that the policies they voted for would also biologically penalize, or fail to benefit, themselves.[7] Research indicated that the instantiation of these "white backlash policies" increased the rate of death for everyone to whom they applied—including whites (Metzl, 8).[8] Where medical treatment is harder to get and pay for, white people (as well as non-white people) suffer worse medical outcomes and die more frequently. "Anti-blackness, in a biological sense, then produces its own anti-whiteness. An illness of the mind, weaponized onto the body of the nation" (Metzl, 15). Public health is diminished not only by the lesser medical outcomes of oppressed people; it is further affected by the misperceptions others have about the oppressed. Public health is undermined by privileged people's misperception that they are immune to the risks of impaired access to health care and expanded access to guns. Ultimately, the choices voters make based on their misperception of their own risk status

[6] Or who were marginalized differently. Some of Metzl's interview subjects who supported policies that were not in their own medical self-interest were economically disadvantaged.

[7] Metzl points out that these voters do actually "benefit" in the sense that they maintain a commitment to their worldview and to their place in a racialized hierarchy. They do not, however, benefit in terms of access to medical treatment, educational opportunity (which is correlated with enhanced health outcomes), or protection from gun violence.

[8] Metzl's analysis also shows that men's anxieties about power and protection generate support for lax gun laws, which then lead to an uptick in death by suicide, particularly by white men. In this matter, "privilege itself becomes a liability, [because] ...the data suggests 'being a white man who lives in Missouri' then emerges as its own, high-risk category" (110).

cause increased risks that accrue to everyone. In the cases Metzl analyzes, privilege, misperceptions of marginalized people, and misperceptions of risk work together to create deadly outcomes.

15.4 Media Coverage of Zika

As we will show, rhetorical examples of news coverage during the 2016 Zika outbreak reveal intersectional obstacles to social equality that undermine disease risk mitigation. Drawing upon the work of rhetorical theorist Kenneth Burke in *A Grammar of Motives*, we have chosen these "representative anecdotes" because they illustrate these intersectional obstacles in form and content. A focus on non-intersectional references to oppression would fail to reveal the ways in which experiences of oppression differ. Brummett notes that an analysis of representative anecdotes provides a framework that allows the critic to "sum up the essence of a culture's values, concerns, and interests in regard to some real-life issues or problems" and reveals familiar patterns and forms of response (1984, 164). The news media is particularly well suited for this approach because its "stories" are dramatic in nature and follow a predictable structure. As Brummett states, "Burke would remind us that media content tells a story, and that critical methods which look for stories are thus appropriate tools for media analysis. To reveal the formal stories being told, and the real or symbolic ills they cure, gives media criticism the ability to move from social commentary to social knowledge" (1984, 174).

Media coverage of an important public issue has been shown to influence public opinion. Both Agenda Setting Theory and Framing Theory demonstrate that news coverage creates viewer priorities and influences the public's understanding of issues (Sell et al. 2018). This influence, taken as a whole, means the news media "function hegemonically" to preserve existing cultural norms and oppressive practices (Crenshaw 1997, 220). The Zika epidemic received a great deal of coverage by the news media in the United States. A content analysis of 25 major news sources and over 800 news stories on Zika found the media engaged in "risk-elevating" or amplification messages that focused on "mosquito-borne transmission, potentially catastrophic birth defects, and potential spread in the United States" (Sell et al. 2018, 2519; 2521). The authors concluded the widespread coverage and the amplification of risk was potentially so alarming that the average person would seek to psychologically distance themselves from those they perceived as "high risk," resulting in "reduced vigilance" (Sell et al. 2018, 2522).

A second study noted that one way the news media influenced this distancing was to make Zika a woman's problem while simultaneously failing to interview women, effectively silencing their voices (Tinga et al. 2018, 2). CNN's reporting directed women to take precautions, including not traveling to affected areas and avoiding pregnancy. On the rare occasion that CNN offered advice to men, reporters suggested that those returning from a trip to an afflicted area "might consider" avoiding sex or using a condom (Tinga et al. 2018). Both of these studies reveal that the news media coverage of Zika could have prompted viewers to mitigate risk by perceiving their privilege as a shield of invulnerability. Men were at the top of that privileged hierarchy and thus could not be asked to participate in a woman's responsibility to protect her unborn children from the potentially devastating effects of the Zika virus. The panic we could have anticipated from the media coverage of Zika did not occur because privilege blinded its holders to their own risk. Media coverage assigned risk to women, including those we would normally consider privileged, socially ostracizing and silencing them.

At this time, intersectionality has only rarely been used as a rhetorical lens. Carrie Crenshaw was among the first to do so, drawing attention to the way in which "the presence of derogatory images and the absence of [women's] experiences" in media presentations reinforces systems of privilege (1997, 220). She examined news reports of women serving in the Gulf War, demonstrating how the lived experience of intersecting identities reveals embedded and reinforced oppressions that a traditional gendered analysis would not. In a similar vein, our analysis will show how media reporting on Zika reinforced systems of privilege that both separated women and undermined accurate perceptions of risk. Our analysis will reveal intersectional oppressions that undermine actual risk mitigation, and reinforce the separation of privileged women from poor women and women of color.

15.5 Trivialization and Erasure of Risk in Media Accounts

While media reports of Zika included information on reducing the risk of transmission, much of the news coverage seemed to downplay the risk, in some cases through the use of humor. A 2016 *Miami Herald* article ("Moms-to-Be Go the Extra Mile to Avoid Zika. Just Ask the One in the Beekeeper Suit") focused on the efforts of pregnant and privileged white women to avoid Zika. A photograph accompanying the article showed the birth announcements of one white Miami couple, the pregnant wife leaning against a palm tree with her smiling husband's hand resting protectively on her rounded abdomen. The mother-to-be, wearing beekeeper's garb to protect herself against Zika-laden mosquitoes, proclaimed her shock at the invasion of this foreign pathogen. "When I saw the first picture of a baby with microcephaly, I burst into tears," she said. "It seemed so far away in the jungle, and then it was in my backyard." The same article featured another white, wealthy mother-to-be who almost never left her house, with "insecticide misters outside and bug zappers inside." On the rare occasion when she had to go outside, she wore full protective gear and "drench[ed] herself in insect repellent." Photos accompanying this article provide farcical illustrations of privileged white women going to extremes in their panic to protect themselves and their unborn babies (Harris 2016). The portrayal of these women as absurd distanced readers, who were encouraged to mock the pregnant women's exaggerated response to the Zika threat. The *New York Times* gave more straightforward advice. In their "Guide to Help Pregnant Women Reduce Their Zika Risk," women were told to "stay indoors," "avoid certain neighborhoods," and "cover-up and use insect repellent." The *Times* story pointed out that relocation was an option, although it added the caveat that "some women" could not afford to do so (Belluck 2016). For those women planning to take a Florida vacation, *Bloomberg News* advised avoiding the beach at peak mosquito time: "Pack the chicest cover-up you own, and wear it often. Request a room on your hotel's higher floors. And don't forget your bug spray" (Ekstein 2017). These stories support Tinga et al.'s conclusion that media reporting made women entirely responsible for protecting their unborn children from Zika. In each case, pregnant women and privileged people were seeking shelter from the risk by avoiding "certain neighborhoods" and even "relocating." Yet these articles offered readers an emotional and metaphorical "relocation," since the news pieces suggest the Zika risk was relevant only for pregnant women. The articles encouraged readers to feel relief that they were exempt from the threat, as well as amusement at pregnant women wearing hazmat suits while fleeing their beautiful Florida homes. What the articles did not encourage was any sense of empathy or concern, instead downplaying and Trivializing the risk.

These articles focused on privileged women, women who lived in beautiful homes and could afford to avoid "certain neighborhoods." But their focus on the privileged distracts us from those not mentioned, the pregnant women who could not afford beekeeper's gear or insecticide misters, women in poor neighborhoods, women who did not have screens on their windows or air conditioning to help them stay inside, many of them women of color. In Texas and Florida, the areas of the country where Zika was most prevalent, poor women and women of color were most at risk because their homes often provided little protection, and "they had little money to spend on insect repellent or protective clothing" (Bond 2017, 853–54). There were very few options available for these erased women. Not only were these women unable to take protective measures that were available to privileged women, but they were also often unable to find late-term abortions when they learned that their babies would be born severely disabled. Late-term abortion has been outlawed in at least one-third of the United States, and in every state where women are at higher risk for a Zika infection (Guttmacher Institute). Ultimately, it was these intersectionally oppressed women who disproportionately faced the prospect of giving birth to profoundly disabled children, children with medical bills costing millions (Cakir 2016). The struggles of these women were obliterated by media Erasure.

The Trivialization of the privileged pregnant makes it easier to overlook the Erasure of intersectionally oppressed women, leading to their "symbolic annihilation," a term coined by media theorist George Gerbner in the 1970s and then used by Barbara Tuchman to underscore how our culture has failed to recognize the needs of women, treating them as either unworthy of attention or reducing them to stereotypical portrayals (Gallagher 2014). Decades later, Gallagher wrote that the same concerns that Tuchman raised in 1978, "issues of power, values, representation, and identity," still arise in contemporary media portrayals of women. "This mediated invisibility, it was argued, is achieved not simply through the non-representation of women's points of view or perspectives on the world. When women are 'visible' in media content, the manner of their representation reflects the biases and assumptions of those who define the public—and therefore the media—agenda" (Gallagher 2014,

23). Even privileged women, slathered in insecticide and attired in hazmat suits, are caricatured for our amusement. As K. Crenshaw and others have written, this Trivialization of women's legitimate concerns erects intersectional barriers that impede the formation of necessary coalitions. The Trivialization and Erasure of women's concerns often renders women unable to see that others face the same obstacles, making it unlikely that they will "collaboratively fight to tear down structural regimes that serve to oppress peoples across multiple axes" (Cho et al. 2013, 803).

The symbolic annihilation of women within these stories does "function hegemonically" to preserve privilege (Crenshaw 1997), and thereby alleviates risk in the minds of those who are privileged and not-pregnant. Privilege creates an illusion of immunity, allowing wealthier white men and women to see themselves as somehow safe from disease. In the case of Zika, those who decided they were not at risk ignored consideration of how they might themselves be vectors for spreading the virus, particularly by sexual contact and by travel to outbreak areas. Their attitudes were reinforced by media reports showing they were not alone. In August 2016, the *Miami Herald* showcased tourists in several stories with little or no worry about Zika because they did not plan to get pregnant. The *Herald* also featured local business owners more afraid of losing income than of spreading Zika (Levin 2016). One male tourist said, "I don't feel too concerned because it only affects pregnant women" (Viglucci and Levin 2016). Another man, vacationing with his wife, was even more blunt. "It doesn't scare me," one coupled-up tourist told the paper. "The chances are slim. The danger is mostly to pregnant women, and we are not pregnant and not planning to be." A local restaurant owner diminished concerns: "The thing to remember is the symptoms are fairly benign unless you're a pregnant woman." An owner of a clothing store promised to protect customers by spraying lavender and incense (Viglucci and Levin 2016). Privilege and denial have always complicated the course of disease. Privileged people deflect risk by blaming the Other, and by failing to see themselves as potential vectors or as morally obligated to help stop disease transmission. As Sihna and Parmet (2016) observe, this type of thinking has persisted throughout history. It is "the Other" that is the risk to be avoided, whether the other has lesser means, a different skin color, or a more "exotic" location (Sihna and Parmet 2016). This stereotype not only separates the privileged from their own culpability in spreading disease, but it creates a fatalistic view of the "others" who perish. "Others" are perceived as "inherently closer to disease: [and are] more deserving of death from it" (Edwards 2014). Our analysis of these representative anecdotes shows that the news media enabled and reinforced this Othering by engaging in the "symbolic annihilation of women," further encouraging the misperception of risk by the privileged.

15.6 Philosophical Analysis of Oppression and Risk Perception

Our argument uses rhetorical criticism to show that the abstract theoretical conclusions reached from a philosophical perspective are in fact concrete concerns for societies facing epidemics; and uses philosophical analysis to show that rhetorical criticism reflects more than the momentary concerns of media professionals and the individuals whose views are portrayed in newscasts and newspapers. Instead, the rhetorical criticism reflects the deeper structures of a society built around oppression, privilege, and pervasive unknowing, demonstrating that the biological vulnerability of the marginalized is mis-portrayed using Trivialization, Erasure, and Othering. Oppression and privilege generate conditions for socially-produced and socially-reinforced ignorance of the experiences of the oppressed, and the patterns of thought revealed by philosophical analysis are *fundamental* to the way humans understand risk under conditions of social injustice. Pandemic crises in socially unjust circumstances will produce three oppression-related misperceptions. The privileged may (mis)perceive oppressed others as the sole bearers of risk (Risk Othering); themselves as immune to risk due to their privilege (Risk Denial); and the risk of marginalized others as unimportant (Disowned Risk). The perverse outcome of any of these risk misperceptions is the intensification and spread risk itself (Risk Reification).

Risk Othering occurs when I believe contagion accrues to those who are Other, based on some characteristic(s) inherent to their Otherness. Only the marginalized are vulnerable, meaning that I am not. Believing myself immune from contagion, I will see no need to engage in my own risk mitigation behaviors. I may travel to a pandemic hotspot, go maskless, or participate in large social gatherings. Such behavior increases my own risk, based on my perception that I am not the sort of person who is vulnerable to contagion. Consequently, I not only increase my own risk; I also increase the likelihood of becoming a contagion vector for my family, friends, and co-workers.

Risk Denial occurs when socially privileged people misrepresent their level of risk based on the mistaken belief that their social privilege is medically protective. This mistake is not baseless, since socially privileged people are also medically privileged. They are able to afford medical care and healthy lifestyles that are out of reach for marginalized people (for instance, most privileged people have access to organic fruits and vegetables while many marginalized people live in food deserts). White Americans consistently experience better health outcomes than BIPOC Americans, an effect that is compounded by attention to socioeconomic status (Stepanikova and Oates 2017). Privileged people are multiply protected from causes of ill-health that oppressed persons experience, including the stress of marginalization. The *New York Times* noted that the effects of discrimination on one's health are cumulative and linked to a variety of poor health outcomes, including "smaller babies and higher infant death rates, a greater risk of cancer, [and] depression" (Khullar 2017). Along multiple vectors, social privilege confers extra biological vigor and longevity, while the effects of lifelong discrimination are "toxic to the cells, organs and minds of those who experience it" (Khullar 2017). Even though medical privilege results from social privilege, neither confers immunity to contagious pathogens. It is Risk Denial to think otherwise. As in Risk Othering, the perception that one is not at risk makes one less likely to attend to one's actual level of vulnerability. This, in turn, increases the likelihood of spreading infection to others.

A third conceptual lens emerges from the heightened medical risk oppressed persons do face in epidemic contexts. Those who are working class, poor, or employed in service industries may find themselves interacting with large numbers of potentially infected people, as, for instance, a grocery store clerk or drugstore cashier would need to do. People with professional jobs are more likely to be able to work from home or otherwise limit their contact with the public; people in lower-paid positions, however, are unlikely to have this luxury. Additionally, managers for customer-service jobs are not always careful about enforcing protocols that would offer some protection to their employees (such as requiring that customers wear face masks). For employees without economic security, it may be necessary to accept these conditions, even though it means a greater risk of infection, and of spreading contagion to family and friends. Marginalized people who live in population-dense cities will also not have the financial means to flee to the country or suburbs when local infection levels rise. Women may be pregnant, which is a potentially serious risk during Zika outbreaks.

This third conceptual lens does not at first glance concern perception. Marginalized persons' level of risk during pandemics is actually greater than is that of privileged persons, and this remains so whether or not it is accurately perceived. Therefore, this point may appear irrelevant to an argument about the role played by *perceptions* of risk in managing contagion. And this would be true if Western societies allocated resources for contagion management proportionally, based on varying levels of actual risk. But we do not extend greater preventative and after-care resources to the poor and to persons of color. Thus, there is a perceptual aspect to our handling of oppressed persons' heightened actual risk in pandemic contexts: we tend to perceive that actual heightened risk as unimportant. We ignore and under-respond to the persons most at actual risk because they are considered less valuable members of society. This Disowned Risk is a misperception of the significance of the actual risk for marginalized people. And it is also a misperception of the truth that preventing infection for any member of our society is protective for all members of our society. Given the many service roles marginalized people play for privileged people, unaddressed contagion in the former population will quickly leak into the latter. Contagion is no respecter of social status.

In each of these forms of risk misperception—Risk Othering, Risk Denial, and Disowned Risk—there lurks a fourth form of risk, Risk Reification. Ignoring the risk to any person, or the risk of any governmental or personal choice, during a context of heightened contagion has the consequence of enlarging the risk of contagion for everyone concerned. Thus, our misperceptions of risk based on our own privilege, others' oppression, and our ignorance about privilege and oppression lead to worse health outcomes for all.

15.7 Relationship Between Rhetorical Criticism and Philosophical Analysis

Our interdisciplinary analysis explains how oppressive societies encourage the misperception of risk, and demonstrates that media messages perpetuate hegemonic structures of privilege under conditions of social injustice. The philosophical categories of Risk Othering, Risk Denial, Disowned Risk, and Risk Reification provide a fundamental explanation of why media messages are likely to engage in Trivialization, Erasure, and Othering. Further examination of the relationship between the abstract philosophical categories and the rhetorical categories shows that they reflect and reinforce one another.

15.7.1 Othering and Risk Categories

Risk Othering provides an account of how persons' and media Othering reflect background systems of privilege, oppression, and socially-generated ignorance. Othering, the media portrayal of privileged persons' response to others' pandemic risk as "not my problem," also exhibits Risk Denial. To the extent that I see another person's risk as solely about that person and her social location, I assume that my privilege protects me from sharing that risk. Finally, media Othering involves Disowned Risk. Once I label another person's risk as "not my problem" I have determined that, from my particular perspective, it is not a problem at all. Together, the Risk Othering, Risk Denial, and Disowned Risk implicated in Othering ensure that I will not attend to the actual risk of contagion, which makes the risk itself larger and more dangerous, producing Risk Reification.

15.7.2 Trivialization and Risk Categories

Trivialization portrays actual risk as minimal, silly, or irrelevant. Here, too, we see Risk Othering, Risk Denial, and Disowned Risk. Risks that are borne by other people are easier to dismiss as irrelevant, which is Risk Othering. My belief that a risk is minimal or irrelevant-to-me is enhanced by my conviction that I am immune based on my privileged status, which is Risk Denial. Media portrayals of actual, substantial risks as merely trivial are Disowned Risk, risk we can safely ignore as a vulnerability for our inferiors. Once again, these risk misperceptions together fuel Risk Reification.

15.7.3 Erasure and Risk Categories

Erasure is perhaps the most diabolical of the risk misperceptions revealed by rhetorical criticism. Ignoring the presence and problems of oppressed persons, particularly the intersectionally oppressed, in pandemic contexts eliminates our ability to attend to those problems, even as this elimination hides its own functioning. The people we don't notice, with their challenges and their risk, simply disappear. Of course, their disappearance in our perception does not eliminate them as potential contagion vectors; rather, their unaddressed vulnerability makes them more likely to spread infection. This is Risk Reification. Erasure is also Disowned Risk, a way of leaving the oppressed to tend to the oppressed. The concerns of the oppressed are most powerfully not my problem, and hence disowned, if I refuse to see those problems in the first place. Less overtly, Erasure is also both Risk Othering and Risk Denial, since it is easier to ignore a vulnerability altogether when the risk is associated with a group other than mine and when I am convinced my privilege renders me immune.

15.8 Conclusion

Our analysis shows that news media reinforce the tendency of privileged people to avoid responsibility for halting transmission during disease outbreaks. The media's portrayal of women during the Zika crisis reinforced the perception that privileged persons had no role to play in blocking contagion. The media's trivialization, erasure, and othering illustrate our broader philosophical argument that risk othering, risk denial, and disowned risk work together to reify risk. Without a fundamental shift in our thinking about social inequality, dangerous ignorance surrounding privilege and oppression will "block inferences, obstruct explanation, and perpetuate problems" (Mills 1997, 7) in every pandemic, and will be "weaponized onto the body of the nation" (Metzl 2019, 15), causing needless suffering and death. Our theoretical approach would provide similar insights when applied to the COVID-19 pandemic. In the United States the pandemic has disproportionately affected BIPOC communities, particularly those who tend to be regarded as "front-line" workers, with the best estimates indicating that BIPOC residents account for 34% of COVID-19 cases, far outstretching their share of the population (Fortuna et al. 2020). Yet the social response to the loss of over 500,000 lives has been uneven at best, with scores of white Americans protesting mask mandates, social distancing, and other measures designed to stop the spread of the disease. This politicized reaction to COVID-19 restrictions is just the latest manifestation of Mills' "racial contract" (1997), leading one writer to observe that "the perception that the coronavirus is largely a black and brown problem licenses elites to dismiss its impact" (Serwer 2020).

Increased globalization means an increased likelihood of epidemics. Ebola, Zika, and COVID-19 will not be our last encounters with highly contagious and dangerous pathogens. Understanding the sources of actual risk and the ways in which we misperceive risk will be crucial for protecting ourselves and our society in the future. Pandemics are, by definition, about seemingly uncontrollable contagion. Contagion can, however, be diminished if each person takes their own vulnerability seriously and if we hold our media and government accountable for responding to actual rather than misperceived risk. Trivialization, erasure, and othering are damaging to public health, as are oppression, ignorance, and the unexamined investment in privilege that produce risk othering, risk denial, disowned risk, and reified risk. In a contagion crisis, social justice and personal health are intimately intertwined.

References

Alcoff L (2007) Epistemologies of ignorance: three types. In: Sullivan S, Tuana N (eds) Race and epistemologies of ignorance. SUNY Press, pp 39–58

Belluck P (2016) A guide to help pregnant women reduce their Zika risk. New York Times. https://www.nytimes.com/2016/08/27/health/a-guide-to-help-pregnant-women-reduce-their-zika-risk.html. Accessed on 14 Mar 2021

Bond J (2017) Zika, feminism, and the failures of health policy. Washington Lee Law Rev Online 73(2):841–885. https://scholarlycommons.law.wlu.edu/wlulr-online/vol73/iss2/13. Accessed on 14 Mar 2021

Brandt A (1988) AIDS and metaphor: toward the social meaning of epidemic disease. Soc Res 55(3):413–432

Brenan M (2020) Roundup of Gallup COVID-19 coverage. Gallup Blog. https://news.gallup.com/opinion/gallup/308126/roundup-gallup-covid-coverage.aspx. Accessed on 14 Mar 2021

Brummett B (1984) Burke's representative anecdote as a method in media criticism. Crit Stud Mass Commun 1(2):161–176. https://doi.org/10.1080/15295038409360027

Burke K (1945) A grammar of motives. University of California Press, Berkeley, 1969

Cakir CT (2016) Is the Zika funding delay rooted in sexism? With women's health at risk, it would seem so. Romper. https://www.romper.com/p/is-the-zika-funding-delay-rooted-in-sexism-with-womens-health-at-risk-it-would-seem-so-11741. Accessed on 14 Mar 2021

Center for Disease Control (2016) Case counts in the United States." https://www.cdc.gov/zika/reporting/2016-case-counts.html. Accessed on 14 Mar 2021

Center for Disease Control (2020) Data and statistics on Zika and pregnancy. https://www.cdc.gov/pregnancy/zika/data/index.html. Accessed on 4 Feb 2021

Cho S, Crenshaw KW, McCall L (2013) Towards a field of intersectionality studies: theory, applications, and praxis. Signs 38 (4):785–810. www.jstor.org/stable/10.1086/669608. Accessed on 8 Jan 2021

Code L (2007) The power of ignorance. In: Sullivan S, Tuana N (eds) Race and epistemologies of ignorance. SUNY Press, pp 213–230

Crenshaw K (1989) Demarginalizing the intersection of race and sex: a black feminist critique of antidiscrimination doctrine, feminist theory, and antiracist politics. University of Chicago Legal Forum, 1989. Article 8. 131–167. https://chicagounbound.uchicago.edu/uclf/vol1989/iss1/8. Accessed on 14 Mar 2021

Crenshaw C (1997) Women in the gulf war: toward an intersectional feminist rhetorical criticism. Howard J Commun 8(3):219–235. https://doi.org/10.1080/10646179709361756

Dann C (2014) NBC/WSJ poll: 71% back mandatory quarantines for Ebola health workers. http://www.nbcnews.com/storyline/ebola-virus-outbreak/nbc-wsj-poll-71-back-mandatory-quarantines-ebola-health-workers-n239576. Accessed on 14 Mar 2021

Ekstein N (2017) Should you still be worried about Zika? Bloomberg News. https://www.bloomberg.com/news/articles/2017-02-08/should-you-still-be-worried-about-zika. Accessed on 14 Mar 2021

Edwards S (2014) From miasma to Ebola: the history of racist moral panic over disease. Jezebel. https://jezebel.com/from-miasma-to-ebola-the-history-of-racist-moral-panic-1645711030. Accessed on 14 Mar 2021

Fortuna L, Tolou-Shams M, Robles-Ramamurthy B, Porche M (2020) Inequity and the disproportionate impact of COVID-19 on communities of color in the United States: the need for a trauma-informed social justice response. Psychol Trauma Theory Res Pract Policy 12(5):443–445. https://doi.org/10.1037/tra0000889

Frye M (1983) Oppression. In: The politics of reality: essays in feminist theory. Crossing Press/Random House, pp 1–16

Gallagher M (2014) Media and the representation of gender. In: Carter C, Steiner L, McLaughlin L (eds) The Routledge companion to media and gender. Routledge, pp 23–31

Harding S (2006) Two influential theories of ignorance and philosophy's interests in ignoring them. Hypatia 21(3):20–36. https://doi.org/10.1111/j.1527-2001.2006.tb01111.x

Harris A (2016) Moms-to-be go the extra mile to avoid Zika. Just ask the one in the beekeeper suit. Miami Herald. http://www.miamiherald.com/news/health-care/article109554377.html. Accessed on 14 Mar 2021

Heldke L (2006) Farming made her stupid. Hypatia 21(3):151–165. https://doi.org/10.1111/j.1527-2001.2006.tb01118.x

Keller J (2016) Why aren't Americans flipping out about Zika? Because they already did over Ebola. Pacific Standard. https://psmag.com/news/why-arent-americans-flipping-out-about-zika-because-they-already-did-over-ebola. Accessed on 14 Mar 2021

Kanen J, Haberkorn J (2016) Is Zika the New Ebola? Politico. https://www.politico.com/story/2016/01/is-zika-the-new-ebola-218362. Accessed on 14 Mar 2021

Khullar D (2017) How prejudice can harm your health. New York Times. https://www.nytimes.com/2017/06/08/upshot/how-prejudice-can-harm-your-health.html. Accessed on 14 Mar 2021

Levin J (2016) South Beach tourists mostly blasé about Zika in Miami Beach. Miami Herald. http://www.miamiherald.com/news/health-care/article96892637.html. Accessed on 14 Mar 2021

Medina J (2013) The epistemology of resistance: gender and race oppression, epistemic injustice, and resistant imaginations. Oxford University Press

Medina J (2016) Ignorance and racial insensitivity. In: Peels R, Blaauw M (eds) The epistemic dimensions of ignorance. Cambridge University Press, pp 178–201

Metzl J (2019) Dying of whiteness: how the politics of racial resentment is killing America's heartland. (Kindle edition). Basic Books

Mills C (1997) The racial contract. Cornell University Press, Ithaca, NY

Mills C (2007) White ignorance. In: Sullivan S, Tuana N (eds) Race and epistemologies of ignorance. SUNY Press, pp 11–38

Ortega M (2006) Being lovingly, knowingly ignorant: white feminism and women of color. Hypatia 21(3):56–74. https://doi.org/10.1111/j.1527-2001.2006.tb01113.x

Outlaw L (2007) Social ordering and the systematic production of ignorance. In Sullivan S, Tuana N (eds) Race and epistemologies of ignorance. SUNY press, pp 197–212

Pateman C (1988) The sexual contract. Stanford University Press

Sell T, Watson C, Meter D, Konk M, Ravi S, Pechta L, Lubell K, Rose D (2018) Frequency of risk-related news media messages in 2016 coverage of Zika virus. Risk Anal 38(12):2514–2544. https://doi.org/10.1111/risa12961

Serwer A (2020) The Coronavirus was an emergency until Trump found out who is dying. The Atlantic. https://www.theatlantic.com/ideas/archive/2020/05/americas-racial-contract-showing/611389/. Accessed on 14 Mar 2021

Sinha M, Parmet W (2016) The perils of panic: Ebola, HIV, and the intersection of global health and law. Am J Law Med 42(2–3):223–255. https://doi.org/10.1177/0098858816658269

State Bans on Abortion Throughout Pregnancy. Guttmacher Institute. https://www.guttmacher.org/state-policy/explore/state-policies-later-abortions. Accessed on 6 Feb 2021

Stepanikova I, Oates G (2017) Perceived discrimination and privilege in health care: the role of socioeconomic status and race. Am J Prevent Med (1 Suppl. 1):S86–S94. Doi: 10.106/j.amepre.2016.09.024

Tinga T, Pruchiewska U, Buozis M, Kute L (2018) Gendered discourses of control in global journalism: women's bodies in CNN's Zika reporting. Fem Media Stud. https://doi.org/10.1080/14680777.2018.1426619

Towers S, Afzal S, Bernal G, Bliss S, Espinoa B, Jackson J, Judson-Garcia J, Khn M, Lin M, Mamada R, Morena V, Nazari F, Okuneye K, Ross M, Rodriguez C, Medlock J, Ebert D, Castillo-Chavez C (2015) Mass media and the contagion of fear: the case of Ebola in America. PLoS One 10(6). Doi: 10.137/journal.pone.0129179

Townley C (2011) A defense of ignorance: its value for knowers and roles in feminist and social epistemologies. Lexington Books

Tuana N (2006) The speculum of ignorance: the women's health movement and epistemologies of ignorance. Hypatia 21(3):1–19. https://doi.org/10.1111/j.15272001.2006.tb01110.x

Viglucci A, Levin J (2016) Zika's tourism scare spreads to South Beach. Miami Herald, https://www.miamiherald.com/news/health-care/article96801697.html. Accessed on 14 Ma 2021

Chapter 16
Multi-sector Situational Awareness in the COVID-19 Pandemic: The Southwest Ohio Experience

David M. Hartley, Andrew F. Beck, Michael Seid, Susan Cronin, Christine L. Schuler, Laura Raney, Muhammad Zafar, Robert Kahn, and Peter A. Margolis

Abstract The SARS-CoV-2 virus emerged and spread quickly, leading to a pandemic within weeks of the first recognized cases. In this chapter, we describe a successful effort to develop and maintain situational awareness of COVID-19 presence and effects across Southwestern Ohio. Healthcare, public health, schools, and businesses organized and worked together to characterize the threat, align on common goals, share data, and spread messages in pursuit of disease suppression. The resulting communication across sector lines was key for understanding and managing the evolving phases of the COVID-19 pandemic in Southwestern Ohio.

Keywords COVID-19 · Situational awareness · Learning system · System design · Rapid response

16.1 Background

Reports of epidemic pneumonia in Wuhan, China, began circulating in the world press in late December 2019 (Zuo et al. 2019; Huang 2020). By early January 2020, news of the outbreak was well-publicized in the US press (Gan 2020; Associated Press 2020; CBS News 2020). Cases were soon observed outside of China (Tan 2020), and the first case in the United States was reported on January 21, 2020 (McNamera 2020). Coronaviruses (CoV) commonly cause human disease, most notably severe acute respiratory syndrome (SARS) and Middle East respiratory syndrome (MERS), as well as less virulent respiratory illnesses including the common cold. This specific virus, SARS-CoV-2, caused coronavirus disease-2019 (COVID-19), including a range of symptoms that can result in death. COVID-19 affected Italy severely (Remuzzi and Remuzzi 2020) in February–March, and in March–April, the impact on New York

D. M. Hartley (✉) · A. F. Beck · M. Seid · S. Cronin · C. L. Schuler · L. Raney · R. Kahn · P. A. Margolis
Cincinnati Children's Hospital, Cincinnati, OH, USA
e-mail: david.hartley@cchmc.org

D. M. Hartley · A. F. Beck · M. Seid · C. L. Schuler · M. Zafar · R. Kahn · P. A. Margolis
University of Cincinnati College of Medicine, Cincinnati, OH, USA

D. M. Berube (ed.), *Pandemic Communication and Resilience*, Risk, Systems and Decisions, https://doi.org/10.1007/978-3-030-77344-1_16

City in the US was intense (PBS News Hour 2020). By early March, Ohio reported its first cases (US News 2020), and by the end of the month, the total number of Ohio cases had grown to over 400 (NBC4 2020). Soon cases extended into the Southwest Ohio region, which includes the Greater Cincinnati metropolitan area, home to more than two million people.

Given stresses to health infrastructure and other systems observed in Asia, Europe, and parts of North America, leaders responsible for preparing Southwest Ohio for COVID-19 needed important questions answered quickly. Hospital decision-makers asked, for example, when a patient surge could be expected, how many patients were anticipated, and whether field hospitals needed to be constructed in order to keep pace with healthcare demand. Public health leaders asked how many contact tracing personnel might be needed and how tracing procedures could be optimized. School leaders considered the ramifications of closing and requirements for reopening. Businesses needed to know how the workforce would be affected and for how long.

One approach to addressing such questions is to develop mathematical models to provide guidance. Such a strategy was undertaken (e.g., OSU COVID-19 Response Modeling Team 2020). However, without well-characterized, locally relevant, empirically validated inputs, uncertainties in the epidemiology of COVID-19 and other factors initially limited model confidence. A second approach, described here, is to gather, analyze, and share data on disease and factors related to healthcare systems, public health, and other associated sectors; collectively, this approach enables decision-makers to maintain awareness of the situation and adapt operations to meet changing conditions and trends.

16.2 Approach

The goal at the onset of our work was to develop the means to provide regional COVID-19 situational awareness (Endsley 1995; Hartley et al. 2019) to stakeholders as rapidly as possible and to improve this capability continually as time passed (Beck et al. 2021). Based on experience in developing and managing rapid learning healthcare networks (Britto et al. 2018), we applied methods for networked organizational design and quality improvement to build a situational awareness and continuous learning system. Broadly, our approach included identifying and organizing stakeholders; facilitating goal setting and planning; assessing execution and performance; and acting on what is learned to improve the system through time (Fig. 16.1). At each stage of the approach, effective communication between stakeholders was key to understanding one another's views and objectives and identifying commonalities and differences.

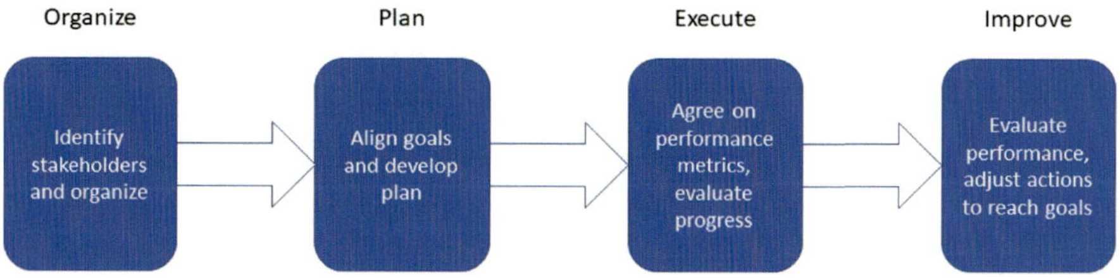

Fig. 16.1 Approach to developing cross-sector COVID-19 situational awareness

16.2.1 Identify Stakeholders and Organize

The first steps were taken in late March 2020, when a Cincinnati-based not-for-profit regional health information organization (RHIO; Adler-Milstein et al. 2009) convened a steering committee composed of senior regional hospital operational leaders representing the 22 regional hospitals. With operational decisions to be made on a near daily basis, the committee initially sought to apply multiple, sometimes conflicting, national or state data and forecasting models with unclear regional relevance. A situational awareness subcommittee, comprised of physicians, epidemiologists, data scientists, programmers, and project managers, was formed to facilitate data acquisition, integration, interpretation, and planning. Initially, both the steering and subcommittees met daily to facilitate rapid progress.

16.2.2 Align Goals and Develop a Plan

Over approximately 10 days, the situational awareness subcommittee worked with healthcare system leaders across four areas (Beck et al. 2021). First, and fundamentally, together we developed and agreed on a shared regional goal: *to suppress regional SARS-CoV-2 transmission to reduce disease burden while maintaining economic productivity*. Second, we worked to specify scope and define populations, geographies, and organizations of interest. This enabled activity to be focused on areas of highest operational and epidemiologic relevance. Third, we worked to establish a shared theory of action by identifying drivers and measures of a successful system (e.g., effective healthcare delivery, public health driven prevention, planning, coordination, and delivery of response). A key driver diagram (i.e., a visual depiction of the team's concept of important contributors to the overall aim or goal of the effort) was developed (Fig. 16.2). This diagram enabled individual stakeholders to understand which drivers they could influence, which could be influenced by other partners, and which required coordinated efforts to achieve the shared goal. It also highlighted the interconnectedness of the stakeholders and built an awareness that uncoordinated actions can have unintended consequences for partners. Fourth, we

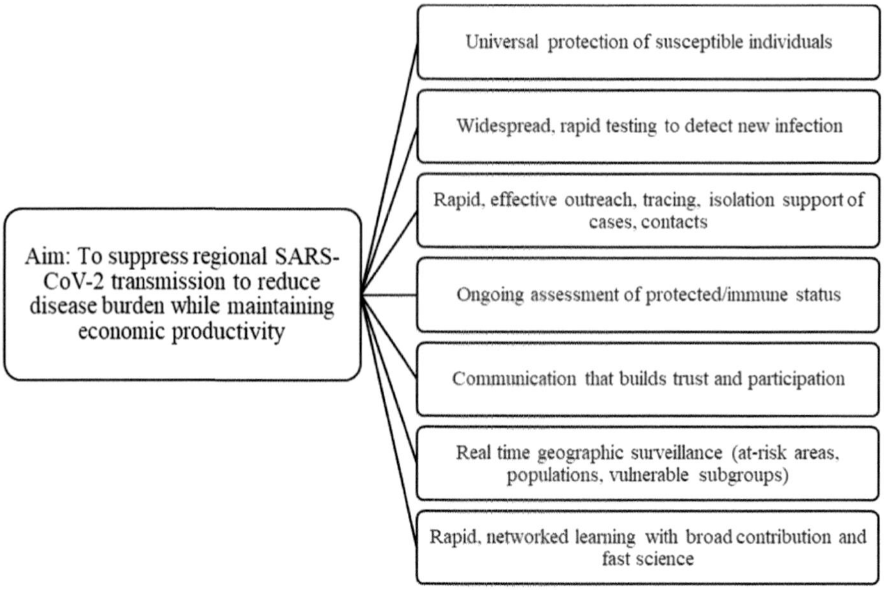

Fig. 16.2 Simplified key driver diagram for COVID-19 situational awareness system for Southwest Ohio, 2020

helped stakeholders iteratively define and test relevant measures, so that progress could be quantified and charted.

The steering committee, which met regularly, reviewed pre-existing agreements for information sharing across healthcare systems and worked with public health partners to ensure all partners understood data sharing limitations and appropriate data use. The steering committee, informed by the situational awareness subcommittee, also communicated with local and state elected officials involved in the pandemic response to understand their actions and identify possible areas of collaboration and synergy.

16.2.3 Identify and Produce Execution and Performance Metrics

The situational awareness subcommittee quickly ramped up measurement and analytic capabilities (Beck et al. 2021). Initial measurement activities began in mid-March, 2020, when SARS-CoV-2 testing was limited and laboratory turnaround times were slow (in some instances, 10 days or longer). The subcommittee, therefore, focused initially on using the most accessible data, which included hospitalizations and emergency visits for influenza-like illness, fever, and cough. We accessed, via the RHIO, hospital specific electronic health record (EHR) data and then scaled this approach across hospitals throughout the region. As testing became less scarce and laboratory capacity improved, we extended this infrastructure to observe additional relevant measures, including COVID-specific healthcare utilization, completed and positive tests for SARS-CoV-2, and time from test to result. Using geocoded address data, we created maps of testing and positive case rates at the census tract level of aggregation which, in Greater Cincinnati, align closely with commonly recognized neighborhood boundaries.

By late April, we were able to integrate additional healthcare and public health data sources to achieve more complete population health situational awareness. Daily numbers of patients with COVID-19 hospitalized, in intensive care units (ICU), and on ventilators were obtained from the Ohio Hospital Association (OHA) Resource Tracker. Numbers of occupied and available hospital beds and ventilators, across all causes, became available through an Ohio Department of Health (ODH)-operated database (SurgeNet; https://ohio.surgenet.org/Registration.aspx). Although Greater Cincinnati encompasses parts of 3 states (Ohio, Indiana, and Kentucky), OHA and SurgeNet databases are Ohio-specific. Still, most regional hospital resources of interest are within Ohio. Separately, cross-state local health departments receive and then report data on positive SARS-CoV-2 tests, COVID-19 cases, and deaths to state departments of health. National outlets like the New York Times and academic institutions like Johns Hopkins University serve as data aggregators and afforded our team electronic access to data for analysis (New York Times 2020; Johns Hopkins University 2020). Although these sources are national in scale, they make disaggregated data available at state and county levels and such data were incorporated into our analyses. Finally, we used a Google-developed depiction of county-level mobility trends to provide an approximate another measure of adherence to nonpharmaceutical interventions beyond mask wearing and hand hygiene (Badr et al. 2020; Google 2020).

We compiled measures into daily "dashboard" visualizations designed to provide a common operational picture (Department of Homeland Security 2008) of the state of the pandemic in the region (Fig. 16.3). Versions of the dashboard were distributed to stakeholders across sectors and made available to the public when possible via email and the web. Each dashboard element was carefully defined, and those receiving the data had easy access to operational definitions, sources, and analytic approaches used. Measures included in the dashboard were chosen to relate downstream effects of the pandemic (e.g., healthcare utilization, death) to disease incidence, the effective reproduction number for the region and for each included county, tests, and interventions (Beck et al. 2021). We were also able to provide assessments across vulnerable sub-populations (e.g., those residing in congregate care facilities such as nursing homes, living in impoverished neighborhoods, and of minority race or ethnicity). Additional measures assessing regional resource requirements (e.g., hospital occupancy, ventilator use, personal protective equipment (PPE) availability) were also included.

As described in Beck et al. 2021, some dashboard measures utilized statistical process control (SPC) methods to depict day-to-day variation in measures and detect important signals (Benneyan et al. 2003; Johnson et al. 2018; Thor et al. 2007). Other

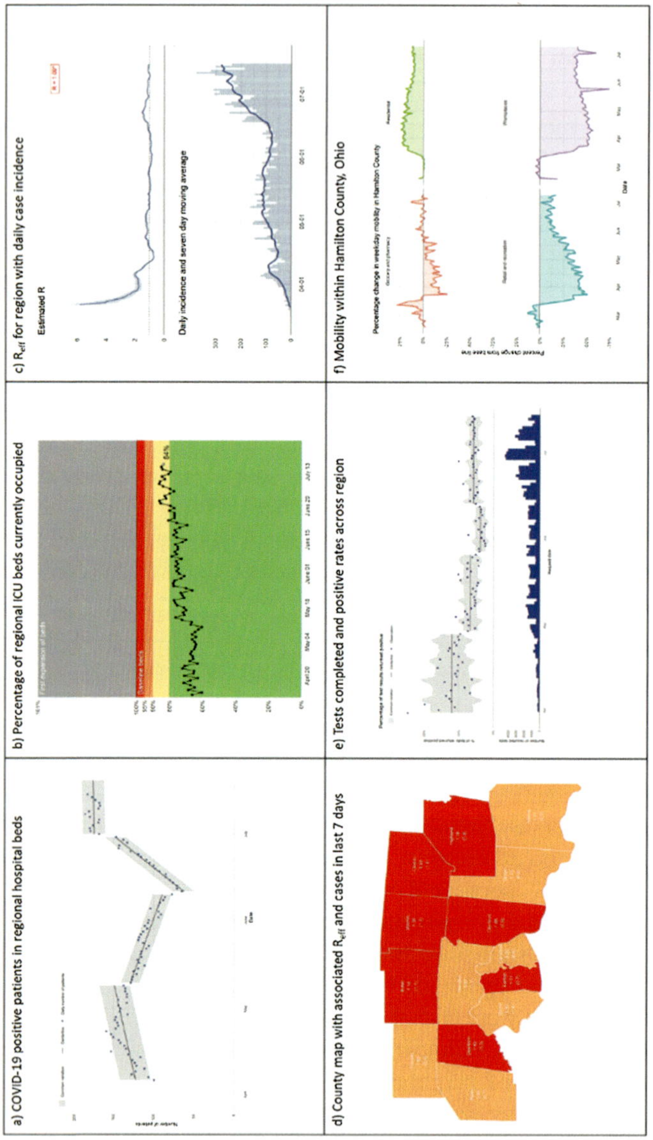

Fig. 16.3 Example of common operational picture of COVID-19 in Southwest Ohio, 2020

measures utilized geospatial analytics and geocoded data when available, to identify potential areas with elevated disease incidence in need of further investigation (Chowkwanyun and Reed 2020; Webb Hooper et al. 2020).

16.2.4 Evaluate and Adjust Activities to Improve

Evaluation and improvement strategies focused on meeting specific decision-making needs of the regional steering committee. Based on those needs, we found data and rapidly designed and prototyped measures thought to be valuable on a small scale. We assessed measure limitations and opportunities to improve data sources and gathered feedback from the committee on the prototyped measures and presentation. We then refined measures according to feedback and scaled them up. Often, this meant taking a measure from a single hospital or healthcare system to all those in the region; or from a single county to all regional 14 included counties. Measure updates were generally incorporated within 24 h. When data gaps or errors were identified, they were either mitigated or acknowledged if mitigation proved impossible. This process of rapid, stakeholder-informed iteration resulted in useful, accurate, and trustworthy measures. It built will with stakeholders, helping to identify ways to bring more individuals and sectors into the fold. As a result, the data collected, analyzed, and promulgated were used broadly to communicate awareness and support decisions across sectors.

16.3 Related Activities

The foundation provided by this work played an important role in many critical activities related to managing the pandemic in Southwest Ohio.

16.3.1 Multi Agency Coalition

Epidemics do not respect jurisdictional boundaries. From the beginning of the response, collaboration and coordination across sectors (healthcare systems, public health, congregate care) was recognized as a necessity and a key challenge. For example, limitations in the availability of personal protective equipment (PPE; e.g., masks, gowns, goggles) led to insufficient supplies in certain skilled nursing facilities, despite the fact that many cases arose there. Further complicating coordination was the number of public health jurisdictions. Many counties had multiple health departments for cities within their geographic boundaries. To address the challenge of cross-sector coordination, Ohio directed its nine public health regions to come together, forming three zones and developing Multi Agency Coalitions (MACs) comprised of representatives from the health care, public health, and congregate care sectors. Each MAC was led by a zone leader who represented regional activities to the state level. As the leadership structure evolved, our team also evolved to support situational awareness and planning for all three sectors represented on the MAC.

16.3.2 Congregate Care

As an example of support, consider congregate care (CC) facilities, which represent a diverse range of living settings with varied degrees of non-medical and medical supervision. These include skilled nursing facilities (e.g., nursing homes, assisted living, hospice, psychiatric facilities), transitional care facilities (e.g., sober living, re-entry facilities), residential facilities (e.g., group homes, senior living), shelters, and correctional facilities or jails. CC facilities are vulnerable to COVID-19 outbreaks due to communal living, interactions of residents, staff and visitors, and limited resources for pandemic response. Residents of nursing facilities are especially vulnerable due to their age, medical comorbid conditions, and underlying frailty. Facility supervision varies from constant supervision for residents in nursing facilities to nearly absent in independent living facilities and shelters. Similarly, the resources and medical training among staff is highly variable. There are over 400 congregate living facilities in Southwest Ohio.

CC facilities are designed to function independently and are not organized for collaborative regional work like that needed for COVID-19 pandemic response. To meet this challenge, a steering committee for CC coordination was formed under the MAC. Representatives from each facility type (i.e., nursing homes, assisted living, group homes, senior living, transitional facilities, shelters, and correctional facilities) were recruited to join the committee. Diverse membership was essential for understanding each context, their challenges and needs. Steering committee members were also instrumental in reaching out to other similar facilities for participation in regional response activities.

For situational awareness among CC facilities, we relied on three data sources. The first two were the publicly available data for skilled nursing and correctional facilities. Both such facilities have mandated data sharing to the Centers for Medicare and Medicaid Services (CMS) (https://www.medicare.gov/nursinghomecompare) and the Ohio Department of Rehabilitation and Correction (https://drc.ohio.gov/), respectively. These data are publicly available and are transformed into visual plots that enable learning over time by our data team. While CMS data provide situational awareness for all nursing homes in the region, the data lag by a two-week period, which reduces its utility for time-sensitive actions such as identification of and response to an outbreak or critical shortage of PPE.

We developed a data acquisition mechanism for real-time situational awareness of all CC facilities in the region. Variables included were selected by consensus with each category of CC facility to ensure that data acquisition required minimal additional effort and variables are meaningful for the facility and region. This data registry is now collected and maintained in REDCap, a secure web application for building and managing online surveys and databases (https://www.project-redcap.org/). The two-step process entailed initial signup and then weekly surveys. At initial signup, details of the facility were obtained. Weekly surveys are sent to all facilities for current data on disease burden among residents and staff and adequacy of PPE supplies.

This work with the CC sector involved engaging diverse, independently functioning facilities into a collaborative effort. Participation was voluntary and not required by any supervising authority. Thus, engagement and participation relied on perceived value by the leadership of each facility. Such facilities were often challenged by workload, staffing, and PPE adequacy; entering data into a weekly survey was an added requirement to an already stressed system. Nonetheless, over time, we had increasing numbers of facilities submitting data weekly. Through real-time dashboards, the CC steering committee provided guidance and direction to facilities with outbreaks and those with critical PPE shortages. The CC dashboard was updated weekly and shared with the CC steering committee. De-identified data were shared with committee work groups for different facility types, as well as with the MAC.

This effort also spawned the development and dissemination of standardized responses to outbreaks, PPE shortages, and testing requests. In addition, these data supported townhalls and educational seminars for residential facilities, shelters, assisted living facilities, and nursing homes.

16.3.3 Public Schools

By July 2020, there was increasing national attention to the impact of COVID-19 on the upcoming school year. An array of potential sources of guidance emerged. The CDC produced early measures and guidance on school reopening, but it was limited in specificity and was criticized as politically influenced. A state dashboard with a color-coded warning system was developed (Ohio Public Health Advisory System 2020), and while adopted by many districts, the measures were a complex bundle and not tailored to inform school opening. Several initiatives, including the Harvard Global Health Institute (Linville-Engle et al. 2020) and researchers at the Johns Hopkins University Center for Health Security (Cicero et al. 2020), produced more specific guidance grounded in emerging COVID-19 measures. The lack of clear federal and state guidance, the dizzying array of potential approaches, and the unfamiliarity of school district administrators with infectious disease epidemiology created a chasm of uncertainty regarding how best to approach school opening.

In late July, members of our situational awareness subcommittee met with leaders from two local health departments to review existing guidance and align our priorities. We focused on the Harvard Global Health Institute guidance and adapted published reports into a simpler guidance document. The document offered a brief background, a parsimonious set of lead measures, and a decision algorithm, along with clear branding of the team that had published the document. The guidance was hosted on a local health department website, with data for the lead measures (drawing from the regional dashboard) updated weekly. Data privacy issues with one measure led to engagement of the RHIO, which took on the role of final distribution to the public health departments.

Importantly, while early guidance was based on theory, school districts accumulated significant actual COVID-19 experience in the fall. The impact of community incidence was mitigated by protective measures in schools. Direct measures of case incidence in schools, including in-school and after-school transmission, and the student and staff absences due to quarantine, became primary considerations over community incidence. Consistent with original intent, we revised the guidance based on emerging experience and evidence.

16.3.4 Public Communication and Messaging

Based in part on the situational awareness provided by the dashboards, the need for public messaging about the COVID-19 pandemic emerged simultaneously with other aspects of the community response. This became especially necessary with a significant increase in community incidence of COVID-19 beginning in late June. Community stakeholders from various sectors, led by a business community task force and the Cincinnati Regional Chamber of Commerce, created the Regional COVID Communications Center (RC3) to provide non-partisan, fact-based, culturally competent, and equitable information about COVID-19 prevention, community spread, testing availability, and other critical news. Leaders recruited a communications consultant with expertise in public relations and communications strategy to run RC3, which was designed to serve as the hub through which pandemic-related messaging would flow, including via key partnerships with hospitals, medical professionals, the business community, and key grassroots and community leaders.

One example of such collaborative communications efforts launched in advance of the July 4th holiday weekend, with RC3 partnering with a local Fortune 100 consumer goods company to develop a #MasksOn campaign. Through this partnership, a simple, clear messaging campaign encouraged Ohioans: *when we're out, masks on.* Signage was paired with bold black and white graphics. Variations on this message with similar color schemes and graphics were created to complement the primary message such as: "Shades on. Mask on.", "Grills lit. Mask on.", "Cups full. Masks on." Prior to a local or state-wide mask mandate, these messages were intended to maximize relevance to the everyday lives of community members and added variety to the messaging while maintaining consistency.

Campaign materials were displayed on billboards and spread via multiple social media sites, radio, and television outlets. Local companies and organizations were encouraged to share and adapt the materials for use with their employees and customers. In addition, a toolkit was created and made accessible to the public online, including materials that could be printed and used on social media platforms. This signage quickly became recognizable across the region and remained well-received given the consistent, simple branding and intentional alignment with messaging from the state level.

Collaboration with researchers affiliated with a local university and academic health center was another core pillar of our pandemic communications strategy. The local National Institutes of Health-funded Clinical and Translational Service Award (CTSA) research hub sponsored a COVID-19 Critical Community Challenge Grant program to support community-based pandemic response efforts. Many funded projects addressed communications, especially messaging to key sub-populations. These researchers were quickly identified as a rich resource for informing regional communications efforts, as they were identifying barriers to COVID-19 mitigation strategies in real time within high-risk populations. Drawing from their expertise and new learnings was key to addressing the nuances of each individual sub-population and ensuring that subsets of the population were not neglected in messaging efforts. Creating relationships and fostering ongoing, open dialogue between communications professionals and these researchers remained a priority as the pandemic waxed and waned and fine-tuning of messages proved necessary.

Beyond local partnerships, external connections were made, and relationships proactively established, with professionals outside our own community. Specifically, we connected with public health professionals in other communities at the city, county, and state level, experts at other academic institutions, and leaders in analogous public health learning collaboratives. Using directed online searches related to communications and other aspects of a comprehensive pandemic response (e.g., testing strategy, contact tracing), our team identified communities using novel, well-developed tools and strategies across the country. This approach was balanced by connecting with public health officials in a neighboring county comparable to our own. Directly connecting with outside experts provided insights about multiple aspects of messaging efforts: funding sources, messaging goals, measures to quantify success, and nuances of various messaging strategies. Learnings from these outside resources facilitated rapid, critical appraisal of our own strategy and identified gaps in our messaging.

16.4 Discussion

We used network organizational design principles and rapid system design and improvement methods to facilitate development of an actor-oriented information infrastructure (Fjeldstad et al. 2020). The resulting infrastructure enabled operational leaders to develop a shared understanding of the interdependent components of the pandemic response system they were simultaneously both a part of and seeking to manage. The focus on cross-sector collaboration in pursuit of shared goals promoted efficient communication within and between groups. As a result, institutions that had not previously interacted with one another began sharing data and coordinating actions in efforts to reduce COVID-19 burden and save lives.

The RHIO, comprised of a coalition of hospitals and healthcare systems that share data and collaborate on planning projects including emergency preparedness, provided essential infrastructure and capability for aggregation of clinical data. Relevant data were identified by users and team members with expertise in epidemiology, medicine, and data science, and visualized on a series of dashboards that formed a common operational picture of the pandemic in the region. The resulting situational awareness allowed users to perceive the system, follow it through time, comprehend the importance of changes and trends, and project the implications of changes for action (Endsley 1995). The awareness infrastructure has operated continuously from early April through the time of writing in late 2020 and provided a foundation for important actions and activities across sectors during the pandemic.

The result was efficient and widespread communication based on near real-time data. The emergence of the Regional COVID Communications Center in response to an awareness of a summertime COVID-19 surge demonstrates that such information can support public health messaging needed to influence human behavior (in this case, mask wearing and other nonpharmaceutical interventions (Hartley and Perencevich 2020)). In addition to the regional messaging campaign, local news media monitored the public facing dashboard and interviewed members of the local medical community several times in the summer and fall when dashboards illustrated intensified disease activity in the region.

Resilience is the ability to prepare for and adapt to changing conditions, including natural threats such as pandemics (White House 2013). Resilient infrastructure must be robust and adaptable to circumstances that may change over time or emerge abruptly. Much public health infrastructure in the United States is not resilient; it is often old and not able to meet requirements placed on it by the COVID-19 pandemic. In the State of Ohio, for example, epidemic-related information is reported to and stored in the Ohio Disease Reporting System, a system now in its second decade. Technical glitches inherent to this system led to delayed reporting of COVID-19 data in the state at least once in the fall, affecting situational awareness at a time when incidence was increasing exponentially throughout the state (Borchardt 2020). The regional capability described here was developed rapidly and responded dynamically to user requirements. Because it can easily be adapted to new circumstances, it possesses a degree of resiliency and redundancy over existing systems.

Around the world, there are hundreds of local responses to COVID-19 like the one described in this chapter. Such initiatives have the potential to accelerate how we learn about designing more flexible and adaptive systems to both respond to epidemic or pandemic threats and improve population health more generally. We believe, for example, that there is merit in building capability that will be relevant to epidemics of social inequities in health, mental illness, opioids, and even sports injuries. By harnessing insights and innovations of adaptive local responses, regions can improve the health of population broadly.

References

Adler-Milstein J, Bates DW, Jha AK (2009) U.S. Regional health information organizations: progress and challenges. Health Aff (Millwood) 28(2):483–492. https://doi.org/10.1377/hlthaff.28.2.483

Associated Press. Mystery bug behind China's pneumonia outbreak identified: what to know. Published 10:31am EST, 9 January 2020 https://www.nbcnews.com/health/health-news/mystery-bug-behind-china-s-pneumonia-outbreak-identified-what-know-n1112891

Badr HS, Du H, Marshall M, Dong E, Squire MM, Gardner LM (2020) Association between mobility patterns and COVID-19 transmission in the USA: a mathematical modelling study. Lancet Infect Dis 20(11):1247–1254

Beck AF, Hartley DM, Seid M, Cronin S, Schuler CL, Raney L, Zafar M, Kahn R, Margolis PA (2021 Apr) Rapid, bottom-up design of a regional learning health system in response to COVID-19. Mayo Clin Proc 96(4):849–855

Benneyan JC, Lloyd RC, Plsek PE (2003) Statistical process control as a tool for research and healthcare improvement. Qual Saf Health Care 12(6):458–464

Borchardt J Ohio daily COVID-19 numbers delayed hours due to 'technical glitch,' 'overburdened' health departments. Cincinnati Enquirer November 18, 2020. Available online at https://www.cincinnati.com/story/news/2020/11/18/ohio-daily-covid-19-numbers-delayed-hours-due-technical-glitch/3770422001/. Accessed 25 Nov 2020

Britto MT, Fuller SC, Kaplan HC, Kotagal U, Lannon C, Margolis PA, Muething SE, Schoettker PJ, Seid M (2018) Using a network organisational architecture to support the development of learning healthcare systems. BMJ Qual Saf 27(11):937–946

CBS News Wuhan pneumonia outbreak in China blamed on new SARS-type virus. Published 9:39 am 9 January 2020 https://www.cbsnews.com/news/wuhan-pneumonia-outbreak-china-mystery-illness-blamed-new-sars-type-virus-today-2019-01-09/

Chowkwanyun M, Reed AL Jr (2020 Jul 16) Racial health disparities and Covid-19—caution and context. N Engl J Med 383(3):201–203

Cicero A, Potter C, Sell TK, Rivers C, Schoch-Spana M Filling in the blanks: national research needs to guide decisions about reopening schools in the United States. Johns Hopkins University. Available online at https://www.centerforhealthsecurity.org/our-work/publications/filling-in-the-blanks-national-research-needs-to-guide-decisions-about-reopening-schools-in-the-united-states. Accessed 25 Nov 2020

Department of Homeland Security (2008) Common operating picture for emergency responders, in system assessment and validation for emergency responders highlight. US Department of Homeland Security 2008. Available online at https://www.dhs.gov/sites/default/files/publications/CommonOpER_HLT_0908-508.pdf. Accessed 25 Nov 2020

Endsley MR (1995) Toward a theory of situation awareness in dynamic systems. Hum Factors 37(1):32–64

Fjeldstad OD, Johnson JK, Margolis PA, Seid M, Hoglund P, Batalden PB (2020) Networked health care: Rethinking value creation in learning health care systems. Learn Health Syst 4(2):e10212

Gan N A mysterious virus is making China (and the rest of Asia) nervous. It's not SARS, so what is it? CNN. Published 2:18am ET, 7 January 2020 https://www.cnn.com/2020/01/06/health/china-pneumonia-intl-hnk/index.html

Google. Global Mobility Report. Available online at https://www.google.com/covid19/mobility/. Accessed 25 Nov 2020

Hartley DM, Mui WL, Corley CD (2019) The role of event-based biosurveillance in biodefense. In: Singh SK, Kuhn JH (eds) Defense against biological attacks. Springer Nature Switzerland AG

Hartley DM, Perencevich EN (2020) Public health interventions for COVID-19: emerging evidence and implications for an evolving public health crisis. JAMA 323(19):1908–1909

Huang K World Health Organisation in touch with Beijing after mystery viral pneumonia outbreak. South China Morning Post. Published: 4:59pm, 1 January 2020 https://www.scmp.com/news/china/politics/article/3044207/china-shuts-seafood-market-linked-mystery-viral-pneumonia

Johns Hopkins University (2020) COVID-19 Data Repository by the Center for Systems Science and Engineering (CSSE) at Johns Hopkins University. https://github.com/CSSEGISandData/COVID-19. Accessed 25 May 2020

Johnson DP, Patterson BL (2018) The natural order of time: the power of statistical process control in quality improvement reporting. Hosp Pediatr 8(10):660–662

Linville-Engler B, Allen D, Siddarth D, Friedhoff S, Sumpter M, Kunau R Pandemic resilience: getting it done. Harvard Global Health Institute. Available online at https://ethics.harvard.edu/files/center-for-ethics/files/ttsi_technical_advice_handbook_2.0_june_30_2020_final.pdf. Accessed 25 Nov 2020

McNamara A CDC confirms first case of coronavirus in the United States. CBS NEWs. Published 6:01 pm 21 January 2020 https://www.cbsnews.com/news/coronavirus-centers-for-disease-control-first-case-united-states/

NBC4. Coronavirus in Ohio Monday update: 442 cases reported, 6 deaths. Published 11:25am EDT 23 March 2020 https://www.nbc4i.com/community/health/coronavirus/coronavirus-in-ohio-monday-update-gov-dewine-dr-amy-acton-to-hold-230-briefing/

New York Times (2020) Coronavirus in the U.S.: latest map and case count. https://www.nytimes.com/interactive/2020/us/coronavirus-us-cases.html. Accessed 25 May 2020

Ohio Public Health Advisory System (2020) Ohio Public Health Advisory System. Available online at https://coronavirus.ohio.gov/wps/portal/gov/covid-19/public-health-advisory-system. Accessed 25 Nov 2020

PBS News Hour. Surge in deaths overwhelms New York's morgues, hospitals. Published 6:48pm EDT 31 March 2020 https://www.pbs.org/newshour/health/surge-in-deaths-overwhelms-new-yorks-morgues-hospitals

Remuzzi A, Remuzzi G (2020) COVID-19 and Italy: what next? Lancet 395(10231):1225–1228. https://doi.org/10.1016/S0140-6736(20)30627-9. https://www.sciencedirect.com/science/article/pii/S0140673620306279?via%3Dihub

Tan SW Thai officials confirm first case of coronavirus outside China. The Washington Times Published 13 January 2020 https://www.washingtontimes.com/news/2020/jan/13/thailand-officials-confirm-first-case-coronavirus

Thor J, Lundberg J, Ask J et al (2007) Application of statistical process control in healthcare improvement: systematic review. Qual Saf Health Care 16(5):387–399

US News and World Report. Ohio confirms first coronavirus cases, continues testing. Published 9:34 am March 10, 2020 https://www.usnews.com/news/best-states/ohio/articles/2020-03-10/ohio-confirms-first-coronavirus-cases-continues-testing

Webb Hooper M, Nápoles AM, Pérez-Stable EJ (2020) COVID-19 and racial/ethnic disparities. JAMA 323(24):2466–2467

White House. Presidential policy directive—critical infrastructure security and resilience. PRESIDENTIAL POLICY DIRECTIVE/PPD-21, February 13, 2013. Available online at https://obamawhitehouse.archives.gov/the-press-office/2013/02/12/presidential-policy-directive-critical-infrastructure-security-and-resil. Accessed 25 Nov 2020

Zuo M, Cheng L, Yan A, Yau C Hong Kong takes emergency measures as mystery 'pneumonia' infects dozens in China's Wuhan city. South China Morning Post. Published: 2:35 pm, 31 December 2019 https://www.scmp.com/news/china/politics/article/3044050/mystery-illness-hits-chinas-wuhan-city-nearly-30-hospitalised

Chapter 17
Coping and Resilience: Reframing What It Means to Have a Good Pregnancy During COVID-19

Ashley Archiopoli

Abstract Ahead of every life-stage, individuals develop a set of expectations for how they imagine the period of life to progress; among these is the value-laden and tenuous stage of pregnancy and welcoming a new baby. As I anticipated on my own foray into parenthood, my set of expectations included prescriptions for how I intended to care for my physical, mental, relational, and social health throughout pregnancy, in short, my vision of a good pregnancy. In January 2020, I found out that I was pregnant with my first child; at the same time, the first official case of COVID-19 was diagnosed in the United States. While it was not evident that January, COVID-19, and its many implications disrupted my notions of how my pregnancy would proceed. This chapter applies the theory of problematic integration to the experience of pregnancy during the COVID-19 global pandemic using a narrative (re)telling examining how problematic integrations were created and managed through communicative practices.

Keywords COVID-19 · Pregnancy · Autoethnography · Narrative · Problematic integration · Resilience

17.1 Introduction

On day five of my newborn's life, I sat on my couch covered in pee, poop, and breastmilk, trying to console my son as we attempted—yet again—to properly latch. While this scene in itself is frustrating, I found comfort in the normalcy of it all. I expected there would be sleepless nights and new challenges in life with a newborn. I knew that it would take time to adjust to one another, him to living in the world, and me in my new role as a mother. This unglamourous moment was an experience for which I had mentally prepared for and expected. It was a welcome reprieve from the previous seven months of fear, grief, and anger created by the all-consuming cultural context of the COVID-19 global pandemic.

A. Archiopoli (✉)
University of Houston-Downtown, Houston, USA
e-mail: archiopolia@uhd.edu

D. M. Berube (ed.), *Pandemic Communication and Resilience*, Risk, Systems and Decisions, https://doi.org/10.1007/978-3-030-77344-1_17

Ahead of every life-stage, individuals develop a set of expectations for how they imagine the period of life to progress; among these is the value-laden and tenuous stage of pregnancy. As I anticipated my foray into motherhood, my set of expectations included prescriptions for how I intended to care for my physical, mental, relational, and social health. In alignment with my partner, we began trying to conceive in late 2019. In January 2020, I found out that I was pregnant with my first child; simultaneously, the first case of COVID-19 was diagnosed in the United States. While it was not evident that January, COVID-19 and its many implications would soon disrupt and challenge my notions of a good pregnancy. This chapter applies the theory of problematic integration to the experience of pregnancy during the COVID-19 global pandemic using a narrative (re)telling examining how problematic integrations were created and managed through communicative practices.

17.2 Theoretical Grounding

This work is grounded in the discipline of communication; everyday activities such as self-talk, discussions with family and friends, giving lectures, information seeking, and the act of entertaining or being entertained constitute the collective that is communication (Carey 1989). Carey (1989) advanced a cultural approach to communication which defines communication as "a symbolic process whereby reality is produced, maintained, repaired, and transformed" (p. 23); central to this definition is the symbolic process. This foundational series of concepts known as symbolic interactionism posit that meaning is created through interactions with the world around us; humans work collaboratively to assign meaning to people, places, and objects through our use of language (Mead 1934). In short, meaning is found in our engagement with others, with ourselves, and with our environments. Then the study of communication investigates the everyday worlds we create and live in (Pearce and Cronen 1980). One such way to investigate lived experiences is through the application of narrative theory. The following section takes a look a narrative theory as a framework for creating meaning and ties it with the theory of problematic integration.

17.3 Narrative Paradigm Theory and Problematic Integration (PI)

Fisher (1984) defined narration as "words or deeds that have sequence and meaning for those who live, create, or interpret them" (p. 2). Narrative is an all-encompassing term that refers to the act of telling stories. The term narrative represents a wide range of forms. To be considered a narrative an artifact must convey meaning through a sequence of events with a beginning, middle, and end; the conclusion has some outcome or significance, and it is conveyed in a memorable or engaging fashion

(Browning 2009). Then, stories are discourse that aid in understanding personal experience and the experiences of others (Fisher 1984). The narrative paradigm offers an alternative to the traditional rational paradigm wherein narrative is the basis for understanding, judgment, and action (Fisher 1984). In this paradigm shift, we draw upon narrative coherence and fidelity to form narrative rationality. Narrative coherence is the degree to which a story makes sense, the probability of the story; interrogation of coherence considers if the story makes sense to the hearer and if it aligns with other similar stories. Then narrative fidelity is the degree to which it feels true, meaning the story is backed by good reasons that resonate with the audience (Fisher 1984).

Similar to narrative paradigm theory is the theory of problematic integration (Babrow 1992, 2001, 2009; Babrow, Kline, and Rawlins 2005) provides a framework for making sense of the world around us. Foundational to the theory are the two widely held orientations toward the world: probabilistic and evaluative. The probabilistic orientation explores the association, expectation, or likelihood of a particular outcome. In general, this is how an experience is expected to unfold. Then the evaluative orientation assesses the value assigned to the expected outcome (Babrow 1992, 2001, 2009).

The theory of problematic integration advances three basic claims. (1) First, the probabilistic and evaluative orientations we make are interrelated; we make sense of the world at the intersection of the two. (2) Integrating probabilistic and evaluative orientations may become problematic. These problems may take different forms, such as when probabilities and values diverge, ambiguity or uncertainty of the situation, ambivalence toward the outcome, or impossibility of the desired outcome (Babrow 1992, 2001, 2009). No matter the form of problematic integration, probabilities and values destabilize each other (Babrow 2001), and one problematic form can transform into another (Babrow 2001). (3) Finally, communication is integral to the creation, management, and transformation of problematic integration.

Babrow et al. (2005) synthesized narrative and PI as complementary approaches to theory and research; they find that the two work together to make-sense of experiences in relation to the nature of being, context, and time. Meaning is found in the configurations of contingencies—in the stories that we live and tell (Babrow et al. 2005). The authors theorize that narrative and PI work together and create meaning in the tensions between the sequencing of events and the exigencies we created. In the following section, I provide a chronological narrative of my pregnancy followed by an analytical (re)telling in which I explore how problematic integrations were constructed and managed and the resulting resilience I found in myself.

17.4 Constructing the Narrative

Russell and Babrow (2011) stated of narrative and PI theory, "PI theory reminds us that stories are often fundamentally provisional, subject to revision, or open-ended. In short, life experience regularly presents us with challenging meanings, when we must make sense of, reflect on, reaffirm, or perhaps reform narrative integrations of causal and moral order" (p. 242). The narratives told here are used to examine the complexities of making-sense, to explore the problems found in integrating probabilistic and evaluative orientations, and further analyze how the problematic integrations were managed. Note, I use the past tense in this analysis for clarity, but as I type these words, the story continues to be (re)written as COVID-19 continues to contextualize our realities, and new information and challenges arise daily.

17.4.1 Frames of Reference

This section offers my narrative using three frames of reference: the expected story, the story lived, and the story in reflection. The expected, or desired, story represents the original probabilistic orientation, the way I anticipated my prenatal period to unfold. From there, I move into the lived account, which is a chronological telling of my prenatal period. Then, during the story in reflection, I apply aspects of PI in order to make sense of the narrative.

17.5 The Expected Story

The construction of this narrative began with writing the story of parenthood with my husband and partner, he as a father and me as a mother. Prior to conceiving, my partner and I worked together to create our narrative vision of what it means to be a parent and who we want to be as parents. My partner and I discussed our parenting styles and the ways we planned to integrate our children into our lives, balancing our own interests, work, and social lives. We discussed how we intended to share the responsibilities of caring for our child through a weekly or daily negotiation in which we prioritize each other's schedules. We discussed the importance of exposing our child to new experiences and imagined including our baby in social outings.

Once this story was well established, I began to write the story of my pregnancy, expressing my expectations and desires, what I deemed to be a good pregnancy. I envisioned using the time to prepare physically, mentally, and socially in various ways. I imagined attending obstetrician appointments with my partner, tracking the milestones of pregnancy together, pampering myself, practicing prenatal yoga, meeting other new moms, and spending time getting ahead on my work. I imagined my ideal pregnancy setting up my new phase of life: motherhood.

Our aforementioned notions were constructed through communicative acts in our relationships with friends and family members; we developed our understanding of what it means to be pregnant or a parent. In terms of problematic integration, we developed our probabilistic orientation, our anticipation of the life event, and our formed expectations, desires, and wishes. Further, we evaluated pregnancy and parenthood as a value-laden event and placed high importance on the pregnancy and our role as parents. The two appeared to be perfectly integrated, feeling secure in our shared narrative, we began trying to conceive. Two months later, we found out that I was pregnant—with an expected due date in early October 2020. At my first prenatal appointment, the pregnancy was confirmed with a heartbeat, and we received three sonogram photos that we promptly placed on our refrigerator. In this provisional story, our desires were fulfilled, and we began laying out additional expectations as we planned for the duration of pregnancy. Then, two months into my pregnancy, our perfectly formed integration was disrupted by the uncertainty associated with COVID-19.

17.6 The Lived Story

While COVID-19 was certainly in the news as I was writing my expected story, it had not yet entered my orbit. In the early months of the pregnancy, my perceived risk of COVID-19 was low due to factors such as being younger, in good health, and no diagnosed cases in my region of the country. My low level of concern was further characterized by the New Orleans trip I took with my partner in early February 2020. We spent the days strolling the crowded streets of the French Quarter, ate in packed restaurants, and waited among seas of people to catch a glimpse at Mardi Gras parades. Instead, my attention and energy were placed on monitoring the standard risk that accompanies pregnancy. In the early weeks, I took daily pregnancy tests, took extra care to avoid trauma to the body, and began optimizing my diet for pregnancy by eliminating alcohol, caffeine, deli meat, raw fish, and the like.

Toward the end of my first trimester in March 2020, my university, my state, and the US began shutdowns in order to mitigate the spread of the virus. My obstetrician's office also began to take precautions. My second appointment with my obstetrician was scheduled for March 19, 2020. Ahead of the appointment I was informed that my partner could no longer attend in order to limit exposure. Further, I was required to go through a screening upon entering the building. In fact, I was screened multiple times prior to my appointment: via phone call before arriving, when I entered the building, when I entered my obstetrician's practice, and when I reached the exam room.

Sitting in the waiting room, I snapped a photo of myself wearing my entrance tag that included the date and the floor of the building I was visiting. I sent the picture to my sister, who was also pregnant, just a few months ahead of me, and we discussed our concerns. We feared things such as contracting the virus and impacting our unborn children or laboring alone, as some news reports indicated was happening at the time in New York City. In the exam room during my fourth COVID-19 screening of the day, tears began streaming down my face as I felt the entire experience of my pregnancy shift. My first appointment was filled with smiles, joy, and excitement. In this new world, smiles were obscured by masks; the excitement was replaced with procedure.

In my meeting with my doctor, I asked about the precautions I should take to keep myself safe. She instructed me to work from home, to avoid crowds in places such as the grocery store and to get groceries delivered if possible, to avoid to-go food because we did not know how the virus spread, and packaging was questionable. Despite all of these limitations, she instructed me to take care of my mental health by getting outside to exercise daily.

It is on this day that COVID-19 became tangible, and the threat was real; yet, at the same time, the uncertainty that surrounded it made it intangible. While the collective working knowledge of COVID-19 was updated daily, data about the impact of COVID-19 on pregnant women were conflicting and incomplete. In an effort to be as informed as possible, I watched 24-h news channels, refreshed COVID-19 case counters and local public health websites, engaged with social media communities, read thought pieces about being pregnant during a pandemic, and dissected academic findings on pregnancy and COVID-19 with other pregnant women in my life. The uncertainty and my communicative interactions fostered fear. In order to manage the unknown, I lived my life in a way that minimized risk and operated only within the boundaries I deemed to be comfortable and safe. In the early days of the pandemic, I avoided all risk. The days passed with a similar cadence, wake-up, handle work tasks as needed, 30-min walk in the afternoon, watch television until bed. Repeat.

17.6.1 April—June 2020

In April, my prenatal appointment was moved to a telemedicine appointment. This consisted of a 10-min phone call in which I was instructed to continue avoiding others but to get exercise and sunshine if possible. It was a disappointing and uninspiring interaction. I was somewhere in between; there was no real medical reason to be seen and no ultrasounds necessary. It made sense that my appointment was virtual, but this began to further dull my excitement and exacerbate my frustrations.

The rest of April 2020 feels like a blur, and it bled into May 2020. Most of May proceeded the same way. However, I distinctly remember the day of my May prenatal appointment. I was ecstatic to have what I deemed a good reason to leave the house. That morning I woke and took a shower with purpose; in that shower, I danced with excitement for the day's outing. When I arrived at my appointment, the same somber tone from the appointment in March remained, but this time I co-signed the necessity of screening and procedure to ensure the safety of the health care providers and their patients. Due to scheduling conflicts, my appointment was rushed, and I was given only a few short minutes with my health care team. I cried again. I was disappointed, frustrated and overall felt dejected. I was upset that my appointment was so quick, but mostly because this was the one activity that I deemed as a worthy event to see someone else in-person, and it did not proceed as I imagined. Not that I anticipated to be greeted like a best friend, or did I? This was a clear sign that the isolation was wearing on my mental health.

The following week I took myself to my 20-week anatomy scan, which is a major scan where the health of the child is confirmed, and many learn the sex of their child. Fortunately, my partner and I learned the sex together over the phone due to an elected blood draw and screen in the first trimester. When I was settled in the exam room I asked if I could video call my partner but was told it was against policy—one not related to COVID-19—but nonetheless, it was another moment that I could not share with my partner. So, I sat in the room by myself as the technician scanned my belly and talked me through what I saw on the screen. Yet another lackluster moment contextualized by COVID-19.

As the months wore on, I pushed boundaries as my burnout and feelings of isolation grew. I began to evaluate every activity in a risk versus reward framework. I asked myself if taking part in a particular activity was worth the risk. Changes happened gradually; I began to integrate activities that I deemed as low risk or worthy of the risk. For example, ordering to-go food, grocery shopping in a low-traffic market, and meeting trusted family and friends in outdoor spaces. In June, I got a pedicure. Beforehand I carefully reviewed the salon's safety precautions and purchased KN95 masks for the occasion. My heartbeat was rapid as I entered the building, and my mind raced with all of the negative outcomes throughout the entire pedicure, but in the end, I was relaxed and happy. I needed the break.

17.6.2 July—September 2020

July proceeded much like the months that preceded it; however, this month included one major milestone, which was a visit to see my sister and her new baby, a 12-h car-ride-away. Two-weeks after my niece was born, my sister and I discussed at length the pros and cons of the visit. We debated the risks for her as the mother of a newborn and for me traveling by car, during COVID, while pregnant. We oscillated on the trip several times. First agreeing that the trip should be canceled due to risk, then hours later deciding that it is important for both of us to see each other. We analyzed the decision endlessly for days over the phone. I finally decided that I would speak with my obstetrician at my upcoming appointment about the trip in order to make my final decision. She was enthusiastic and told me that it would be good for my mental health. Then with approval and reassurance from my obstetrician and negative COVID-19 tests, my partner and I were off to Kansas City.

The trip exposed me to how different areas of the country were handling the pandemic with varying levels of seriousness. For example, Oklahoma did not have a mask mandate at the time, which made me anxious at every stop in the state along the way. I declined a door being held open for me by a mask-less assistant; and I shuddered as a mask-less woman coughed as she passed me in the bathroom. We arrived at my sister's house a few hours later, but the scene in the bathroom replayed in my head. Had I been exposed? I took a shower upon entering their house and told my sister about the bathroom incident. We immediately rationalized and determined that this did not count as an exposure. It was a splendid few days together; I was happy to be able to care for my sister, who, up to that point, had no visitors due to COVID-19 restrictions. She cried when we left, and so did I. The trip brought me joy, and it helped to ease my isolation. But, in the corners of my mind, I continued to replay the bathroom incident and counted down the two weeks we would need to wait until we could confirm that my partner and I did not expose my sister, brother-in-law, and their new baby. On day 13, my sister reported a fever—a COVID-19 symptom. I was distraught at the thought that I might have exposed my sister. She, too, was distraught and wore a mask around her newborn until she was able to seek medical care. They were able to evaluate her and determine that she was experiencing fever due to postpartum complications, which was a relief, as the issue was resolved quickly. This also sparked internal conflict, being that the postpartum complication was the favorable outcome.

Toward the end of August and the beginning of September, I made the decision to treat myself to a series of outings that I would not do once my child arrived. This is where I really began to push on the walls that confined me. In the span of one week, I had my hair colored and cut, got another pedicure, and celebrated my pregnancy with friends at a physically distanced baby shower. The COVID-friendly event took place outdoors, and chairs were spaced at least six-feet apart. We used a karaoke machine to play games and chat with each other. It was really a wonderful night, contextualized by 2020. Yet, I was so uncomfortable around my friends, people that I trusted and that I share reciprocal care because of the unknown that characterizes COVID-19. The entire event in my honor, I was on guard. I did not want to get too close to my friends because I did not know for certain they did not carry COVID-19. I felt guilty and ashamed that I treated my friends like this but also relieved when the party was over, and I could retreat to my controlled space.

Around the same time, my partner and I began taking every parenting class offered by our hospital. Because he was not allowed at any prenatal appointments or ultrasounds this was a way for us to bond and prepare for our baby. We felt the crush of COVID-19 in more unique ways as we slogged through our series of courses. Our virtual hospital tour was a voiced-over PowerPoint; we took an online birth class, an online CPR and safety class, an online breastfeeding class, online sleeping and schedule class. Most memorable was when we were instructed to have a virtual car seat check. At the crux of irritation and humor, my partner and I joked, will the birth also be virtual? It wasn't.

On October 3, 2020, we welcomed our son into the world. I felt a sense of relief when he arrived. I was tested for COVID-19 upon my admission into the hospital, and a few hours later, I was informed that my results were negative. Thus, I was allowed to labor unmasked. Everyone in the room remained wearing masks, including my partner. After the birth, we were transferred to a mother and baby room, and my fear of COVID-19 faded into newborn bliss. However, I was promptly reminded that I welcomed my son amidst a pandemic when a nurse came to run a diagnostic on my son. As she pricked his foot, I asked if I should stand by him for comfort, and her reply was, "No ma'am, we are still in COVID." This simple statement shocked me back into the present.

17.7 The Story in Reflection

In this section, I spend time making-sense of my narrative; the story expected contrasted with the lived story. In terms of problematic integration, at the outset of my pregnancy my probabilistic and evaluative orientations were in perfect alignment. I placed a high value on the transitional period of life, and I planned to prioritize my needs to ensure my ideal pregnancy. However, I did not anticipate the context of COVID-19. More so, my experience was dominated by my internalization and subsequent enaction of the dominant public health narratives: "fourteen days to slow the spread," "flatten the curve," "stay six feet apart," "safer at home," "outside is safer than inside."

In the COVID-19 landscape, I still held to my probabilistic notion of an ideal pregnancy. However, new evaluations were introduced: (1) my pregnancy and child hold the most value in my life and are to be sheltered from the threat of COVID-19, (2) to be infected with COVID-19 is bad, and to borrow from narrative paradigm theory there was no *good reason* for me to contract it as my work did not necessitate public-facing work at the time, and (3) contracting COVID-19 would be born out of a selfish act—meaning that if I were to contract COVID-19 while pregnant, I chose my personal needs above the needs of my unborn child. This is an early indicator of indoctrination into motherhood culture and the accompanying motherhood guilt. These evaluations served to guide my actions throughout the months of pregnancy. I worked to make-sense of the information available to me at the time I made choices about my physical, mental, and social health that fit within my boundaries of comfort and safety. Often prioritizing being risk avoidant over caring for myself and my mental health. It is in the divergent integration of the probabilistic and evaluative orientations that I lived my life in those early months.

Foremost, I felt grief. I mourned the loss of the pregnancy I anticipated. I grieved the time lost to COVID-19 including, the trips that were canceled, tampered celebrations, and date nights that never occurred. I felt isolated from family, friends, acquaintances, colleagues, and students. And I grieved the loss of everyday social interactions in public spaces. Much of my pregnancy took place in isolation—though it was—and continues to be—a privilege to work from home. I envied those that were able to retain a sense of routine by reporting to work. Grief was coupled with frustration—this was found in isolation. That is, the three values kept me contained in isolation. To betray one of these evaluations would invite risk. At best, this meant feelings of shame and guilt. At worst, it meant contracting COVID-19. The feelings of grief and frustration were later joined by the emotion of anger. I was not merely disappointed by the context of COVID-19, but I also became angry that I put my life on hold in March to "slow the spread," yet cases only continued to grow. I was angry that others selfishly went about their daily activities like the threat did not exist or even denied its existence. My personal sacrifices felt futile.

The problematic integrations I experienced gave rise to grief, frustration, and anger. As the pandemic raged on the challenge was the uncertainty of my situation. Pregnancy alone is a highly uncertain time. Sassi Matthias and Babrow (2007) stated, "…although pregnancy and childbirth are natural and typically healthy, the enormous values at stake magnify the importance of uncertainties that arise in these processes about the health of mother and child" (p. 62). Further, the uncertainty of pregnancy can manifest in a variety of ways, such as anxiety, depression, obsession, or fear (Fairbrother et al. 2016; Frías et al. 2015; Menkedick 2020). In fact, the theory of PI has been applied previously to various aspects of pregnancy and the postpartum period. This includes the process of selecting and receiving care during pregnancy (Matthias 2009), due dates (Vos, Anthony, and O'Hair 2014), prenatal and genetic testing (Kirkscey 2017), breastfeeding (Koerber, Brice, and Tombs 2012), and contraceptives and preventing pregnancy (Sundstrom et al. 2017).

Then contextualizing an already uncertain period in one's life within the cultural context of a global health crisis produced even greater mental burden, which demanded the introduction of my fourth evaluation—that is to take on some level of risk as needed for my mental health. One of the defining characteristics of the COVID-19 pandemic is the uncertainty associated with the virus. Applying PI's concepts of epistemological and ontological uncertainty, we can unpack how the feelings of grief, frustration, and anger were created. (Babrow 1992, 2001); Babrow (2009) differentiates two types of uncertainty: epistemological uncertainty and ontological uncertainty. Epistemological uncertainty occurs when knowledge of the situation is incomplete, overwhelming, or even inconsistent, while ontological uncertainty appears in the causal structures, for example, how one event impacts another.

The cultural context of COVID-19 is wrought with epistemological uncertainty. How does the virus spread? What is the role of asymptotic carriers? What is the efficacy of masking? Are pregnant women and their unborn children more vulnerable to COVID-19? What are the best therapeutics for the virus? What are the long-term effects of the virus on the body? And questions about vaccination development and rollout as well as the efficacy of the vaccination as a preventative and protective measure. In the early stages of the pandemic, we did not have these answers, and my first three evaluations were born out of epistemological uncertainty. Though our collective knowledge of COVID-19 grew throughout the course of my pregnancy, it remained incomplete, which introduced doubt. This doubt provided an opening for ontological uncertainty.

Ontological uncertainty surfaced most often in questions about how another's behaviors might impact me and my health. We are subject to the choices of other people that we encounter and negotiate and manage the uncertainties associated with those interactions, such as COVID-19 exposure. This boiled down to the ways that epistemological uncertainty presented and impacted relational aspects of life—namely, my interactions with family members and friends. In the later months of my pregnancy, e.g., July—September, I introduced a fourth evaluation to prioritize my mental health in the high-stress time, (4) there is worth in assuming some level of risk in order to care for myself and my mental health. The dilemmas I experienced cultivated difficult feelings, and those feelings transformed as the pandemic wore on and worsened daily. The fourth evaluation that emphasized my mental health more closely aligns with ontological uncertainty. Instead of acting risk-avoidant, I made calculated risks using my risk versus reward metric.

Though my city and state had mask mandates and limited gathering sizes, these policies only governed behavior at a public level. Individuals make their own risk assessments unique to them or their family unit. These are shaped by our adopted understanding of COVID-19 and accompanying safety precautions. Then, it was up to my partner and me to maintain our agreed upon boundaries. This became especially important as public restrictions in my state began to loosen. This sometimes lead to difficult conversations with my family members and friends or even avoidance of loved ones to maintain peace of mind. The story of my baby shower is a prime example of this. I wanted nothing more than to revel in the celebration, but questions of ontological uncertainty plagued my consciousness, and I remained guarded and distant toward loved ones. Some family interactions proved trickier and lead to the decision to isolate from those family members or ask for negative COVID tests ahead of seeing one another—all of this leading to further feelings of discomfort and perpetuated isolation.

As I retold the story of my pregnancy, I reflected on the chronological events, highlighting impactful moments in time, and revisited the general anguish felt throughout; but I also see the ways that I grew as an individual. Though I faced immeasurable loss and it wore on my mental health, I also reflect upon and celebrate the good that happened; notably, I met the big goals, I gave birth to a healthy and happy baby, I avoided COVID, my marriage is thriving, and not to mention I was awarded promotion and tenure in 2020. As events unfolded, and I was faced with a series of problematic integrations that challenged my expectations and notions of what it means to have a *good pregnancy*. Despite my expectations and desires not being fulfilled and the many challenges presented by these problematic integrations, I see how enduring these events created resilience in me. I see how this tremendous challenge impacted me as an individual and, in turn, as a mother. Complementary studies also find resilience among pregnant women during COVID-19 (Farewell et al. 2020; Preis et al. 2020), and it is my hope that this account adds to the literature on this topic.

17.8 Conclusion

This (re)telling examines the meaning that was made about what it means to live through COVID-19 during the vulnerable stage of pregnancy. Daily acts of communication—worked to construct my narrative, and it is through the application of problematic integration that I was able to make-sense of the damage and further redefine and reclaim. My narrative is characterized by uncertainty understood in the flux of time. It should be noted that at the time of writing, the COVID-19 pandemic continues to grow, and the long-term effects on the physical, mental, and social health of individuals and communities are still largely unknown.

This chapter represents ideas and reflections that are temporal, a snapshot in time from the beginning of the COVID-19 pandemic through the first nine months in the United States. In these months, I embodied the experience as a pregnant woman living in the United States during the global pandemic. Today, I still hold many of these same boundaries, with my circle of trusted family and friends. Now I co-exist with COVID-19 and with the epistemological and ontological uncertainty. However, I find the fear to be less overwhelming and remain hopeful for the future.

References

Babrow AS (1992) Communication and problematic integration: understanding diverging probability and value, ambiguity, ambivalence, and impossibility. Commun Theory 2(2):95–130

Babrow AS (2001) Uncertainty, value, communication, and problematic integration. J Commun 51(3):553–573

Babrow AS (2009) Problematic integration theory. In: Foss SLK (ed) Encyclopedia of communication theory

Babrow AS, Kline KN, Rawlins WK (2005) Narrating problems and problematizing narratives: linking problematic integration and narrative theory in telling stories about our health. Narratives Health Healing Commun Theor Res Pract 31–52

Browning L (2009) Narrative and narratology. In: Littlejohn KAFSW (ed) Encyclopedia of communication theory, vol 1. SAGE Publications, Inc., pp 673–677

Carey JW (1989) Communication as culture: essays on media and society. Unwin Hyman, Boston

Fairbrother N, Janssen P, Antony MM, Tucker E, Young AH (2016) Perinatal anxiety disorder prevalence and incidence. J Affect Disord 200:148–155. https://doi.org/10.1016/j.jad.2015.12.082

Farewell CV, Jewell J, Walls J, Leiferman JA (2020) A mixed-methods pilot study of perinatal risk and resilience during COVID-19. J Prim Care Community Health 11:2150132720944074

Fisher WR (1984) Narration as a human communication paradigm: the case of public moral argument. Commun Monogr 51(1):1–22. https://doi.org/10.1080/03637758409390180

Frías Á, Palma C, Barón F, Varela P, Álvarez A, Salvador A (2015) Obsessive-compulsive disorder in the perinatal period: Epidemiology, phenomenology, pathogenesis, and treatment. Trastorno Obsesivo-Compulsivo Durante El Periodo Perinat Epidemiología Fenomenología Etiopatogenia y Tratamiento 31(1):1–7. https://doi.org/10.6018/analesps.31.1.168511

Kirkscey R (2017) Patient decision aids for prenatal genetic testing: probability, embodiment, and problematic integration. Health Commun 32(5):568. Retrieved from http://ezproxy.uhd.edu/login?url=https://search.ebscohost.com/login.aspx?direct=true&db=edb&AN=121727068&site=eds-live&scope=site

Koerber A, Brice L, Tombs E (2012) Breastfeeding and problematic integration: results of a focus-group study. Health Commun 27(2):124. Retrieved from http://ezproxy.uhd.edu/login?url=https://search.ebscohost.com/login.aspx?direct=true&db=edb&AN=71115763&site=eds-live&scope=site

Matthias MS (2009) Problematic integration in pregnancy and childbirth: Contrasting approaches to uncertainty and desire in obstetric and midwifery care. Health Commun 24(1):60–70. https://doi.org/10.1080/10410230802607008

Mead GH (1934) Mind, self and society, vol 111. Chicago University of Chicago Press, Chicago

Menkedick S (2020) Ordinary Insanity: fear and the silent crisis of motherhood in America

Pearce WB, Cronen VE (1980) Communication, action, and meaning: the creation of social realities, Praeger

Preis H, Mahaffey B, Heiselman C, Lobel M (2020) Vulnerability and resilience to pandemic-related stress among US women pregnant at the start of the COVID-19 pandemic. Soc Sci Med 266:113348

Russell LD, Babrow AS (2011) Risk in the making: narrative, problematic integration, and the social construction of risk. Commun Theor (3):239. Retrieved from http://ezproxy.uhd.edu/

login?url=https://search.ebscohost.com/login.aspx?direct=true&db=edsgea&AN=edsgcl.261885445&site=eds-live&scope=site
Sassi Matthias M, Babrow AS (2007) Problematic integration of uncertainty and desire in pregnancy. Qual Health Res 17(6):786–798
Sundstrom B, Ferrara M, DeMaria AL, Baker-Whitcomb A, Payne JB (2017) Integrating pregnancy ambivalence and effectiveness in contraceptive choice. Health Commun 32(7):820. Retrieved from http://ezproxy.uhd.edu/login?url=https://search.ebscohost.com/login.aspx?direct=true&db=edb&AN=122658277&site=eds-live&scope=site
Vos SC, Anthony KE, O'Hair HD (2014) Constructing the uncertainty of due dates. Health Commun 29(9):866–876. https://doi.org/10.1080/10410236.2013.809501

Part VI
International Case Studies: Experiences and Resiliency

Chapter 18
Pandemic Resilience: What We Can Learn from a Rural Liberian Village's Response to Ebola

Crystal D. Daugherty

Abstract From 2014–2016 Ebola ravaged the West African country of Liberia. With an already weak health infrastructure, Ebola quickly impeded Liberia's health and medical communities. Exploring narratives from rural Liberian villages this chapter explores the impact a pandemic can have on communities and individuals. Throughout the lived experience themes of local knowledge, network and relationships, mental outlook, and communication emerge in Liberian narratives. Ultimately these narratives lead to acknowledging the importance of identifying challenges and understanding the role of communication in developing resilience before rather than during pandemics.

Keywords Liberia · Ebola · Communication · Resilience · Communities

18.1 Ebola Pandemic

The streets are empty and the market silent. As you walked by houses, there were no families in the yard, no greetings to be heard, only a watchful and suspect gaze from a window. There are no community gatherings, the local churches are shuttered, and friends have not spoken to each other in weeks. The older men who loiter in the community spaces have gone home where it is safe.

This scene sounds all too familiar to millions of people across the globe. The COVID-19 pandemic has disrupted every aspect of our daily lives. However, this scene is not from the current COVID-19 pandemic; these are notes from my field journal detailing life in rural Liberia during the 2014–2016 Ebola outbreak. Rural Liberians are no strangers to the harsh realities of life. From 1989 to 1997, and again from 1999 to 2003, the country of Liberia was ravished by back-to-back civil wars. In the aftermath of both wars the economic, social, and physical infrastructure was destroyed. Eleven years after the end of the second civil war, Liberians found themselves picking up the pieces yet again. While still rebuilding physical infrastructure

C. D. Daugherty (✉)
Northern Kentucky University, Highland Heights, KY, USA
e-mail: daughertyc3@nku.edu

D. M. Berube (ed.), *Pandemic Communication and Resilience*, Risk, Systems and Decisions, https://doi.org/10.1007/978-3-030-77344-1_18

such as road, hospitals and clinics, Liberians were met with the world's largest and deadliest health concern, Ebola.

During the 2014–2016 Ebola outbreak in West Africa, there were more than 28,652 cases and 11,325 deaths (CDC 2019). Libera accounted for 10,678 cases, and 4810 deaths (CDC 2019). Ebola alone would have been problematic. However, the Ebola epidemic was complicated by a struggling health system and inadequate understanding of cultural practices by outside organizations (WHO 2015). While the numbers stated above seem small compared to the number of coronavirus cases in the United States, the similarities between the two crises are uncanny. Through the health narratives of rural Liberians, this chapter attempts to explore the question what we can learn about communication and resilience from Liberia/Liberian's response to Ebola and how might it help during current and future pandemics?

18.2 Resilience and Communication

Defining resilience, as is evident in a simple search, has proven to be complicated. Many scholars differ in how they approach and define this particular phenomenon. Some scholars view resilience at the national level stating that national resilience "evokes solidary with the victims of tragedy, as well as shared feeling of resoluteness in the face of anxiety, uncertainly, and despair" (Bean 2018, p. 23). For some, resilience is defined through a community, stating that resilience is a "community's capacity to bounce forward following an adverse event such as a disaster or crisis" (Houston 2018, p. 19).

Others approach resilience through a cultural lens and suggest that culture serves as a resource for building resilience (Fleming and Ledogar 2008). Many researchers turn to the American Psychological Association definition of resilience, which is "the process of adapting well in the face of adversity, trauma, tragedy, threats, or significant sources of stress—such as family and relationship problems, serious health problems, or workplace and financial stressors" (APA 2012; Southwick et al. 2014).

It is important to note that despite the varying definitions, resilience is a process that attempts to use resources to sustain well-being, either at the individual, community, national, or global level (Southwick et al. 2014). Because resilience is a process and has so many factors, especially cultural factors, it is difficult to define and generalize. However, by studying how various cultures or communities go through the process, we can see examples of successes or failures and then adapt those for future problems. Southwick et al. (2014) suggest that resilience can be enhanced before trauma through flexibility, promoting healthy development, and supporting adaptive systems such as families or communities. However, this type of work cannot be accomplished unless there is another process in place, effective communication.

While scholars hold different views on defining resilience, there is explicit agreement that communication is critical to developing resilience (Salzarulo et al. 2015). The scholarship supports the notion that resilience requires communication. Afifi (2018) states that "resilience is created through communication in close relationship"

(p. 7). Researchers such as Beck (2016) and Buzzanell (2010) also back the claim that resilience is directly connected to communication. For both scholars communication is key to the resilience process. Beck explores the role of communication in close relationships stating that partners communicated resilience and that patterns of resilience were promoted via interpersonal communication. Buzzanell also suggests that resilience is "created, sustained, and enhanced through discourse" (Afifi 2018, p. 7; Buzzanell 2010).

Individuals and communities who can effectively communicate their experiences and needs during traumatic events are more likely to be stronger in the future. Because "resilience is a socially constructed term, and as such, it's creation and maintenance can only occur through communication and communication technologies" understanding the importance of communication is paramount (Salzarulo et al. 2015, p. 1). Houston (2018) joins researchers in claiming that centralizing communication is vital because of its connection to interpersonal networks, community systems, and resilience producing processes. In reviewing this scholarship, it became abundantly clear that Liberian narratives offered a unique perspective of how communication and resilience work. With this literature as a guide, I propose the following research question: What elements of resilience are present in rural Liberians' narratives?

18.3 Method

This chapter is part of a larger research project exploring coping, health literacy, and health narratives in rural Liberia. With institutional review board approval, I collected 46 interviews (21 females, 25 males) from individuals and healthcare workers around the rural village of Flehla in Liberia, West Africa. Using a modified McGill Illness Narrative Interview (MINI), which is used to "elicit illness narratives in health research" protocol, I was able to conduct semi-structured interviews about health structures, narratives, and beliefs (Groleau et al. 2006, p. 671). Interviewees ranged in age from 18–80 and represented a diverse tribal affiliation. Eighty percent of participants identified as Kpelle tribal members. While there are over sixteen tribes in Liberia, the Kpelle tribe is the largest, representing roughly twenty percent of Liberia's population. The majority of individuals who live in and around Flehla are Kpelle. For members of the Kpelle tribe there is a strong expectation to follow social rules and demonstrate a collectivist culture's characteristics (Murphy 1980).

To analyze the interviews, I used directed qualitative content analysis. Directed content analysis provides more structure than traditional content analysis by using a theory or prior research as a framework for coding and analyzing data (Assarroudi et al. 2018; Hsieh and Shannon 2005). For this chapter, I used Patel et al. (2017) review of community resilience as the framework for coding. The study resulted in nine elements that are present across resilience definitions. Table 18.1 shows the nine elements with their sub-categories.

Table 18.1 Elements of community resilience definition review

Element	Sub-categories
Local knowledge	Factual knowledge; training and education; collective efficacy and empowerment
Community networks and relationships	Connectedness; trust; cohesion
Communication	Effective communication; risk communication; crisis communication
Health	Health services; physical and mental health
Governance and leadership	Infrastructure and services; public involvement and support
Resources	Natural, physical, human, financial, and social
Economic investment	Economic programming and development
Preparedness	Individual, family, and government
Mental outlook	Hope; adaptability; acceptance

18.4 Findings

Of the nine elements, there were four that consistently appeared in the narratives of both individuals and healthcare providers. The four were communication, local knowledge, community networks and relationships, and mental outlook. Health, resources, and preparedness also deserved to be mentioned and were often coupled with one of the predominant elements. However, governance and leadership, and economic investment were never clearly identified or coded in the data. I found that over eighty percent of the coded data was assigned to the elements below. The relationship between these four elements offers exciting insight into how individuals might process and conceptualize resilience.

18.5 Local Knowledge

If a community can know and understand what is happening during a disaster, trauma, or in this case, pandemic, then they are situated to address better community weakness (Patel et al. 2017). Within this theme, the sub-categories proved to be prevalent in Liberian narratives. Early during the Ebola pandemic, many rural community members did not know what was happening. One man shared his experience of his wife showing symptoms: "*We did not know. The following day she started with a runny stomach. Then her eye became red. We did not know- there was no awareness.*"

A woman shared that many Liberians did not have factual knowledge or awareness about Ebola during the early stages and the early symptoms were often misdiagnosed as malaria:

> From past experience we knew that people had runny stomach with malaria. But they didn't die from it. This they die from it. We begin to understand from this experience that it was Ebola because it continued to kill. It continued to kill.

In addition to local knowledge, it is worth noting that many individuals and healthcare providers shared how misinformation spread at a higher rate during the early days, often resulting in mistrust of the government. First, stories began to circulate that this unknown virus was caused by witching. For many Liberians, witching is a spiritual practice caused by angry neighbors or ancestors who are upset with a family member. A healthcare provider who practices traditional bush medicine shared: *"They did not believe what was happening. At first, they thought they had been witched. When they came to me, I could not help them but they were afraid."*

Another form of misinformation came in the form of stories about the government intentionally killing Liberians. The government launched a campaign to report any suspected Ebola cases. Individuals could call a hotline or report a sick family member to a nearby clinic. However, in many rural areas, people did not trust this process. They would report a family member, and an ambulance would show up with people dress in personal protection equipment (PPE). The use of PPE meant that ambulance drivers and healthcare workers were covered in strange white suits. For some, especially those in remote areas, only bush devils concealed their identities. The cultural practice of traditional medicine, cultural and spiritual traditions around bush devils meant that local individuals did not trust the people in white suits.

Additionally, once a family reported their loved one, that person often never returned from the ambulance ride. Stories began to spread that the government was intentionally killing Liberians (Daugherty and Young 2019). A nurse at a local government ran clinic echoed the traditional bush medicine provider's statements:

> Yea- they did not trust us. When a person comes to you with Ebola they will mostly die. It is hard to cure. So now families think you are killing. When the government tell you to call and report a sick member- yeah they will not do it because every time they do that family member does not come home.

It was not until later in the outbreak that villagers began to receive more knowledge about the virus and the required PPE for healthcare workers. Having factual information about what a community is facing is crucial to being able to move through the overcoming and resilience process. Through community based medical teams' knowledge was spread via networks such as churches and small local clinics.

18.6 Community Networks and Relationships

For a community to be resilient, scholars agree that there is a deep need for community members to be connected and, if possible, form a cohesive unit (Patel et al. 2017). Another way to view networks and relationships is social capital. Social capital is a "combination of community participation and social cohesion that arises when such participation is frequent and supportive" (Hanson-Easey et al. 2018, p. 622). In the rural villages there were no community connectedness and cohesion in relation to the government workers. This lack of connectedness proved difficult in providing outlets for help, information, and mental support. Along with communication, this element of resilience was discussed the most by participants. For healthcare providers community networks and relationship were what helped curtail some of the Ebola outbreak.

A local medical doctor shared:

> It was important for us to try and get word out to the churches. To the markets. People needed to stay home. DO NOT visit the sick. Even if it is your pastor do not visit with them because you could die and your family could die." This particular doctor also shared her frustration with how information was being spread. "Why would you send an outsider? Send me, they will see me and say, "oh I know her. I believe her. But even during the early times some did not believe me. Their belief was that their ancestors had them witched. But they could not be cured in the bush so they came back.

Overwhelmingly, trust was a significant factor throughout the narratives. It was hard to trust individual workers who came to remove dead bodies or those who needed medical attention. Those workers wore all white personal protection equipment (PPE), and you could only see their eyes. For many Liberians, who hold strong spiritual beliefs, the PPE clad workers could be mistaken for evil, much like the tribal bush devils who also cover and disguise their identities. In addition to strong spiritual beliefs held by community members, many of the government workers did not have social capital in the communities where they were trying to work.

Through community networks, factual information began to spread, and villagers started taking precautions such as limited contact with others and increased hygiene. An older man, Eli, shared:

> We did not sit under the market tree anymore. Once the doctor told us this would kill us, we would sit outside our houses but your neighbors did not come to your house. You might greet them from the path but they did not come to your home. Even if a person was bringing you something-for me the market vendor would send me fruit- you would lay your money in the mixture [a sanitizing mixture that was also used for handwashing]. They would give you food and you would sit inside until they left. You didn't touch anything.

Without the concentrated effort to spread factual information, many Liberians would have continued to demonstrate distrust of the government. Eli continued:

> Yes many people would hide their sick. They would not call because the government was killing everyone who went to the Ebola clinics. But then we found out that was not true. The local medical teams told us about the disease and we could know that it was not the government.

Eli also shared how the community network also served as a way to boost morale and influenced the mental outlook for many Liberians who were isolated during Ebola.

18.7 Mental Outlook

Patel et al. (2017) suggest that mental outlook be defined as "attitudes, feelings and views when facing the uncertainty that typically occurs after a disaster or when contemplating a future one" (p. 12). The listed subcategories suggested were hope, adaptability, and acceptance. Throughout interviews Liberians mentioned their adaptability to their surroundings. When asked how they coped with the unknown during this time, an older woman shared: "*We should not be afraid. We have come through these things before. In the time before the war with Ebola there was another war. We can come through this war as well.*"

Others did not share the same sentiment as this particular woman. A young mother shared her concerns: "*Yes, it is bad for me. I was worr[ied] about that [Ebola]. Worry about work. But I found small jobs so I don't have to worry about that.*"

Other participants shared just how worrisome this time was. John, a community member, shared:

> People were dying. People die in front of us. The ambulance would come and take them and they would go away. It was dark days. So, then we would be worrying about the sickness because when it grabs the person you can't survive. It was very serious for us in Liberia.

When prompted, another woman shared how her community connections helped her to stay strong:

> I felt strong then. Because of the interaction [with community members], even during the time of Ebola, I feel comfortable. I feel secure because if you do get sick or cannot get something then your friends or the people from your church will come to help. They will come and encourage you about your life.

For many Liberians, as is true across cultures that experience resilience producing events, mental outlook is constantly evolving based on how many factors such as the ones mentioned above. However, the ability to communicate the various attitudes, values, and beliefs during uncertain times helps to establish a narrative precedent for what community members can do in the future.

18.8 Communication

Across the board, communication, specifically effective communication was valued as an essential part of resilience (Patel et al. 2017). Communication was the element that had the most codes and appeared most frequently with other elements. Many of the individuals who were interviewed expressed frustration with the lack of effective communication.

When discussing the lack of factual information, one community member shared: "*Why didn't they tell it to us plain. If they had told us early then maybe many would have survived. But then again it is hard tos get information to the communities. It is hard to spread that news.*" Later in their interview, this individual shared when they found out the level of risk that faced the community: "*Yeah we were traveling and there was a new check point. That is when we were told 'My man, you cannot be traveling this way. It is very dangerous and you can contract the virus.*"

Communication was also central to how community networks and relationship maintained their connections with each other. Ellen, an elderly woman, shared how church prayer group retained connection during this time:

> At first it was very lonely. But then sisters and brothers from the church would come visit. Oh, they did not come into your house and you did not go to church. But they might walk down the path by your house and say hello. They might give you a smile and a word from there. It made me feel like I was not alone.

Later in our interview, Ellen also shared how various community networks served as a channel of information:

> That is how I found out. Yes! When they were walking by, they said, 'Ma Ellen, it is safe to go to the clinic now. But first you wash your hands.' And that was true. The clinic up the road was open again but now you were checked three times before you got there. Once on the road, once outside the gate, and again inside. You had to wash your hands each time.

The interviews demonstrated that the community had created a shared narrative. This narrative, which was shared in the retelling and circulation of experiences, explained the importance of communication, knowledge, networks, and mental health in this community. Even as I received each individual's stories, I could sense the significance of what I was hearing. This shared knowledge, the weaving together of stories, and re-telling of them to an outsider (me) served as a way to validate the experiences of the community. It became apparent even during data collection that time and time again, the country of Liberia and its people were resilient.

18.9 Communicating Resilience Moving Forward

Both communication and resilience are complicated processes that vary depending on multiple factors. Communicating for resilience looks like helping others understand, prepare, and co-exist with traumatic events, including pandemics. In examining the data through the lens of Patel et al. (2017), communicating resilience in Liberia was about local knowledge, networks and relationships, mental health, and communication. While there may be some differences among other cultures, Patel' et al. review of resilience and the nine elements and subcategories offers a robust framework for examining what resilience means.

Through the analysis, there are two significant implications paramount to advancing the scholarship on communication and resilience: identifying challenges and focus on communication. First is understanding the challenges scholars and practitioners face during pandemics. As was evident in the Ebola crisis, trust plays a vital role in communication, the networks that are used, and in the validity of factual knowledge. Early in the pandemic, many rural Liberians did not trust the government or the government workers who were spreading accurate information about

Ebola. Establishing trust by using the networks and relationships available may have shortened the Ebola crisis or at least elicited a faster community response early in the pandemic.

An underlying concern here, especially in countries like Liberia, once colonized, is the historical narrative and implications of trusting outsiders and, in this case, the networks outsiders use. The colonization of modern-day Liberia and both civil wars are connected to power struggles from groups who considered themselves "in" and groups who were considered to be "other". Often these influences go unnoticed in the moment, especially during a traumatic health disaster like Ebola. Looking at the historical implications of colonization and its long-lasting impact on a country or culture, it is worth noting that the distrust of outside groups could also be connected to colonization. From this, others can see the value of establishing and maintaining trust *before* a pandemic begins. In the US, we are seeing this play out in real-time. Distrust of media and government has been growing over the past four years. However, now that the US is amid of a global pandemic, many citizens do not trust the media or government's attempts to share information and preventative methods.

Another challenge to consider, tied directly to trust, is the networks and relationships used to disseminate information and resources. Utilizing small local networks such as community centers, places of worship, local clinics should be the primary method of information dissemination. Rural Liberians did not trust the outsiders who came to their villages sharing information. Looking at the historical implications of colonization and the long last impact it has on a country or culture, it is worth noting that the lack of network relationships from outside groups could also be connected to colonization. In this case, it has looked like an outside group swoops in tells locals what to do, and then once the crisis is over, they leave just a quickly. Using this approach there is no time or emphasis on building network relationships. In Libera, the outsiders, often European or American organizations, were not familiar with the established networks and had not built a sustainable relationship with community members. Other communities can learn from this by identifying critical networks and relationships that would help share resources, knowledge, or services before a crisis happens.

The final challenge that faces communities trying to build and communicate resilience is resources. This element was not overly present in the interviews with rural Liberians, but the absence may be due to long term lack of resources in many rural villages. For many Liberians, the lack of resources did not compound the Ebola outbreak. The lack of attention or mention of resources was very intriguing. It is hard to say why individuals did not mention resources or lack thereof. However, from my own experience in Liberia and previous casual conversations, I believe this is because rural Liberians have learned how to be resilient despite limited resources. This is a fascinating concept that warrants additional in-depth study. Other communities should note how adequate resources, whether financial or structural, contribute to the resilience process.

The second implication of this research is the importance of focusing on communication. As previously stated, like resilience, communication is a foundational process that is directly tied to each of the elements present throughout the research.

As with identifying challenges, communication is foundational in creating trust. For Liberians, the government waited until the country was in shambles to try and develop local trust. A long history of civil unrest and government mistrust was not easily forgotten. Communicating and building trust with community members before pandemics will serve at-risk communities well.

When looking at the rural villagers' response to the Ebola pandemic, we see an exciting link between trust and networks. While there was distrust of outsiders and government workers, the networks and relationships within the communities, especially those that were longstanding, demonstrated the important role of communication as prevention and resilience. Many of the networks and relationships had established strong communicative ties prior to the pandemic. Because of the strong ties once the outbreak began those networks and relationships adapted. Gone were the days of sitting at the market or meeting in houses of worship. It is this demonstration of strong communicative practices and adaptability that other communities should mirror.

Finally, communication is critical in identifying, sharing, and expanding resources. As stated earlier the fact that Liberians did not specifically mention resources in detail does not mean that moving forward, we should discount the role resource availability plays in creating resilience. Communities that are able to identify and share, or have access to, resources are creating an environment that fosters resilience (Patel et al. 2017). Communicating about resources that are available prior to pandemics or traumatic events will encourage communities to seek out the resources they need to become resilient. The process of communicating and fostering resilient behavior prior to an event is important to all communities.

It was difficult to imagine empty streets and a silent market as I sat under a massive tree with a community member who agreed to be interviewed. His description of the 2014–2016 Ebola outbreak was vastly different than my experiences in Liberia prior to 2014 and later in 2016. People were packed in the market that day. Just three months shy of being declared Ebola-free, community life was returning. I sat and listened as the older man shared how the community survived and why he was not worried about future outbreaks or traumatic events: "*Oh we will be fine, yeah. Mama Liberia is strong. We always find our way again.*"

Acknowledgements The United States colonized modern-day Liberia in the 1800s. To this day we can easily identify the ripple effect of colonization during periods of civil unrest and in modern day global affairs. As an ethical practice, I acknowledge my race and nationality's role on my privileged experiences in rural Liberia. With that being said, my most profound respect and gratitude is extended to the people of Liberia, specifically the residents of the rural villages where I conducted interviews. Your hospitality, resilience, and joy are indescribable. Thank you for allowing me to share your experiences and stories in such a meaningful way.

References

Afifi TD (2018) Individual/relational resilience. J Appl Commun Res 46(1):5–9

American Psychological Association (2012) Building your resilience. https://www.apa.org/topics/resilience. Accessed 27 Feb 2021

Assarroudi A, Nabavi FH, Armat MR, Ebadi A, Vaismoradi M (2018) Directed qualitative content analysis: The description and elaboration of its underpinning methods and data analysis process. J Res Nurs 23:42–55

Bean H (2018) National resilience. J Appl Commun Res 46(1):23–25

Beck G (2016) Surviving involuntary unemployment together: the role of resilience- promoting communication in familial and committed relationships. J Fam Commun 16(369):385

Buzzanell P (2010) Resilience: talking, resisting, and imagining new normalcies into being. J Commun 60:1–14

Center for Disease Control and Prevention (2019) Ebola (Ebola Virus Disease). https://www.cdc.gov/vhf/ebola/history/2014-2016-outbreak/index.html. Accessed 27 Feb 2021

Daugherty CD, Young AJ (2019) "If I die, who will tell their stories?" Emerging health legacies following the 2014–2016 Ebola epidemic. In: Kellet P (ed) Narrating Patienthood: engaging diverse voices on health, communication, and the patient experiences. Rowan, London, pp 47–60

Fleming J, Ledogar RJ (2008) Resilience, an evolving concept: a review of literature relevant to Aboriginal research. Pimatisiwin 6(2):7–23

Groleau D, Young A, Kirmayer ULJ (2006) The mcgill illness narrative interview (MINI): an interview schedule to elicit means and modes of reasoning related to illness experience. Transcult Psychiatry 43(4):671–691

Hanson-Easey S, Every D, Hansen A, Bi P (2018) Risk communication for new and emerging communities: The contingent role of social capital. Int J Disaster Risk Reduction 28:620–628

Houston JB (2018) Community resilience and communication: dynamic interconnections between and among individuals, families, and organizations. J Appl Commun Res 46(1):19–22

Hsieh HF, Shannon SE (2005) Three approaches to qualitative content analysis. Qual Health Res 15:1277–1288

Murphy WP (1980) Secret knowledge as property and power in Kpelle society: elders versus youth. Africa 50:193–207

Patel S, Rogers MB, Amlot R, Rubin GJ (2017) What do we mean by "community resilience'? A systematic literature review of how it is defined in the literature. PLoS 9:1–27

Salzarulo A, Mundorf N, Sakar J, Terui M, Lei W (2015) Communication as a tool for empowerment: a model for resilience. China Media Res 11(4):78–87

Southwick SM, Bonanno GA, Masten AS, Panter-Brick C, Yehuda R (2014) Resilience definitions, theory, and challenges: Interdisciplinary perspectives. Eur J Psychotraumatol 5(1):25338

World Health Organization (2015) Factors that contributed to undetected spread of Ebola virus and impeded rapid containment. https://www.who.int/news-room/spotlight/one-year-into-the-ebola-epidemic/factors-that-contributed-to-undetected-spread-of-the-ebola-virus-and-impeded-rapid-containment. Accessed 27 Feb 2021

Chapter 19
The Role of Scientific Output in Public Debates in Times of Crisis: A Case Study of the Reopening of Schools During the COVID-19 Pandemic

Gabriela F. Nane, François van Schalkwyk, Jonathan Dudek, Daniel Torres-Salinas, Rodrigo Costas, and Nicolas Robinson-Garcia

Abstract In exceptional circumstances such as pandemics, the expectation is for policy to be supported by science. However, the lack of scientific consensus during the COVID-19 pandemic places strain on decision making. In this chapter, we focus on COVID-19 effects on children and the public debate around the reopening of schools. The aim is to better understand the relationship between policy interventions and the subsequent use of scientific information by the public. We combine information from scientific articles and preprints with their appearance in (social) media. First, we investigate different related COVID-19 scientific areas. Second, we identify news and social media attention around this scientific output, focusing on three countries: Spain, South Africa, and the Netherlands. We then analyze the activity in (social) media and news outlets and conclude by discussing how scientific publications, media and policy actions shape public discussion in the context of a health pandemic.

Keywords Children · Schools · Science-society interaction · Altmetrics · Public policy · Public health measures

---

G. F. Nane (✉) · J. Dudek · N. Robinson-Garcia
Delft Institute of Applied Mathematics, Delft University of Technology, Delft, The Netherlands
e-mail: g.f.nane@tudelft.nl

F. van Schalkwyk · R. Costas
DST-NRF Centre of Excellence in Scientometrics and Science, Technology and Innovation Policy, Centre for Research On Evaluation, Science and Technology, Stellenbosch University, Stellenbosch, South Africa

J. Dudek · R. Costas
Centre for Science and Technology Studies, Leiden University, Leiden, The Netherlands

D. Torres-Salinas
Departamento de Información y Comunicación, Universidad de Granada, Granada, Spain

D. M. Berube (ed.), *Pandemic Communication and Resilience*, Risk, Systems and Decisions, https://doi.org/10.1007/978-3-030-77344-1_19

19.1 Introduction

The COVID-19 pandemic has turned the world upside down. While it rapidly became clear that certain population groups are more at risk, with the elderly and adults who have underlying health conditions being at higher risk of developing severe illness from COVID-19, much uncertainty remained surrounding children (Rajmil 2020). The uncertainty characterized both the infection rates among children as well as their range of symptoms. As empirical evidence accumulated at tremendous pace, fewer infection rates have consistently been reported in children compared with adults, as have the milder symptoms (Cruz and Zeichner 2020; Götzinger et al. 2020; Goldstein et al. 2020). Very few cases in children have been linked to severe symptoms, such as multisystem inflammatory syndrome and Kawasaki-like disease (Viner and Whittaker 2020; Webb et al. 2020).

Nonetheless, the infection rate in children is biased, given the testing policies in many countries. For example, in the Netherlands only those (mildly) symptomatic are eligible for COVID-19 testing. And whereas everyone with cold-like symptoms could be tested over the summer, from September 26 2020 up until the time of writing (December 2020), children of 12 years old or younger have been placed under special testing rules and were only allowed to be tested following the presentation of serious symptoms. Adults and children older than 13 years of age could still be tested if they presented cold-like symptoms. The special testing rules stipulated being sick or being in contact with someone who had tested positive for September 26 2020.

A substantial body of work about COVID-19 and children has focused on the role of children in spreading the virus. The role of children in the transmission of the virus was questionable from the onset of the pandemic, and the topic is still under debate. To illustrate: "Children may play a major role in community-based viral transmission" according to Cruz and Zeichner (2020) whereas Ludvigsson (2020) reports that "children are unlikely to spread the coronavirus". Media reporting reflects this debate. A recent news article in *Nature* (Lewis 2020) reports that young children are unlikely to spread the virus, whereas an article in *The Conversation* (Hyde 2020) states that "children may transmit coronavirus at the same rate as adults".

Despite the ongoing debate about the role of transmission in children, schools received distinct attention early on. The enclosed and crowded environment, prone to poor ventilation and where children and educators spend 6 to 8 h daily, creates the conditions for a high-risk environment. Hence, strict measures have been necessary in the face of scientific uncertainty and closing schools was among the first measures taken worldwide to reduce the spread of the virus. China and Mongolia were the first countries to close schools in the middle of February 2020, followed by some schools in Italy and San Marino at the end of February 2020. By March 31st, schools in 193 countries were closed due to COVID-19.

The reopening of schools has been part of the first steps in the easing of lock-down restrictions. Whereas 43 countries reopened schools partially and another 40 countries fully reopened schools by June 15th, many other countries in the northern hemisphere postponed the reopening of schools until after the summer holidays. In a considerable number of countries, school reopening has been perceived as being safe, given the low levels of community spreading. Nonetheless, when community spread levels were not low, hot spots of infections were reported at a school and at a school camp (Stein-Zamir et al. 2020; Torres et al. 2020; Szablewski et al. 2020). An ECDC report on COVID-19 in children and the role of school settings in COVID-19 transmission concludes that "there is limited evidence that schools are driving transmission of COVID-19 within the community, however there are indications that community transmission is imported into or reflected in the school setting" (ECDC 2020). Macartney et al. (2020) has reported low transmission rates in the New South Wales educational system during the first wave of the COVID-19 pandemic. Even though most schools remained open in Australia during the first wave, class sizes were reduced during the peak. For some parts of the country (e.g., Melbourne) with high community viral spreading, schools did close.

There is thus no scientific consensus reached so far from the empirical research on children's role in the transmission of the coronavirus. Nonetheless, decision-making concerning school reopening and closure could not and cannot wait for greater scientific consensus.

This chapter presents an exploratory attempt at tracing social discussion around a scientific topic under debate during a global pandemic, combining quantitative and qualitative methods. We focus on the case of COVID-19 and its effects on children which inform the public debate around the reopening of schools. We do so to better understand the relationship between policy interventions during an uncertain and rapidly changing knowledge landscape and the subsequent use of scientific information in public debates related to the policy intervention during a crisis. Our approach is to combine scientific information with their appearance in the popular media, including social media.

We investigate the reopening of schools in three different countries (Spain, South Africa, and the Netherlands), each of which introduced different policy measures, with the aim of analyzing the societal reception of scientific findings in three different national and political contexts. In the case of the Netherlands, after an initial lockdown, schools reopened in May, at quite an early stage of the first wave of the outbreak and have been opened ever since (except for summer holidays). In South Africa, the outbreak took place in March and schools closed until early June, just to close again a month later due to the rise of infections and reopen again in August. Finally, Spain has been one of the European countries with the most restrictive measures at the early stage of the pandemic. Schools did not open until after the summer holidays. For each of these countries, we retrieved information related to these policy interventions as well as the dates on which announcements were made.

19.2 Conceptual Framework

Our study is situated at the intersection between science, politics, and society. We take as our starting point the fact that public communication about science is inherently political and adopt Scheufele's (2014) conceptualization of science communication as political communication. Such a conceptualization considers the broader political contexts in which science–public interactions occur, how stakeholders compete for attention in the political sphere, and how publics interact with the scientific information they encounter in the media—information which may often be contradictory as well as overwhelming in complexity and volume. Moreover, the information is usually presented via multiple traditional and online channels and may change rapidly during times of heightened uncertainty such as the COVID-19 pandemic.

The COVID-19 pandemic presents a unique case of science communication as political communication because the threat posed by the virus is both immediate and global. The consequence is a simultaneous and rapid response to the pandemic by scientists, politicians and the public alike. Science responds to the crisis by conducting research aimed at understanding the behaviour of the virus and to develop effective responses to containing its spread. To share new truth claims with other scientists with the objective of accelerating the discovery of effective responses to the pandemic, findings from scientific research on the novel coronavirus are fast-tracked for publication in preprints and scientific journals. These findings are also communicated to policymakers and to the public, either indirectly via the media or directly via briefings, press releases and social media.

While science advances understanding of the coronavirus, political decisions are taken to control the spread of the virus to protect society. These political decisions constitute policy moments in response to the pandemic. Political decisions are informed by the local context, including the progression of the outbreak and the prevailing political climate, and will varying degrees be influenced by the available science. The degree of influence that science exerts over political decision- making is not only likely to depend on context but is likely to change over time as the perception of the threat changes and as other political issues such as the socio-economic impact and the infringement on citizens' constitutional rights begin to challenge the legitimacy of the measures taken to control the pandemic.

The social response to policy moments during the pandemic is reflected both in the mainstream and in the social media as citizens process and debate both the available scientific information and the political decisions taken. At the same time, scientists themselves seek to popularize or make more accessible the latest scientific information about the virus. Scientists are also attentive to the media leading to the 'medialization' of science, that is, to increasing links and interplay between science and the media (Weingart et al. 2012).

19.3 Data and Methods

Our point of departure is the scientific output generated around the COVID-19 pandemic as it relates to children. This corpus of literature includes scientific papers with a broad scope of topics, including the mental and social effects related to policy interventions, effects of the lockdowns, the closure of schools, and medical issues related with the infection, transmission, and diagnosis of COVID-19 in children. From this set of publications, we trace signals of discussion in social media and news media platforms, to establish a link between the scientific and societal realms.

19.3.1 Data Collection

The data collected for this study was extracted from a variety of sources: scientific publications, news outlets, and social media discussions and policy interventions. Since the outbreak of the pandemic, different community- and organization-led initiatives have been conducted to make scientific publications on COVID-19 openly accessible. In this study, we made use of the COVID-19 Open Research Dataset (CORD-19) and the World Health Organization (WHO) COVID-19 Global literature on coronavirus disease database. These two databases are of special interest due to the combination of sources they include, containing not only studies published in scientific journals but also preprints from the main global repositories (e.g., BioRxiv, MedRxiv, SSRN, etc.).

We downloaded the two complete databases on October 15, 2020. Table 19.1 shows some descriptive values of the size of the database at the time. We searched within the title and abstract fields for documents containing the words 'children' and 'schools'. After merging the 'Pubs children' documents of both datasets, a total of 5713 publications were retrieved. This is our final set of scientific publications from which we t race t heir (social) media reception.

Table 19.1 Descriptive values of the publication databases used in the study

Database	Pubs	Pubs in = 2020	% DOI = in 2020	Pubs children	% DOI children
WHO	113,105	103,084	26.57	4434	30.42
CORD-19	314,001	220,251	50.87	9380	52.03
Merged final set (via DOI)	–	–	–	5713	100

Note Number of total publications by database [Pubs], publications in 2020 [Pubs in 2020], share of publications with Document Object Identifier (DOI) [%DOI in 2020], number of publications related to children and schools [Pubs children] and share of publications with a DOI related to children and schools [%DOI children]

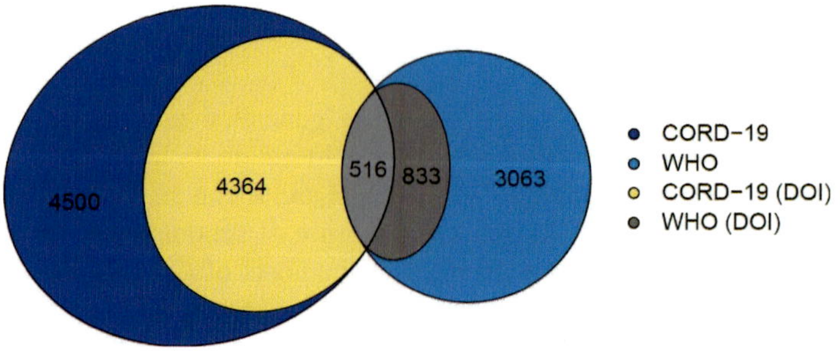

Fig. 19.1 Number of publications from 2020 related to children and schools, indexed in the CORD-19 and WHO databases

The two databases (WHO and CORD-19) do not represent distinctive sets of publications, having quite substantial overlap. To avoid duplicates, the two databases were merged and cleaned. For a reliable merging of the two databases, as well as for the further tracing of the (social) media reception of the publications, it was necessary to count with unique document identifiers (e.g., PubMed Identifiers, Digital Object Identifiers, etc.). Particularly Digital Object Identifiers (DOI) are commonly assigned to scientific publications to univocally identify scientific documents across databases and the web-at-large. The main inconvenience of using DOIs is that we can only identify and combine publication data for half of the papers included in the CORD-19 database and a third of those included in the WHO database (Fig. 19.1).

A final number of 5713 publications along with their DOIs have been collected in our final dataset of scientific output. We proceeded to identify news outlets and social media discussions around the scientific publications in our dataset. News media items mentioning a DOI in our set were identified with data from Altmetric.com, retrieved in October 2020. From a total of 19,922 news items found globally for the set of DOIs in our database, 424 news articles could be identified as originating from the Netherlands, Spain, or South Africa. This was done by matching the URLs of the news outlet coming from Altmetric.com with the URLs of Dutch, Spanish, and South African national newspapers and broadcasting services, as extracted from Wikipedia and other websites listing news outlets. The final list of news outlets from each country was verified and curated manually. We identified 200 news items from Spain, which referenced 81 distinct DOIs. In South Africa, 79 news pieces referenced 72 distinct DOIs and in the Netherlands, 145 news items referenced 83 distinct DOIs. The titles and short abstracts of the news articles (where available in the data from Altmetric.com) were analyzed manually for our study.

Twitter data on mentions of publications were also collected from Altmetric.com in October 2020. This included any tweet identified by Altmetric.com that refers to a DOI in our set of publications. More detailed Twitter data was rehydrated directly from Twitter (using the Twitter API) on December 2, 2020. This resulted in a total number of 540,615 tweets, covering 66.7% (3811) of the publications. The first identified tweet was on January 14, 2019, and the last recorded one on October 24, 2020. From our tweet data, 65.8% (182,548) of the 277,419 distinct Twitter users provided geolocation information. This allowed us to link tweets to the three countries selected in our study. We identified 16,548 tweets with a Spanish geolocation, which referenced 932 distinct DOIs. Much less Twitter activity was captured in the cases of the Netherlands and South Africa. In the Netherlands, 1478 tweets were collected, linking to 229 distinct DOIs, whereas in South Africa, 1062 tweets could be linked to 290 distinct DOIs.

Lastly, data on policy interventions regarding the closure and reopening of schools was retrieved from the UNESCO Institute for Statistics, which includes daily global information on the state of schools since the outbreak of the pandemic. Regarding the announcements and specificities of the measures, we manually searched national news media platforms.

19.3.2 Semantic Analysis

We explored the general semantic configuration of the publications selected by means of co-word maps, extracted from the titles of the articles identified. We employed natural language processing tools to extract noun phrases from titles. We then created a binary co-occurrence matrix to build and visualize the final network. This network provided a baseline to visually inspect the whole body of literature related with COVID-19 and children. Moreover, it also offered a baseline on which to overlay Twitter and news media activity. This was done by coloring the nodes of the network (noun phrases) based on the number of mentions received by the papers containing such terms. Thus, terms with a higher intensity in the color grading belong to papers that are highly tweeted by users of a given country. Figure 19.2 provides a visual aid to help the reader better understand and interpret the contents of such maps.

The data collected on news outlets and tweets were used to investigate the activity and topics covered in the (social) media when reporting scientific outputs, for each country. Overlay maps were used to show those topics. We include overlay maps for tweets in this chapter. We mention that single noun phrases "covid" and "children" were removed from the maps since they are redundant for our analysis. The overlay maps for news outlets are uploaded on Figshare. For illustrative purposes, we selected examples to study differences in the reporting of scientific literature in the news media, by individually reading and analyzing the contents of selected scientific and news articles, and the tweets text. The specific findings will be reported for each country in the following section.

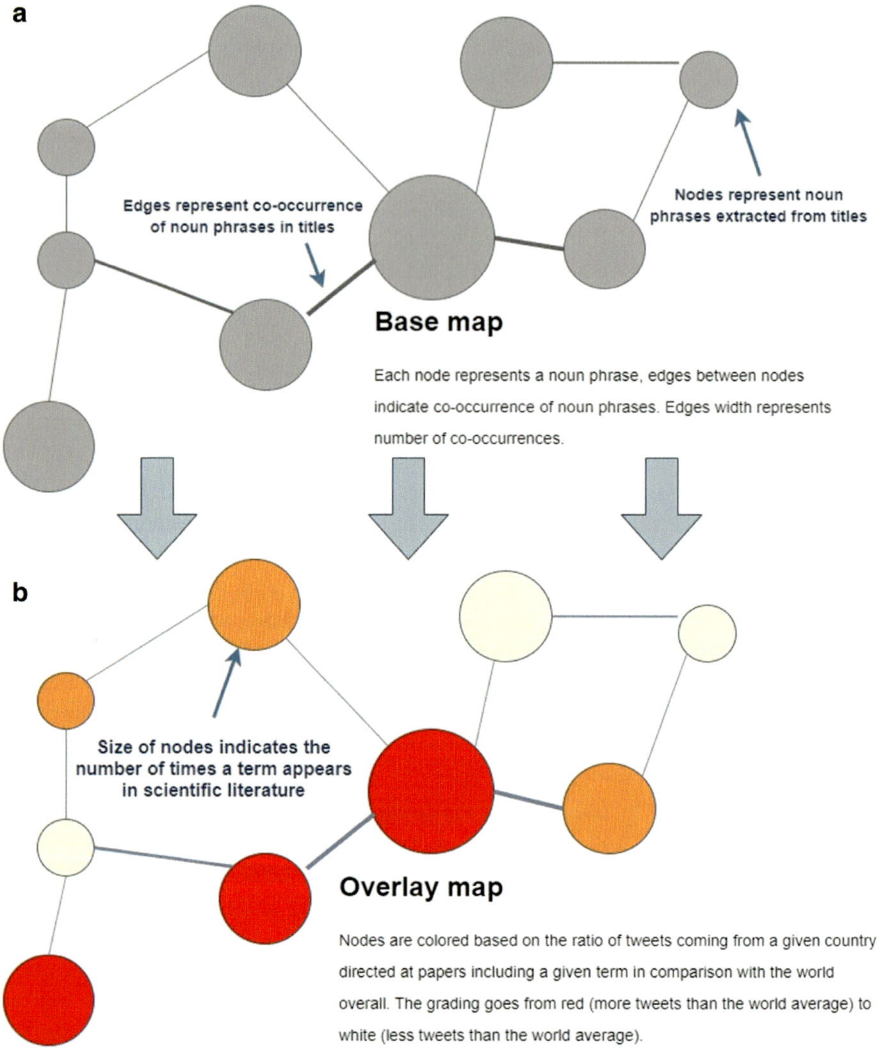

Fig. 19.2 Graphical representation explaining overlay maps and how to interpret them correctly. Graph **a** shows the base map constructed with the complete set of scientific literature, while Graph **b** overlays tweet mentions to publications by coloring nodes based on the intensity of the tweets

19.4 Results

19.4.1 *Semantic Structure of the Research on Children, Schools, and COVID-19*

From a literature review and a clustering analysis, it became apparent that the scientific output consisting of 5713 publications since the start of 2020 focused on the risk of infection, as well as on the development of mild or severe illnesses from COVID-19 in pediatric cases. The large number of asymptomatic cases, as well as the role of children in spreading the virus also received considerable attention. Despite the focus on the physical health of children during this pandemic, attention was given to the psychological effects of quarantine and lockdown during COVID-19 (Orgilés et al. 2020; Idoiaga Mondragon et al. 2020). The vulnerability of children during this pandemic was also researched (Haffejee and Levine 2020; Fouche et al. 2020). Finally, inequality in home-schooling during the pandemic also received attention from scientists (Bol 2020).

Reopening schools does not only involve the health and well-being of children but also of adults who come in close contact with children. In this regard, research focused on the risk of severe COVID-19 illness among teachers and among adults living with school-aged children (Gaffney et al. 2020).

Broad themes such as infection, development of severe symptoms, transmission, and social and psychological impacts of school closures were identified as dominant in the scientific literature about children and COVID-19. The underlying map is available on Figshare.

A manual search identified research output on children's health and school reopening within the three countries. The focus on the scientific output in the three countries varies. In Spain, significant attention has been given to hospitalization and to severe cases of children with COVID-19. Tagarro et al. (2020) reported on the early screening and severity of coronavirus in Madrid by investigating data from 365 tested children in the first two weeks of March 2020. Attention has also been given in Spain to the psychological effects and well-being of children (Orgilés et al. al. 2020; Idoiaga Mondragon et al. 2020).

In the Netherlands, research has focused on the transmission of the virus by children. A study on 54 households "suggest lower point estimates for transmissibility of infection to close contacts from children aged under 19 yrs, and higher point estimates for adults aged over 70 yrs when compared to persons aged 19–69 yrs" (RIVM 2020). Alsem et al. (2020) reported on the effects of the pandemic on pediatric rehabilitation, whereas Bol (2020) investigated the inequalities in home-schooling.

In South Africa, research has been conducted on child protection and resilience (Fouche et al. 2020; Haffejee and Levine 2020) and well-being (van Bruwaene et al. 2020), indicating a focus on the social aspects of school closures and lockdown. Additionally, attention has been given to multisystem inflammatory syndrome in children in South Africa (Webb et al. 2020).

The scientific output results in mixed evidence of infection and transmission as they pertain to children. The limitations of the scientific studies and the consequent levels of uncertainty were conveyed when reporting findings. This is, however, not a unanimous approach. For example, a viewpoint in the Archives of Disease in Childhood (Munro and Faust 2020), is entitled "Children are not COVID-19 super spreaders: time to go back to school". The title appears to be inflated by the urgent need for policy decisions. The authors write "At the current time, children do not appear to be super spreaders. Serosurveillance data will not be available to confirm or refute these findings prior to the urgent policy decisions that need to be taken in the next few weeks such as how and when to reopen schools." They continue "Governments worldwide should allow all children back to school regardless of comorbidities. Detailed surveillance will be needed to confirm the safety of this approach, despite recent analysis demonstrating the ineffectiveness of school closures in the recent past (Viner et al. 2020b). The media highlight of a possible rare new Kawasaki-like vasculitis that may or may not be due to SARS-CoV2 does not change the fact that severe COVID-19 is as rare as many other serious infection syndromes in children that do not cause schools to be closed". The title suggests no uncertainty regarding the role of children in the transmission of coronavirus. A more uncertain approach is taken on the severity of symptoms, where a rare disease "may or may not be" attributed to the novel virus.

19.4.2 Analysis Per Country of Policy, News Outlets and Social Media Response

We investigated the (social) media response to the policy actions related to school closures and reopening. For this, we monitored news outlets and Twitter activity and investigated the extent to which they overlapped temporally with the policy actions. Moreover, we analyzed the topics covered by tweets and in news articles using the overlay maps. For exploratory purposes, we manually inspected the titles of news articles, as well as the full text of selected news items, and the content of selected publications. The analysis is presented for each of the three countries in our study.

Spain

While in the Netherlands and South Africa schools reopened after around two months of closure, in Spain school reopening was delayed until after the summer holidays. Figure 19.3 depicts the announced and implemented measures, in chronological order, both at the national, as well as the regional levels. The policy measures registered no difference between primary and secondary schools. The figure also includes the timeline distribution of the news items and tweets in our database, which have been identified as originating from Spain. A total of 188 news articles and 15,603 tweets were identified between the beginning of February and the end of September 2020. News articles on the topic registered brief appearances before the school closure

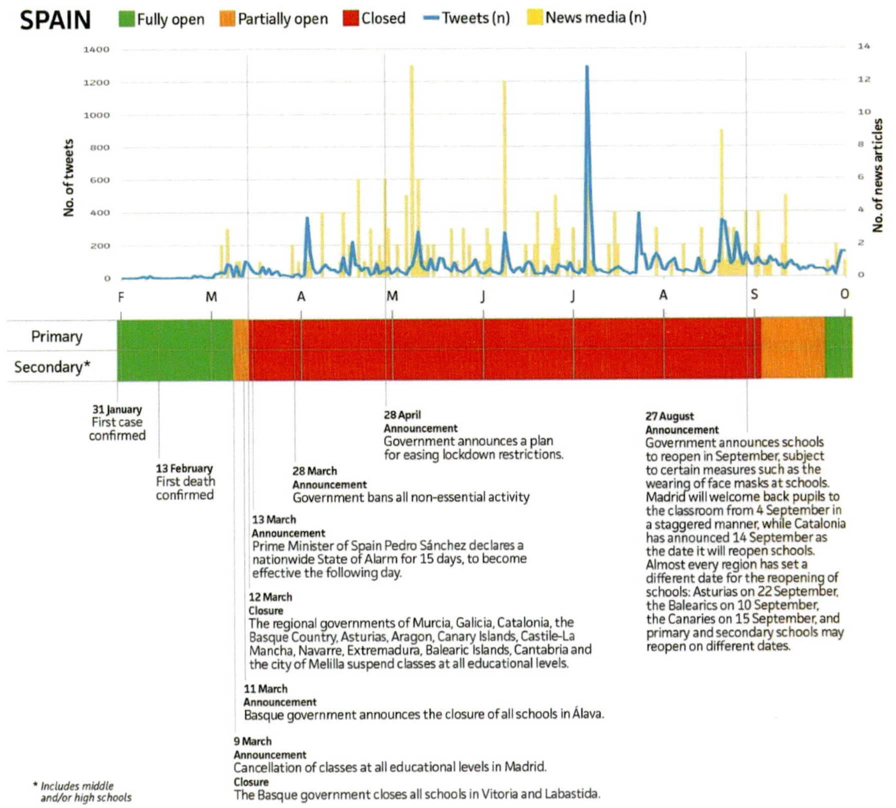

Fig. 19.3 Timeline of announcements and implementation of school closure and reopening in Spain, along with the distribution of tweets (on the left y-axis) and news items (on the right y-axis) mentioning scientific articles

in March, as well as more consistent appearances around the reopening of schools in September. As for tweets, we can observe small peaks around the time of the announcements in March, as well as shortly before and after the schools reopening in September. Further activity has been registered during the school closure, with peaks around the end of April, when the government announced a plan for easing lockdown restrictions, as well as in July and August, when no other policy intervention has been announced nor occurred.

The highest number of tweets in early July is the result of mentions received by a nationwide, population-based seroepidemiological study (Pollán et al. 2020) on the prevalence of SARS-CoV-2 in Spain (ENE-COVID). A total of 1906 tweets followed soon after the article was published in the *Lancet*, at the beginning of July. The second-highest tweeted article (739 tweets) reports on pediatric severe acute respiratory syndrome (Yonker et al. 2020) and received distinct attention before the school reopening, as well as in September. Similarly, attention was paid before the school reopening to the safety of reopening (primary) schools during pandemic (Levinson et al. 2020; Mallapaty 2020).

In the case of news mentions, we do not observe the same activity pattern. There are almost no mentions in March, with sustained but low activity between April and July, and recurrent peaks on specific days at the beginning of May, June, and July. This is followed by more constant activity in the media at the end of August, prior to the reopening of schools. The most mentioned article (20 mentions) received most of the attention in April. The article presents findings on the potential impact of the summer season on slowing the pandemic (Jüni et al. 2020) weighed against an alternative hypothesis that school closures account for such slowing. The two next most mentioned papers have 17 mentions each. In one case, Pollán et al. (2020) discuss the results of a nationwide screening undertaken in Spain between April and May. This paper concludes that at the time of the survey, there was a seroprevalence of around 5% with lower figures for children (<3.1%) and a third of the positive cases being asymptomatic. The third study focuses specifically on children tested positive (Yonker et al. 2020) aiming at "understanding the potential role children play" (p. 45) in the pandemic.

An overview of the topics covered by tweets from Spain is depicted by a VOSviewer map in Fig. 19.4. The nodes in the map present the co-occurrence of the most relevant keywords identified from the titles of the 5713 articles in our data set. The color coding of the map reflects the prevalence of mentions of those articles in tweets from Spain relative to the worldwide collected tweets: The darker the color, the more focus on the keywords relative to the worldwide tweets on the topic.

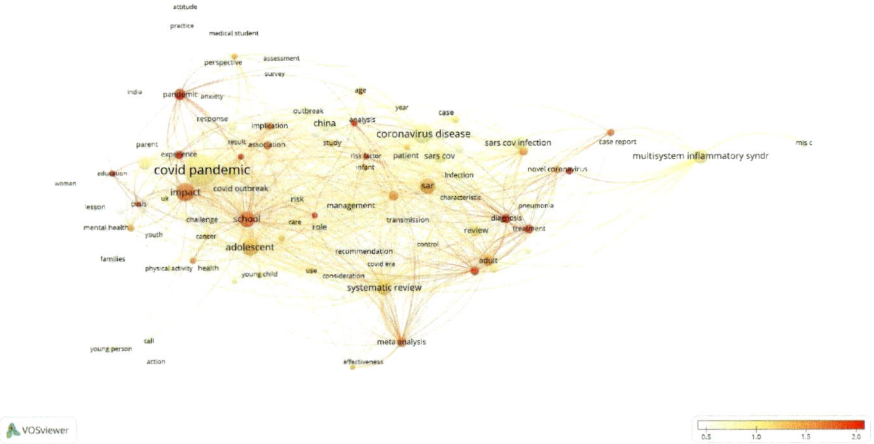

Fig. 19.4 VOSviewer map of the topics covered by tweets in Spain relative to the overall tweets on children and school closure/reopening. A value of 1 (=yellow nodes) denotes similar focus on the topic as for the overall tweets, nodes in the colors orange-red indicate above average focus. Only top terms are included for visualization purposes (the noun phrases "covid" and "children" were removed since they are redundant)

The map in Fig. 19.4 depicts more focus than is the case worldwide on topics such as school and education, but also on adolescence, experience, and impact, which suggests a focus on the social implications related to school closure. Furthermore, an enhanced focus on infection and risk factors, diagnosis, and treatment are visible on the right-hand side of the overlay map. The severity of symptoms, i.e., multisystem inflammatory syndrome appears to be at the same level as for the rest of the world, but at higher levels than for the other two countries. The news overlay map (not shown but uploaded on Figshare) shows less activity around the severity of symptoms and infection, but similar activities for transmission and school, schoolclosure and impact.

South Africa

Figure 19.5 provides the timeline for the school closure and (partial) reopening in South Africa. We note the differentiation between the reopening of secondary (high) schools and primary schools in July and August. Primary schools reopened later than secondary schools. A total of 74 news articles and 992 tweets were identified. Of the 74 news articles, 33 news articles were identified during the first school closure and only 18 in September. The most mentioned studies in the news are those by Yonker et al. (2020) and Hsiang et al. (2020), on large-scale anti-contagion policies, each with 3 mentions. Despite the comparatively low number of identified tweets, we observe Twitter activity around the announcement of school closure in March, and immediately after the first confirmed death, which registers the highest Twitter

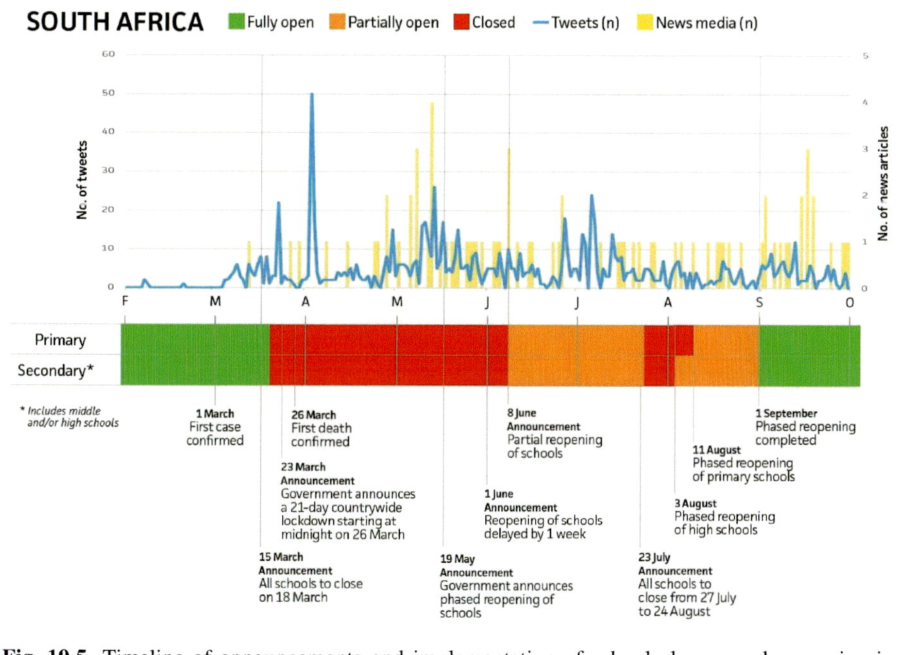

Fig. 19.5 Timeline of announcements and implementation of school closure and reopening in South Africa, along with the distribution of tweets and news items mentioning scientific articles

activity. The highest number of tweets (92 tweets) was in fact in relation to an author correction concerning respiratory virus shedding in exhaled breath and efficacy of face masks (Leung et al. 2020), and concerned the wearing of face masks. This illustrates a textbook example of how correlation between two apparent events does not imply causation.

The second peak of Twitter activity registered in May around the announcement of a phased reopening of schools reveals discussions gravitating either around the Kawasaki-like disease or transmission of the virus by children. Similarly, the discussions from the end of June and beginning of July signal the first Europe-wide study of children (Götzinger et al. 2020) which suggests mild disease in children and rare fatalities and relates to the seroepidemiological study in Spain (Pollán et al. 2020). The Spanish study represents the second highest tweeted article, with 39 mentions.

Figure 19.6 depicts the topics covered by the tweets in South Africa, relative to the total tweets on the topic. Despite the relatively low number of tweets on specific topics, we note that most tweets focused on social aspects (*impact*, *lesson*, *adolescents*), which are depicted on the left-hand side of the map, as well as on *infection*. The focus given to the severity of symptoms was less than in the case of Spain. The news map reflects more attention to the severity of symptoms than captured by the tweet map, similar attention to infection and transmission, as well as to the topic focusing on social aspects, and less attention to 'impact'.

The Netherlands

Like Spain and South Africa, the Netherlands' government decided to close schools in the middle of March. Unlike South Africa, primary schools were reopened first,

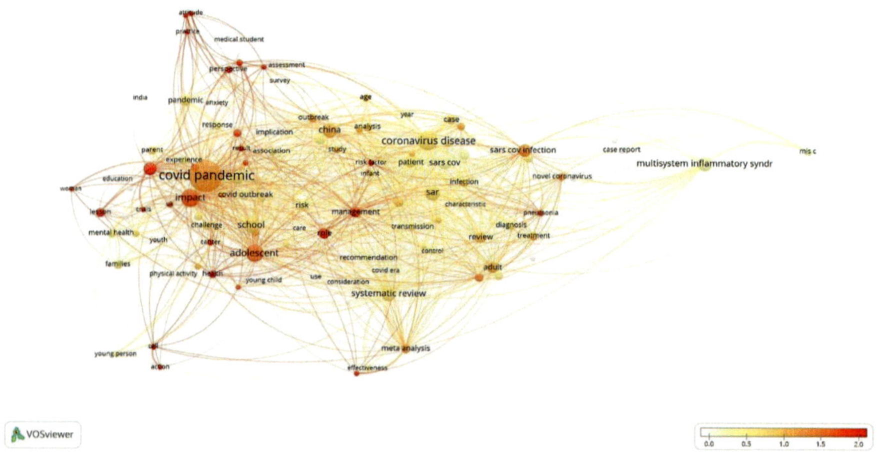

Fig. 19.6 VOSviewer map of the topics covered by tweets in South Africa relative to the overall tweets on children and school closure/reopening. A value of 1 (=yellow nodes) denotes similar focus on the topic as for the overall tweets, nodes in the colors orange-red indicate above average focus. Only top terms are included for visualization purposes (the noun phrases "covid" and "children" were removed since they are redundant)

at half capacity, whereas secondary schools reopened three weeks later, also at half capacity. Additionally, the partial opening was short (on June 8 and June 15 respectively), after which primary and secondary schools were fully open. Finally, it should be noted that the second school closure coincided with the summer holiday in the Netherlands. Figure 19.7 provides the graphical illustration of the measures, along with the announcements.

The news items and twitter activity over the period is also included. A total of 133 news articles and 1277 tweets were collected. Most news media attention was dedicated to the topic during the school closure and the partial reopening—38 news mentions were identified in April and May. Fewer mentions in the news were observed during the summer months, and 21 news mentions were counted in September. The study by Leung et al. (2020) was mentioned 13 times in the news, the highest number of mentions to a scientific article in the Netherlands, supporting the discussion around wearing masks. The authors report that "surgical face masks could prevent transmission of human coronaviruses […] from symptomatic individuals". This finding introduces an interesting observation of the discussion around the face

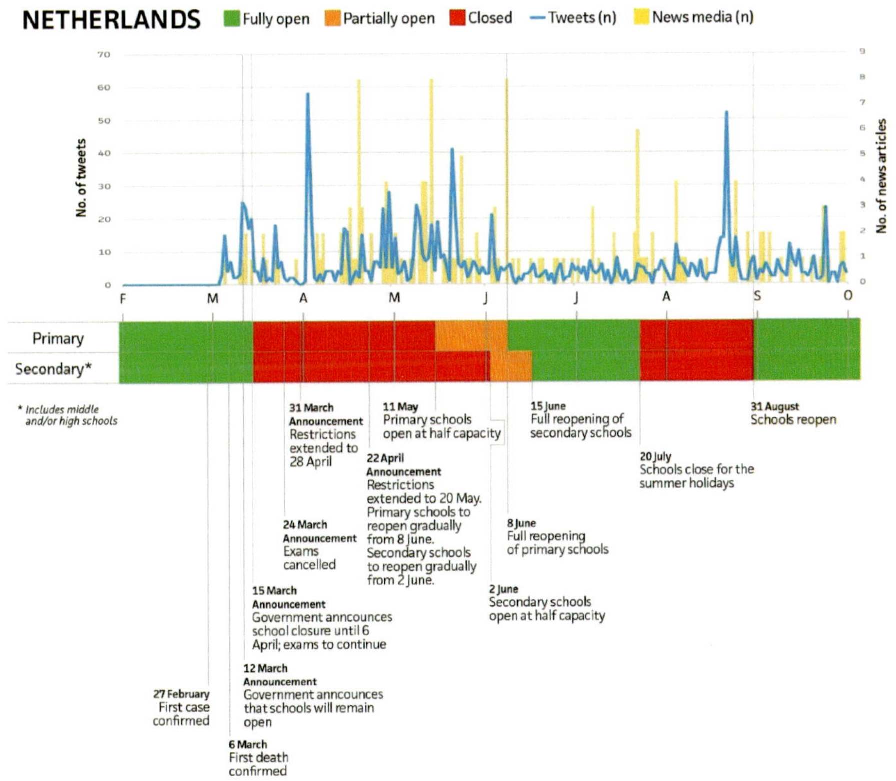

Fig. 19.7 Timeline of announcements and implementation of school closure and reopening in the Netherlands, along with the distribution of tweets and news items mentioning scientific articles

masks in the Netherlands, where face masks only became mandatory in December 2020. In Spain and South Africa, where mask wearing was made mandatory earlier in the pandemic, we do not observe a reaction to this publication, with only one news mention in both Spain and South Africa. A report on coronavirus disease in children from the United States published by the CDC COVID-19 Response Team (Covid et al. 2020) also received 13 mentions in the Dutch news media. The news items referencing this article vary in message. One is titled "Children appear to be less susceptible to corona [virus]", another reads "The number of children with COVID-19 has risen dramatically over the last five months" and another one "Mounting research paints a bleak picture for schools trying to reopen. Most large schools can expect coronavirus cases within 1 week".

Netherlands registers a relatively modest Twitter activity like South Africa. The first sustained discussion on Twitter was stirred up by the announcement on March 12 that schools will remain open. The tweets debated the government decision and the message transmitted by RIVM that children are less susceptible to become infected. The highest tweet activity was generated by the study by Leung et al. (2020) in early April, with 96 tweets. In the middle of May, a systematic review (Lecrerc et al. 2020) on the available literature on examples of SARS-CoV-2 clusters linked to indoor activities, caught the attention of tweeters in the Netherlands. The majority of 70 tweets focused on the need of outdoor sports (for children), as supported by the lack of empirical evidence for outdoor settings. Finally, the second-most tweeted (79 tweets) article mentioned pediatric SARS-CoV-2 (Yonker et al. 2020) and third-most tweeted output (with 71 tweets) provides scientific support for the wearing of face masks (Peeples 2020). Even though face masks are not the topic of our study in this chapter, it is interesting to note again that in the Netherlands, face masks were not mandatory at the time of the Twitter discussion.

Figure 19.8 illustrates the topics covered by tweets in the Netherlands relative to the overall tweets on the topic. Unlike in Spain and South Africa, there is little focus on the social topics, which may reflect the early school reopening in the Netherlands. Most tweets focus on infection and transmission, which are expected topics when schools are open. The news overlay map reflects less activity related to infection and slightly less to severe symptoms, but more to transmission and to school closure than the Twitter overlay map. Similarly, little focus is given to other social aspects.

19.4.3 Overall News Outlets Analysis

The news outlets collected for the three countries were manually scrutinized further. An overall topic analysis based on the titles of the news articles reveals that *infection* appeared in 8 articles in Spain, 12 in the Netherlands, and 2 in South Africa. Kawasaki or other (severe) diseases were mentioned in 7 news articles in the Netherlands, 4 in Spain and 5 in South Africa. *Transmission of the virus* registered the most attention, with 38 news articles in the Netherlands, 29 in Spain and 12 in South Africa. Similarly, *school closur*e was covered by 21 news articles in the Netherlands, 6 in South Africa,

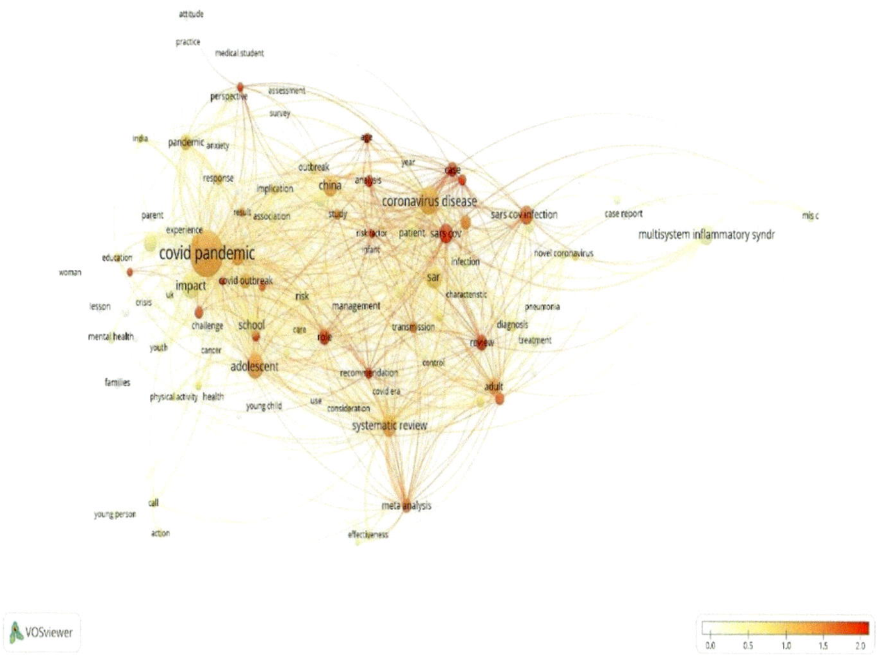

Fig. 19.8 VOSviewer map of the topics covered by tweets in The Netherlands relative to the overall tweets on children and school closure/reopening. A value of 1 (=yellow nodes) denotes similar focus on the topic as for the overall tweets, nodes in the colors orange-red indicate above average focus. Only top terms are included for visualization purposes (the noun phrases "covid" and "children" were removed since they are redundant)

and 5 in Spain. This is remarkable given that the Netherlands registered the quickest time to school reopening, which suggests that the news items generally supported the policy measures for schools reopening.

It is also notable that the news items in the three countries do not appear to have covered the scientific output of the national researchers. The only notable exception is the ENE-COVID national study (Pollán et al. 2020), which was mentioned in 20 news articles.

Overall, the most referenced article in the news items of the three countries is the study on pediatric SARS-CoV-2 (Yonker et al. 2020), which is also the second-most tweeted article in the Netherlands. The article was referenced in 25 news articles across all three countries.

Our exploratory analysis of the news articles also revealed cases of miscommunication in reporting the scientific findings in the news media. A first example comes from a recent systematic review and meta- analysis focused on the infection among children compared with adults (Viner et al. 2020a). Their conclusion was that "There is weak evidence that children and adolescents play a lesser role than adults in transmission of SARS-CoV-2 at a population level. This study provides no information on the infectivity of children." A *Nature* news article (Lewis 2020) reported on the findings as follows: "Researchers suspect that one reason schools have not become COVID-19 hot spots is that children—especially those under the age of 14—are less susceptible to infection than adults, according to a meta-analysis (Viner et al. 2020a) of prevalence studies".

The study on pediatric SARS-CoV-2 (Yonker et al. 2020) concludes that "children may be a potential source of contagion in the SARS-CoV-2 pandemic despite having milder disease or a lack of symptoms" (p. 45). After finding that regardless of the viral load most children are either asymptomatic or have milder disease, they warn "that it would be ineffective to rely on symptoms or temperature monitoring to identify SARS-CoV-2 infection" (p. 51). While the paper clearly emphasizes the potential risk of children spreading the disease, it does not present evidence of higher transmission rates from children. However, a news item from the Netherlands mentioning the article is entitled "Children with mild symptoms from coronavirus appear to be very infectious". Four other Dutch news items report the findings of the study, which are not in line with the guidelines from the National Institute for Public Health and Environment (RIVM) that children play a modest role in the transmission of the coronavirus. Spanish media presented a quite different perspective on the paper's findings, warning in their news articles titles that asymptomatic children are spreaders of the disease.

A final example is linked to a Dutch news item entitled "Children 'do transmit COVID-19' to adults, says researcher whose study was 'misunderstood' as evidence that kids cannot spread coronavirus" published on April 30, 2020. The mentioned study is a review of 78 studies, from which one reports that no child under 10 years of age was found to transmit COVID-19. The article also reports that the German virologist Christian Drosten "pointed out that the findings of the Dutch study from 54 households (RIVM, 2020) which was used as evidence that children do not play a big role in spreading COVID-19, were not statistically significant".

19.4.4 Overall Twitter Analysis

An overall Twitter analysis on the topics identified in the scientific output reveals further insights. *Infection* has been the most popular topic in the tweets in the three countries, with 3031 mentions in Spanish tweets, 218 in Dutch and 216 in South African tweets. However, comparatively few tweets mention *community transmission* in relation with children as a vector of transmission. In Spain, we identified 156 mentions, South Africa 41 and in the Netherlands 12 mentions.

By the same token, the transmission of coronavirus among children has received appreciable attention, with 2170 tweets in Spain, 125 tweets in the Netherlands and 97 in South Africa. In particular, the study by Stein-Zamir et al. (2020), which reports an outbreak in a high school in Israel, was referenced in 714 tweets in Spain—the third-highest tweeted article. The study points out that community transmission is replicated in a school setting and hence the risk of school reopening while community spread is high. This suggests that despite the strict measures on school reopening in Spain, the public debate accounted for the risks of less strict measures. The severity of COVID-19 symptoms in children was captured by 550 tweet mentions of *Kawasaki* and *multisystem inflammatory syndrome* in Spain, 47 in the Netherlands and 31 in South Africa. The tweets in the latter countries promoted a recent publication (Viner and Whittaker 2020).

Munro and Faust's (2020) paper "Children are not COVID-19 super spreaders: time to go back to school" received appreciable attention in the social media. The article metrics, as observed on November 12, 2020, report that this article had been picked up by 45 news outlets, referenced in 2 policy sources of the Scottish government, and tweeted by 5430 accounts. Despite this, the article received modest attention in the three countries in our study: 202 mentions in Spain, 15 mentions in the Netherlands and 3 in South Africa.

We also investigated the profiles of tweeters whose tweets were collected in our dataset. We investigated the share of tweeters who had tweeted about science (i.e., tweeted an academic publication) before the pandemic, that is, tweeters from our sample present in Altmetric data from 2019. We found that 59.3% in Spain, 60.6% in the Netherlands and 65.1% in South Africa had mentioned other scientific articles in their tweets prior to the pandemic. From the total of 8597 distinct tweeters identified in all three countries, 5141 had already referenced scientific output before the COVID-19 pandemic. Moreover, we attempted to determine the professions of the 8597 tweeters.

We identified 740 researchers, 741 health professionals, and 296 journalists based on terms found in the user descriptions of the tweeters. We note a possible overlap between the groups, as someone can be both a health professional and, e.g., hold a Ph.D. (one of the indicators for being a researcher). Given the limited available information on Twitter and the limitation of our search algorithms, we expect that these results are underestimating the true presence of those professions in our dataset.

19.5 Discussion

This study explored the relationship between scientific advances and their societal reception in the context of a global pandemic. We focused on the COVID-19 pandemic and on the specific case of its effects on children. We investigated the publication of scientific findings related to COVID-19 and children, and how this scientific output has been used in news and social media communication following the government measures related to the closure and reopening of schools since the onset of the pandemic until late October 2020. Despite the broad scientific coverage of the topic, both in the news and social media, we observed locally driven communication in the context of a global pandemic. Our illustrative exploration also identified a disconnect between the policy (political) timeline and the resultant communication in the social media, suggesting that the (social) media reaction to policy moments is not supported by frequent referencing to scientific output on the topic of children and school closure during the C OVID-19 pandemic.

Our analysis of the scientific output revealed differences in the science published by researchers in Spain, the Netherlands, and South Africa. Despite a global effort to advance knowledge on the COVID-19 virus to contain its spread, differences at the national level in terms of the focus of the research published was observed. In other words, our findings suggest that different scientific priorities may emerge during a global pandemic as determined by different socio-economic contexts as in the cases of Spain, the Netherlands, and South Africa.

The available scientific output about the role of children and schools in the COVID-19 pandemic has not been picked up in the social media in the three countries in our study to the same degree. We found that only 17.9% of the publications in our database have been tweeted about in the three countries; this is much less than the coverage of about 63.0% of all attention for CORD-19 publications as overall captured by Altmetric (Colavizza et al. 2020). A total of 932 DOIs (16.3% of the scientific output) has been mentioned in the Spanish tweets on the topic. In the Dutch tweets, only 4% of the scientific output (229 articles) has been mentioned, whereas in South Africa 2 89 articles (5%) have been mentioned.

Furthermore, as the timelines reveal, Twitter activity as it relates to mentions of scientific papers did not mirror notable policy events, apart from weak evidence in the Netherlands. For most cases, where activity was noted, it was triggered by a scientific event (the publication of a paper). This observation tallies with the finding that, based on an analysis of Twitter profiles in the sample, approximately one in five of the Twitter accounts was identified as belonging to a researcher or a health professional. It is conceivable that social media is mirroring science given the number of health, research, and other professionals in our sample. Moreover, we were able to identify differences in responses to the scientific output on the topic in the three countries. As in the case of differences in scientific focus between the three countries, a difference in focus in terms of social media content was observed, reinforcing the contextual nature of attention, particularly during a pandemic.

The expectation that scientific papers will be mentioned in the media is not unreasonable given high levels of fear and scientific uncertainty, and since we know that social movements can seek to mobilize action or advocate for a particular ideological position by referring to scientific information in the social media (e.g., anti-vaccination movement) (van Schalkwyk 2019). In our study, we found no evidence during the pandemic of a pro- or anti-schooling movement on Twitter. This claim is based on the lack of a universal or popular hashtag related to the topic and which would point to a specific community of attention on Twitter. Further support is provided by the fact that there are not highly active or dominant Twitter accounts to be found in the data. One would expect to find highly active Twitter users in cases where the platform is used to amplify messaging. In the data, only two accounts (both from Spain) were found to have tweeted more than 100 times during the eight-month period. One of the accounts belongs to a pediatrician while the other to the Spanish Society for Pediatric Infectious Diseases. In the absence of an ideologically-motivated group, movement or collective, and some evidence that scientific rather than political activity is the driver of social media activity, Twitter is not in this case being used as a communication platform to amplify messaging about the risks or benefits of children attending school during the COVID-19 pandemic.

A more speculative explanation for the apparent absence of ideologically motivated groups is the demise of Twitter as a platform for politicized science communication in favour of other platforms and/or closed communication channels such as WhatsApp or Facebook groups, especially given the push by Facebook towards private communities as central to the future of the social network (Dwoskin 2019), and the appeal of private groups to users (Holmes 2018).

Like Twitter, the news media around scientific publications exhibited independent activity from policy moments. Nonetheless, we identified several cases of miscommunication of the scientific output in the news media, which illustrate that despite the apparent absence of politically motivated communities, the intricacies of scientific studies are still not always accurately communicated in the media, and this may lead to distorted messages. These preliminary findings point to the need for further systematic research into the politicization of science communication in specific contexts and across multiple communication platforms during a global pandemic.

Acknowledgements The authors would like to thank Altmetric for providing access to data on news and Twitter mentions. Rodrigo Costas is partially funded by the South African DST-NRF Center of Excellence in Scientometrics and Science, Technology, and Innovation Policy (SciSTIP). Jonathan Dudek is partially funded by TU Delft COVID-19 Response Fund.

References

Alsem M, Berkhout J, Buizer A (2020) Therapy needs and possibilities in paediatric rehabilitation during the COVID-19 lockdown in the Netherlands. Child Care Health Devel

Bol T (2020) Inequality in homeschooling during the Corona crisis in the Netherlands. First results from the LISS Panel

Colavizza G, Costas R, Traag VA, Van Eck NJ, Van Leeuwen T, Waltman L (2020) A scientometric overview of CORD-19. BioRxiv

Covid CDC, COVID C, COVID C, Bialek S, Gierke R, Hughes M, McNamara LA, Pilishvili T, Skoff T (2020) Coronavirus disease 2019 in children—United States, February 12–April 2, 2020. Morb Mortal Wkly Rep 69(14):422

Cruz AT, Zeichner SL (2020) COVID-19 in children: initial characterization of the pediatric disease. Pediatrics 145(6)

Dwoskin E (2019) Facebook says private groups are its future. Some are hubs for misinformation and hate. *The Washington Post*, July 5. Facebook says private groups are its future. Some are hubs form misinformation and hate

ECDC (2020) COVID-19 in children and the role of school settings in COVID-19 transmission. European Centre for Disease Prevention and Control, Stockholm

Fouche A, Fouché DF, Theron LC (2020) child protection and resilience in the face of COVID-19 in South Africa: A rapid review of C-19 legislation. Child Abuse Negl 104710

Gaffney AW, Himmelstein D, Woolhandler S (2020) Risk for severe COVID-19 illness among teachers and adults living with school-aged children. Annal Intern Med

Goldstein E, Lipsitch M, Cevik M (2020) On the effect of age on the transmission of SARS-CoV-2 in households, schools and the community. medRxiv.
Götzinger F, Santiago-García B, Noguera-Julián A, Lanaspa M, Lancella L, Carducci FIC, Gabrovska N, Velizarova S, Prunk P, Osterman V, Krivec U (2020) COVID-19 in children and adolescents in Europe: a multinational, multicentre cohort study. Lancet Child Adolesc Health 4(9):653–661
Haffejee S, Levine DT (2020) 'When will I be free': lessons of COVID-19 for child protection in South Africa. Child Abuse Negl 104715
Holmes R (2018) Are closed (and secret) Facebook groups the future of social media? Medium, 14 Nov. https://invoker.medium.com/are-closed-and-secret-facebook-groups-the-future-of-social-media-25870c0ceeb1
Hsiang S, Allen D, Annan-Phan S, Bell K, Bolliger I, Chong T, Druckenmiller H, Huang LY, Hultgren A, Krasovich E, Lau P (2020) The effect of large- scale anti-contagion policies on the COVID-19 pandemic. Nature 584(7820):262–267
Hyde Z (2020) Children may transmit coronavirus at the same rate as adults: what we now know about schools and COVID-19
Idoiaga Mondragon N, Berasategi Sancho N, Dosil Santamaria M, Eiguren Munitis A (2020) Struggling to breathe: a qualitative study of children's wellbeing during lockdown in Spain. Psychol Health 1–16
Jones TC, Mühlemann B, Veith T, Biele G, Zuchowski M, Hoffmann J, Stein A, Edelmann A, Corman VM, Drosten C (2020) An analysis of SARS-CoV-2 viral load by patient age. medRxiv
Jüni P, Rothenbühler M, Bobos P, Thorpe KE, da Costa BR, Fisman DN, Slutsky AS, Gesink D (2020) Impact of climate and public health interventions on the COVID-19 pandemic: a prospective cohort study. C MAJ 192(21):E566–E573. https://doi.org/10.1503/cmaj.200920
Leclerc QJ, Fuller NM, Knight LE, Funk S, Knight GM, CMMID COVID-19 Working Group (2020) What settings have been linked to SARS-CoV-2 transmission clusters? Wellcome Open Res 5(83):83
Leung NH, Chu DK, Shiu EY, Chan KH, McDevitt JJ, Hau BJ, Yen HL, Li Y, Ip DK, Peiris JM, Seto WH, Cowling BJ (2020) Author correction: respiratory virus shedding in exhaled breath and efficacy of face masks. Nature Med 26(6):981–981
Levinson M, Cevik M, Lipsitch M (2020) Reopening primary schools during the pandemic
Lewis D (2020) Why schools probably aren't COVID hotspots. Nature
Ludvigsson JF (2020) Children are unlikely to be the main drivers of the COVID-19 pandemic–a systematic review. Acta Paediatr
Macartney K, Quinn HE, Pillsbury AJ, Koirala A, Deng L, Winkler N, Katelaris AL, O'Sullivan MV, Dalton C, Wood N, Brogan D (2020) Transmission of SARS-CoV-2 in Australian educational settings: a prospective cohort study. Lancet Child Adolesc Health 4(11):807–816
Mallapaty S (2020) How schools can reopen safely during the pandemic. Nature 584(7822):503–504
Munro AP, Faust SN (2020) Children are not COVID-19 super spreaders: time to go back to school. Archives of disease in childhood
National Institute for Public Health and the Environment (RIVM), the Netherlands. Children and COVID-19. 2020. Available from: https://www.rivm.nl/en/novel-coronavirus-covid-19/children-and-covid-19
Orgilés M, Morales A, Delvecchio E, Mazzeschi C, Espada JP (2020) Immediate psychological effects of the COVID-19 quarantine in youth from Italy and Spain
Peeples L (2020) Face masks: what the data say. Nature 586(7828):186–189
Pollán M, Pérez-Gómez B, Pastor-Barriuso R, Oteo J, Hernán MA, Pérez-Olmeda M, Sanmartín JL, Fernández-García A, Cruz I, de Larrea NF, Molina M (2020) Prevalence of SARS-CoV-2 in Spain (ENE-COVID): a nationwide, population-based seroepidemiological study. The Lancet 396(10250):535–544
Rajmil L (2020) Role of children in the transmission of the COVID-19 pandemic: a rapid scoping review. BMJ Paediatr open *4*(1)

Scheufele DA (2104) Science communication as political communication. Proc National Acad Sci 111(Supplement 4) 13585–13592. https://doi.org/10.1073/pnas.1317516111
Stein-Zamir C, Abramson N, Shoob H, Libal E, Bitan M, Cardash T, Cayam R, Miskin I (2020) A large COVID-19 outbreak in a high school 10 days after schools' reopening, Israel, May 2020. Eurosurveillance 25(29):2001352
Szablewski CM, Chang KT, Brown MM, Chu VT, Yousaf AR, Anyalechi N, Aryee PA, Kirking HL, Lumsden M, Mayweather E, McDaniel CJ (2020) SARS-CoV-2 transmission and infection among attendees of an overnight camp—Georgia, June 2020. Morb Mortal Wkly Rep 69(31):1023
Tagarro A, Epalza C, Santos M, Sanz-Santaeufemia FJ, Otheo E, Moraleda C, Calvo C (2020) Screening and severity of coronavirus disease 2019 (COVID-19) in children in Madrid, Spain. J AMA Paediatrics
Torres JP, Piñera C, De La Maza V, Lagomarcino AJ, Simian D, Torres B, Urquidi C, Valenzuela MT, O'Ryan M (2020) SARS-CoV-2 antibody prevalence in blood in a large school community subject to a Covid-19 outbreak: a cross-sectional study. Clin Infect Dis
van Bruwaene L, Mustafa F, Cloete J, Goga A, Green RJ (2020) What are we doing to the children of South Africa under the guise of COVID-19 lockdown? S AMJ: South Afr Med J 110(7):1–2
van Schalkwyk F (2019) The amplification of uncertainty: the use of science in the social media by the anti- vaccination movement. In Weingart P, Joubert M, Falade B (eds) Science Communication in South Africa: reflections on current issues. African Minds, Cape Town, pp 170–212. https://doi.org/10.5281/zenodo.3557216
Viner RM, Mytton OT, Bonell C, Melendez-Torres GJ, Ward J, Hudson L, Waddington C, Thomas J, Russell S, Van Der Klis F, Koirala A (2020a) Susceptibility to SARS-coV-2 infection among children and adolescents compared with adults: a systematic review and meta-analysis. JAMA Paediatrics.
Viner RM, Russell SJ, Croker H, Packer J, Ward J, Stansfield C, Mytton O, Bonell C, Booy R (2020b) School closure and management practices during coronavirus outbreaks including COVID-19: a rapid systematic review. Lancet Child Adolescent Health
Viner RM, Whittaker E (2020) Kawasaki-like disease: emerging complication during the COVID-19 pandemic. The Lancet 395(10239):1741–1743
Waterfield T, Watson C, Moore R, Ferris K, Tonry C, Watt AP, McGinn C, Foster S, Evans J, Lyttle MD, Ahmad S (2020) Seroprevalence of SARS-CoV-2 antibodies in children-A prospective multicentre cohort study. medRxiv
Webb K, Abraham DR, Faleye A, McCulloch M, Rabie H, Scott C (2020) Multisystem inflammatory syndrome in children in South Africa. Lancet Child Adolesc Health 4(10):e38
Weingart P, Rödder S, Franzen M (2012) Dimensions of medialization: Concluding remarks. In: Rödder S, Franzen M, Weingart P (eds) Sociology of the sciences yearbook 28. The sciences' media connection—Public Communication and its Repercussions. Springer, Dordrecht, pp 363–373
Yonker LM, Neilan AM, Bartsch Y, Patel AB, Regan J, Arya P., Gootkind E, Park G, Hardcastle M, John AS, Appleman L (2020) Pediatric severe acute respiratory syndrome coronavirus 2 (SARS-CoV-2): clinical presentation, infectivity, and immune responses. J Pediatr

Chapter 20
Emotions, Morals and Resilience: The Consumption of News in Ibero-America During the Covid-19 Pandemic

Edson Capoano and Pedro Daniel Rodrigues Costa

Abstract The survey "Perception about information generated during Covid-19" aimed to investigate how the new Coronavirus pandemic changed the audience reception of news. Of the 365 responses collected mostly in São Paulo, Brazil, but also with other Ibero-American respondents, 64.8% felt stunned due to the excess of information consumed, and a set of negative emotions was detected in the act of consuming information, namely with emotions associated with sadness, fear, anger and shame, in about 57% of respondents.

Keywords Emotions · Morality · Moral foundation theory · Health news · Brazil · Covid-19

20.1 Introduction

SARS-COV-2 has impacted the world since the beginning of 2020, from the moment it emerged in Wuhan Province in the previous year. Spreading across the planet at great speed, the new Coronavirus forced changes in the behavior of individuals from personal hygiene habits to social distancing and isolation, altering routines and generating uncertainties in the population as to the configuration of the future. Covid-19 has called for unprecedented global mobilization in terms of the impact on individual, governmental and business agendas, to the point that thinkers of global society assess the pandemic of the new coronavirus as the "world's first fear", according to Lévy (2020, p. 15).

Similarly, for Agamben (2020) the discharge of fear during the pandemic generated anesthetised the collective, removing it's critical capacity over the remaining phenomena. This dynamic will have generated, alternatively, a morality to decide, almost exclusively, in favor of sanitation. Fear was identified as the most used tool for decision-making on how to proceed during the pandemic, which will have established, from the perspective of the Italian philosopher, a vision of the world

E. Capoano (✉) · P. D. R. Costa
Communication and Society Research Centre (CECS) of the University of Minho, Minho, Portugal
e-mail: edson.capoano@ics.uminho.pt

D. M. Berube (ed.), *Pandemic Communication and Resilience*, Risk, Systems and Decisions, https://doi.org/10.1007/978-3-030-77344-1_20

from the perspective of medicine, relegating the other consequently generated crises (economic, environmental, political, social) to the background. According to Lévy, the "exaggeration of the pandemic" (2020, p.14) could only have as its response a growing focus on such crises, which will follow the current period.

The *new normal*, an expression that refers to a series of procedures of individual and collective protection by the sanitarian perspective, was established in a process of collective acceptance, in which rational (the need to contain contamination), emotional (the fear of being contaminated), and moral (the responsibility to protect the collective) were mixed. The fear of evil and morality that encourages being good would have placed the postmodern subject in a typically premodern situation—if Bauman's definition (2007) is taken into account.

In the actual case of coping with a previously unknown virus, there was a worldwide social agreement for the choice of medical sanitation as a single solution, while there was no better solution. With no alternative to decide what to do, nor enough information to safely take decisions, dealing with the first months of the pandemic of the new Coronavirus implied dealing with the pressure of a new normal in our lives. We have often done so in an anxious, depressed, and emotionally exhausted condition for trying to reconcile the socializing demands of the new normal (wearing masks, isolation and social distancing) with our individual demands (desire for security, freedom, survival, financial sustainability and sanitation). The social contract that demands morally correct attitudes during the pandemic made our lives a pendulum between the commitment to the collective dynamics, while fighting for individual sanity (Agamben 2020; Lévy 2020).

But while the decisions to contain the coronavirus were thought out in accordance with the commitment to public health, we still had the right and duty to make certain individual decisions, even though of secondary importance. We had to choose how we faced the pandemic, regarding the decision to wear a mask, the adoption of social distancing, the choice to vaccinate, and even believing whether or not there was a pandemic. Besides having to deal with SARS-COV-2 pandemic in 2020, we also had to face *infodemic*: an excess of information, accurate or not, that make it difficult to distinguish reliable sources and trustworthy guidance from misinformation or fake news (Paho 2020). Therefore, the following questions arise: will the consumption of news about Covid-19, (dis)trust in the press, the presence of fake news in the information process, and the phenomenon of infodemic have generated rational, emotional or moral effects? How were such attributes organized for news readers? What rational, emotional or moral triggers the messages the press generated about the pandemic stimulated in the recipients? In the most problematic moments of the pandemic, how did the news audience keep themselves resilient to persist informing themselves about SARS-COV-2?

Such phenomenon refers to a large increase in the volume of information associated with a specific subject, which can multiply exponentially in a short period of time due to a specific event, such as the current pandemic. The infodemic aggravated the pandemic by making it difficult for reliable sources and trustworthy guidelines to be found and chosen as a reference by citizens in decision-making. As sources of information about the pandemic came from different origins (applications, scientific

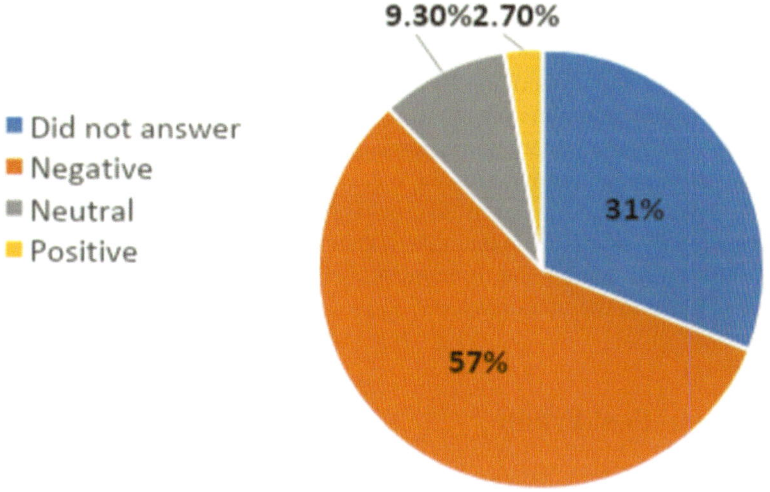

Fig. 20.1 Metropolitan Region of São Paulo (dark spot, on the large map above) and the State of São Paulo (dark spot, inserted in the map of Brazil, on the left). *Source* Wikipedia

institutions, websites, blogs, influencers), there was no way to control what was true, false, incorrect or incomplete about Covid-19 (Costa et al. 2020).

Brazil, the country from which the data in this article came from, suffered from both epidemics. The country has been in the world's leading positions for months in number of cases and deaths. In the period of collection of responses (June 2020), the state of São Paulo had 28,000 confirmed cases of the new Coronavirus (19% of all cases in the country) and was approaching 15,000 dead (24% of all cases in the country) It was the beginning of the first wave of the epidemic in the country, which at that time recorded 1,448,753 cases and 60,632 deaths (Fig. 20.1).[1]

Without a unified sanitary policy for the entire territory, the country was divided between the federal discourse of relativization of the virus hazard and the measures from the states of the federation, which made contingency plans to assist people infected with Covid-19. In addition, the Brazilian population with access to news was bombarded with information about SARS-COV-2, coming from reliable or unreliable sources. A study[2] (AVAAZ 2020) found that 94% of respondents claimed to have

[1] «Coronavirus Brazil». *Ministry of Health. Consulted on 1 July 2020.*

[2] The international organization AVAAZ (2020) conducted a study conducted in Brazil, Italy and the United States with people from the three countries between 9 and 15 April 2020. It selected nine statements about coronavirus and presented them to the interviewees only in text format: two were statements based on correct information and seven were fake contents delied by independent fact-checkers. Six identical false alternatives were presented to all participants in the three countries and, for each country, a false "ad hoc" statement was included that was circulating specifically in local languages.

seen at least one of the fake news about Covid-19 presented by the survey[3]; 73% believed that at least one of the misinformation contents was true or probably true[4]; 46% believed that friends and family were victims of fake news.[5] According to the same study, digital platforms played a major role in information, as well as in the spread of misinformation: 59% of respondents saw at least one of the uninformative content about the new Coronavirus on WhatsApp, and this platform was the most cited source for four of the seven fake news presented to respondents. In relation to Facebook, 55% saw at least one fake content on this social network. WhatsApp and Facebook were among the three most cited sources by Brazilians for all false statements shown in the investigation.

20.2 Theoretical Frame

It is known that individuals are convinced to consume news not only by the rational decision-making process, but also by emotional and moral domains. The media industry has already used this dynamic, through the use of journalistic narratives with emotional or moral triggers (Becket and Deuze 2016). Emotions can be conditioned by an "emphatic media" (Bakir and McStay 2018) or the beliefs of recipients can be used to convince them to consume news (Pennycook and Rand 2019).

There would also be a level of engagement triggered specifically by emotions: intimacy (Siqueira and Bronsztein 2015). After the involvement steps (with the superficial contact of the consumer with the media content, such as the act of accessing the website of a company) and the interaction (search for additional information about the media product, such as the production of comments in social networks), it would take a greater effort on the part of the public, achieved by intimacy, which would encompass the affective connection with the public. Thus, the engagement "involves the feelings and opinions that individuals cultivate or weave in relation to a brand and are usually represented by the words used when referring to them and by the nature of the generated contents" (Siqueira and Bronsztein 2015, p. 148). Finally, one could reach the influence, "this advanced stage of engagement embraces brand awareness, loyalty and the possibility of repurchase" (2015, p. 148).

According to a constructivist perspective, emotions would be socially constructed, since what people feel is conditioned by socialization, through culture and participation in social structures. That is, although emotions always involve a biological component (the nature of emotion and its biological intensity), emotions are conditioned and channeled by sociocultural contexts (Berger and Luckman 2004; Bourdieu 2001). For the sociology of emotions, there are four primary emotions: happiness, anger, sadness and fear. Primary emotions share the same syntax, that is, they are

[3] About 80 percent saw two or more of the fake news; almost 60 percent saw at least three; and 6 percent of Brazilians saw all seven news presented by the study (AVAAZ 2020).

[4] 65 percent of Americans and 59 percent of Italians (AVAAZ 2020).

[5] 41 percent of Italians and 26 percent of Americans (AVAAZ 2020).

perceived in the face of cultural differences. Only one is positive: happiness. Fear, anger and sadness are negative in scope (Stets and Turner 2008).

Emotions played an adaptive role throughout the evolution of people and groups. Negative emotions, such as fear, anger and sadness, make a strong contribution to the generation of social solidarity, leading individuals to sanction socially inappropriate behavior and functioning as a reason to repair damage, both in social relations and in the violation of moral codes (2006). In turn, positive emotions work more in a long-term logic, promoting critical frameworks for the construction of a life that seeks well-being. Positive emotions are the ones that allow us to seize opportunities, create new bonds or strengthen old ones, to the extent that interacting in cooperation reinforces the appearance of positive emotions (Goleman 1995).

Emotions generate feelings, such as the case of hatred and anger, which "are variations of complex bodily feelings caused by a specific object and referred to it" (Damásio 2010, p. 104), driven by consciousness until they are transmuted into feelings (Tiedens 2001). Thus, emotions and feelings are closely intertwined, although the latter are not always the reflection of the former. While anger focuses more on psychobiological processes, hatred is manifested more through the sociological expression of these processes. The masses would be more sensitive to negative emotions as these along with sadness, anger or fear, than to positive emotions, such, as happiness (Tiedens 2001).

For neuroscience, the emotions and moral inclinations of individuals would have innate origin, developed throughout the evolution of the human species. Although this interpretation is viewed with reservations by the Human Sciences, studies of neuroscience on emotions and morality can bring more elements to a wider comprehension of the process of human decision-making, since rationalizations would not contemplate all fundamental components that form individuals' sense of judgment (Mangus et al. 2015); that even brief information about morality—such as whether someone is reliable or not—may shape subsequent individuals' sense of judgment (Delgado et al. 2005); that discourses with moral attributes would make individuals more or less prone to communication, triggering neural stereotyping procedures (if the narrative is suspicious) or would reduce the activation of stereotypes (if narratives have a high burden of empathy).

Specifically on communication, cognitive neurosciences and their media neurosciences subfield seek to understand media consumption and its effects on receptors, and in their studies would have found a relationship of how moral narratives communicate with the neural substrates of reasoning and decision-making (Mangus et al. 2015), as through the activation of brain regions[6] when there are moral judgments, more than other forms of mental operations (Zahn et al. 2011).

Psychology, the science that most investigates emotions, uses various methods such as studies on personality and temperament to evaluate emotions as natural

[6] They are Cortex frontopolar, prefrontal cortex ventromedial, prefrontal cortex and dorsolateral, ATL, later STS, amygdala, hiand the basal ganglia (Mangus et al. 2015).

Table 20.1 Eysenck's table with inventory of emotions and their respective personalities

Psychoticism	Extroversion	Neuroticism
Aggression	Sociability	Anxiety
Antisociality	Assertiveness	Shame
Impersonality	Dominance	Emotion
Egocentrism	Activity	Tension
Dislike	Independence	Irrationality
Determination	Insecurity	Low self-esteem
Coldness	Persecution	Fault
Creativity	Insurrection	Depression
Impulsiveness	Excitement	Melancholy

Source (Bates 1989, in Beatty et al. 2001)

triggers or constructed aspects in each individual,[7] studies that dialogue with those of communication to the extent that such mechanisms may affect the consumption of information, since emotions stimulate personality traits, possible responses and tendencies of reaction to assimilated emotions. (Beatty et al. 2001, p. 69). Or when news can trigger emotions whose personality traits stimulate extroversion, neuroticism and psychoticism, according to Eysenck's model of three factors, *supertraits* or big three (in Beatty et al. 2001) (Table 20.1).

In other words, the emotions contained in the three factors, *supertraits* or big three from the columns above, can trigger or suppress personality reactions: hostility or aggression (characteristics of psychoticism), cooperation or sociability (extroversion), avoidance fearful or resilience over fears. Psychoticism is a trait that denotes the lack of self-control or restriction in the dimension of impulsivity. Extroversion, on the other, is a trait that defines the internalization or externalization of emotions by social situations of interaction. Finally, neuroticism is the trait of anxiety expression. (Beatty et al. 2001, p. 87).

This model makes it possible to understand how individuals interact in situations of inhibition, which fits with this study because the Sars-Cov-2 pandemic has been the largest global social inhibition event of the twenty-first century so far. The *communibiological* paradigm (Beatty 2008), for example, has a model for communicative apprehension with two structures, one responsible for neuroticism, the other for introversion that, combined, produce a list of behaviors known as communication apprehension (Beatty et al. 2001).

[7] (Bates 1989, in Beatty et al. 2001) defines temperament as "individual differences rooted biologically in behavior trends that are present early in life and are relatively stable in various types of situations and over time" (p.4) According to Bates, temperament is conceptualized on three levels: Temperament is expressed behaviorally; (b) temperament is related to individual differences in neurological function and anatomy; and (c) temperament is hereditary (Bates 1989, in Beatty et al. 2001 pp. 60–61). Temperament and Psychobiology indicate that our behavior and communication are not only overruled by conscious control. We believe that hypothetical explanations about human behavior are only necessary when the biological question cannot respond to human communication (Bates 1989, in Beatty et al. 2001).

The field of psychology also focused on morality to understand human actions and reactions when individuals interpret the world around them. Haidt et al. (2011) go beyond the two traditional moral foundations of the field (justice and damage) and credited five moral foundations (justice, damage, loyalty, authority and purity) as responsible for the narratives that individuals assess as right and wrong, and that depending on the social group analyzed, there would be predominance of a foundation over others. Together, the attributes compose the Moral Foundations Theory (MFT) (Graham et al. 2013), which explains the origins and variations of human moral reasoning based on innate and modular foundations and how these alter the rational decision on the consumption of information in different cultures.

There would be four basic precepts for the MFT: nativism (the innate ability of the human being to solve social problems, result of the evolution of the species); cultural learning (the predominance of one foundation over others, what cultural learning will define according to the social formation of the individual); the primacy of intuition (the existence of unconscious and automatic mechanisms that precede rational judgment); and plurality (social problems during human evolution have selected cognitive mechanisms for its resolution). Together, these precepts would have defined so-called "moral" foundations for human decision-making: (1) care/damage (concern for the suffering of others); (2) justice/cheating (preference for reciprocity and justice); (3) loyalty/betrayal (concern for the common good and prejudice against outsiders); (4) authority/subversion (reference to domain and hierarchy); (5) holiness/desecration (concern for purity).

Researchers in this field have already investigated the moral attributes within political discourses (Graham et al. 2009), participation in religious events (Haidt 2009) and even the moral motivations inserted in the discourses of environmentalist Greta Thunberg (Weber 2019, in Tassof, 2019). In the study on the narratives of the environmentalist, for example, Greta's language is related to care/damage (such as "growth", "future" and "suffering") and justice/cheating (such as "right", "solutions" and "consequences").

There are methods to measure this phenomenon, such as the Moral Foundations Questionnaire MFQ methods (Graham et al. 2011). The methodology of the questionnaire, equipped with questions inspired by the MFT, was proven through a sample of more than 30,000 people, in which they found six previously mentioned factors that, when related from the binomial individuality/collectivity, allowed better interpretable results. With several confirmatory stages of the results, the MFQ clarifies which moral attributes have the most factorial burdens, hence, they are the most important for the moral evaluation of the respondents.

20.3 Methodology

Our study entitled "Perception of the information generated during Covid-19" aimed to investigate how the consumption of news about the new Coronavirus affected the public during the Sars-Cov-2 pandemic and under the influence of *infodemic*. The

365 responses (227 valid, mostly from the city of São Paulo, Brazil, but also with some other respondents around Ibero-America) were collected during the month of June 2020. The online form was hosted and made available on the Google Forms platform, with invitations through Facebook platform by the FB Message Sender tool to interest groups with the keywords journalism and journalists.

We obtained 365 responses, of which 227 were valid, originated in most of the metropolitan region of São Paulo[8] (58%), 5.4% from the rest of Brazil and 1.9% from the Americas and Europe. About generations (Dimock 2019), half of the sample is composed of Millennials (53.2% between 23 to 28 years-old): a quarter of the sample is composed of Generation Z (18 to 22 years) has 12.1% of the sample; There is no significant difference between females (48.2%) and males (48.5%); Three quarters of the sample knew someone who had contracted Covid-19 up to that time.

The 13 questions of the survey have semi-structured format (with closed and open answers) and in the Likert scale (1 to 5 points between disagreement and agreement). They were generated with the support of theoretical reference on fake news (Tandoc et al. 2018) and *infodemic* (PAHO 2020); the Eysenck (1986, in 2001) studies on emotions and personality; and the Moral Foundations Theory (Graham et al. 2011). With such questions, the goal was to understand whether news readers who were exposed to fake news or excessive information, therefore, to infodemic, and whether and how such a state led to certain emotions and motivations to continue consuming news. Thus, the following questions were generated about emotions and motivations:

> What motivates you the most when consuming newssabout Covid-19? (semi-structured, open response, with the following options already established:
>
> Prevent health hazards?
>
> Sympathize with other affected people?
>
> Recognize those responsible for crisis management and their powers?
>
> Know the new routine and what life will be like in the future?
>
> Follow authorities and reference figures?
>
> Concern for life on Earth and the environment?
>
> Attention to restrictions on your freedom?
>
> (Capoano and Costa 2020).

20.4 Results

Regarding questions about the consumption of news and fake *news,* we most respondents consumed news on TV and social platforms. About 75.3% of respondents considered themselves well or very informed about Covid-19 and 57.3% increased news consumption. About 64.8% mentioned feeling stunned by the news and 39.7%

[8] The "great São Paulo" is the conurbation of the city of São Paulo with 10 other cities around it and brings together 21.5 million inhabitants, one of the ten most populous regions in the world. Alone the city of São Paulo has 12 million people, and the Brazil has not total 214 million inhabitants (source: Brazilian Institute of Geography and Statistics - IBGE: www.ibge.gov.br).

Graph 20.1 Graph with results of question 11. "The emotions generated by the consumption of news about Covid-19 are mostly…" (results as a percentage). *Source* own authorship. N = 365. NR = Not responding

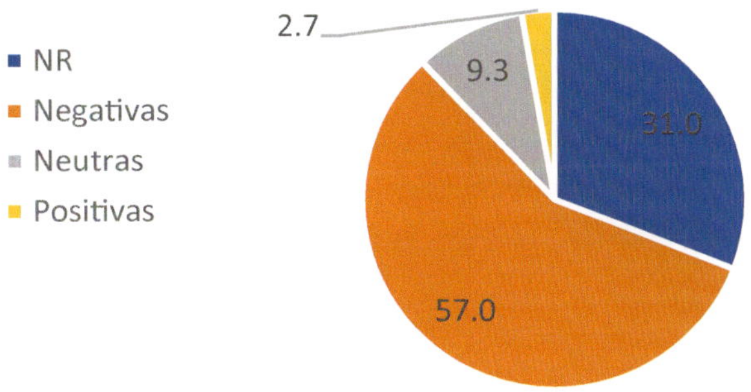

consumed news mainly about Covid-19. In general, most of the news they consumed was based on specialist/scientist, medical/health professional sources, WHO and government. In addition, about 58.1% consider the work of the press in the coverage of the pandemic good. However, 81.9% of respondents identified fake news while being informed. Furthermore, the most consumed fake news format was text and audiovisual, and the most common methods were lies, information out of context, altered or falsified data, and editing the original content.

Specifically on emotions, about 57% consider that most emotion aroused by news consumption was negative, while about one in ten (9.3%, respectively), considered the emotion neutral. Only 2.7% of respondents considered the emotion positive with regard to the pandemic (see Graph 20.1).

About the most recurrent emotion experienced while consuming news during this period of the pandemic, the responses were organized according to the first emotion chosen. We consider the first choice among the list of emotions the one with the highest notoriety. Thus, we analyze the emotion that was mentioned in the first place. The data revealed that in about 15.3% of the cases, the emotion anger was the most chosen by the respondents. Fear (12.3%) was the second choice. Sadness (9%) and disgust (8.8%) was next among the options, closely followed by shame (6.8%) and surprise (5.5%). The remaining emotions chosen in the first place are between 0.3% and 3% (Graph 20.2).

Regarding the moral motivation in the consumption of news about Covid-19, the majority of the answers were concentrated in the option "avoid health hazard" (17.8%) and "know the new routine and how life will be in the future" (15.9%). In 7.9% of the answers, respondents focused on "Concern for life on Earth and the environment", and in 7.1% of cases in "Recognizing those responsible for crisis management and their tasks" and in "other combinations" (Graph 20.3).[9]

[9] The item "other combinations" it comprises a set of combinations with the following options: "Avoid damage to health", "know the new routine and how life will be in the future", "follow authorities and reference figures", "Concern for life on Earth and the environment", "Recognize those responsible for crisis management and their tasks" and "general interest in such disruptive social phenomenon".

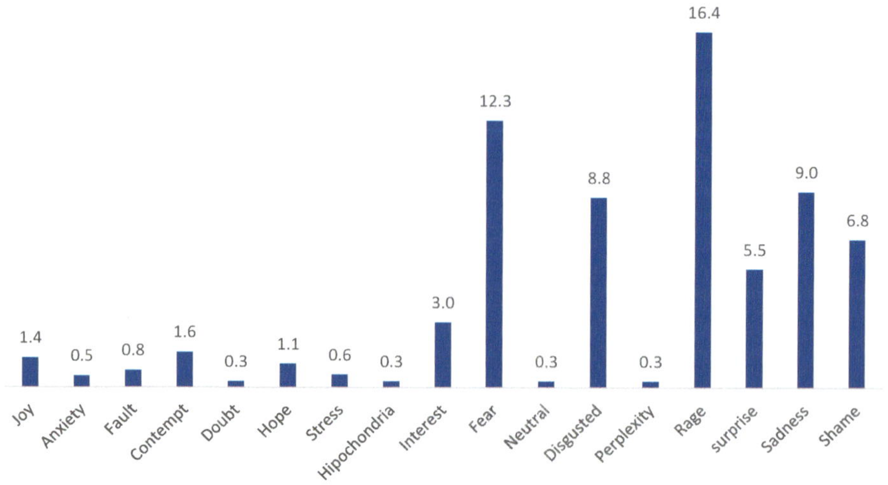

Graph 20.2 Graph with results of question 12. "What is the most common emotion generated by your country's press coverage on Covid-19"(results as a percentage). *Source* own authorship. N = 365

20.5 Final Considerations

As relevant crossover between the meaning of emotions (positive, negative and neutral) and the moral motivation in the consumption of news about Covid-19, about a quarter of the emotions considered negative are related to "avoiding damage to health" (27.4%) and with "knowing the new routine and how life will be in the future" (22.6%). The emotion associated with "avoiding health hazards" is negative in 87.7% of cases, as well as in 81% of cases when the option is "to know the new routine and how life will be in the future".

The emotion associated with "concern for life on Earth and the environment" is negative in 12% of cases, followed by "recognizing those responsible for crisis management and their tasks" (10.1%). On the emotions considered neutral, the only highlighted case is the item "knowing the new routine and how life will be in the future", which represents 23.5%. For positive emotions, the data is close to zero, which means that there is no relationship between emotion and the reason for consumption of news.

If the emotions stated in the responses of our survey are interpreted according to the constructivist framework, we can assume that the biological factors that may have driven emotions of survival, such as fear, sadness and anger, have been combined with the cultural context that sought answers to the pandemic slightly known beforehand, therefore, it encouraged respondents to inform themselves about avoiding health hazard and knowing what life would be like in the future. But as in June 2020, when they responded to the survey, there was still no positive news about the pandemic—such as approved vaccines or case control, the emotional drive that may have spurred

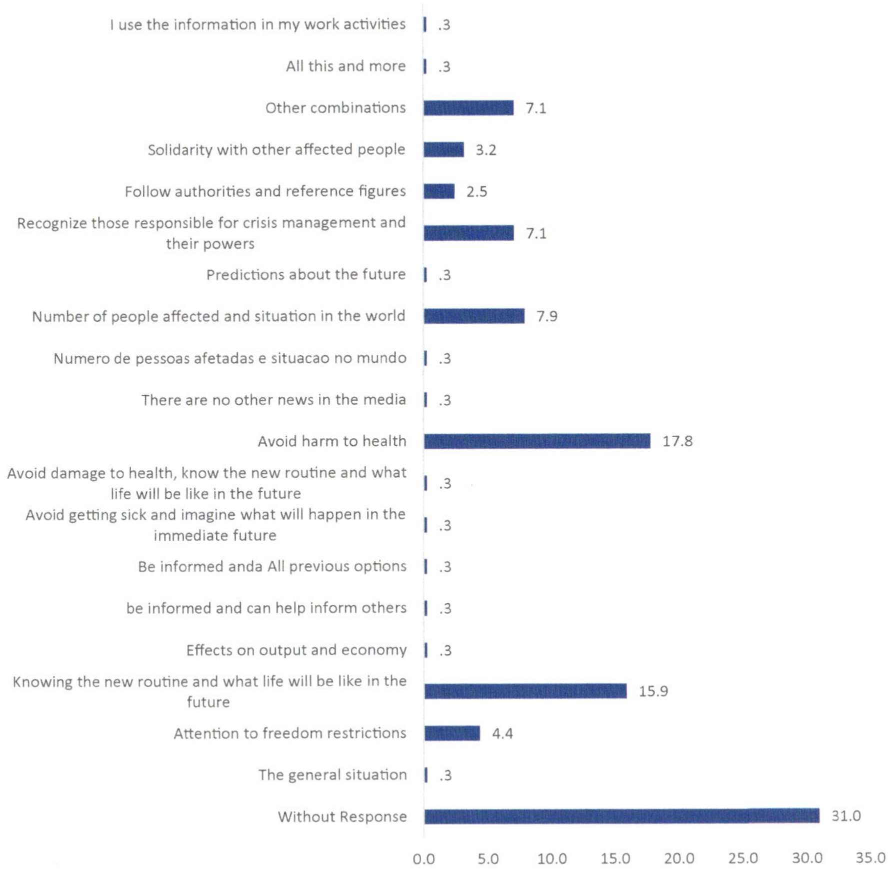

Graph 20.3 Graph with results of question 13. What motivates you the most when consuming news about Covid-19 (results as a percentage)? *Source* own authorship. N = 365. NR = unresponsive

them to consume news was fed back by the fear and anger they felt, the first and second emotions most chosen by the respondents, and generating others yet, such as sadness and shame, the third and fourth emotions most selected by the respondents, since the social context portrayed by the Brazilian press presented the social unruly social scenario, such as lack of use of masks, social distancing and quarantine not adopted by most Brazilian citizens, as obstacles to prevent the advance of Covid-19 in Brazil. It is also understood that the emotions anger and shame may have been generated by a posture of social surveillance in favor of collective health, which may have been carried out both in person, in everyday life, as in the symbolic, in the consumption of news that portrayed appropriate or inappropriate conduct scans for the moment.

From negative emotions, however, one may have drawn energy to resist the moment and seek a solution, in facts recorded in the present or expectations for the future. Therefore, we assume that the search for better social contexts may have been another motivation to read the press during the period, either by altruistic initiatives, such as volunteers to collect clothes and food for the needy, or of health professionals, who served as models of heroism and sacrifice to the rest of society, either by speculation of a solution to the pandemic, such as the discovery of vaccines, or by the projected scenarios of a post-Covid-19 world. In other words, positive emotions were expected to be achieved in the press.

If we adopt the previously presented concepts of neuroscience to interpret emotions and decision-making to continue consuming news during the pandemic, it is clear that rationality was not enough to find the answers to what we should do during the pandemic to be safe and peaceful. Hence, our brain may have sought the answer in the emotions and instinctive interpretations (the moral foundations of neuroscience) of what we were living to complete the puzzle. Thus, news that brought narratives that triggered fear, anger, sadness and shame, may not have been the motivators of emotions, but may have been the rational justification for feeling them.

It is not new for the press to use emotions and moral judgments to attract public interest, as presented in the theoretical frame of this paper. However, in this moment of unprecedented social fragility, it is pondered whether the effort of the media to inform society may have potentialized psychological disturbances of the receiving public.

The *infodemic* environment, taking into account the responses "true", "false", "outdated" and "out of context news", suggests positively. If we frame the emotions of the answers in the personality *supertraits*, they dialogue with the emotions contained in the traits of neuroticism, extroversion and psychoticism. It is inferred that, by following the opposite direction of the Eysenck model—from responses/reactions to the consumption of news about Covid-19 of the survey and relating such emotions generated with the original three-factor model emotions, one safely arrives at the personality *supertraits* stimulated by the news. In conclusion, the survey's responses infer that the consumption of information during the pandemic stimulated individuals' personality *supertraits* in the sample, especially neuroticism (Table 20.2).

Eysenck (1986, in 2001) understands that neuroticism encompasses both calm and emotional stability, as well as nervousness and irritability. The brain of neurotics

Table 20.2 Table with emotions related to the big three of EYSENCK's inventory (Bates 1989, in Beatty et al. 2001)

Respondents' reaction to the news about Covid-19	Emotions related to those of the three-factor model	Personality supertrait
Sadness	Depression and melancholy	Neuroticism
Fear	Insecurity	Extroversion
Anger	Aggression	Psychoticism
Shame	Shame	Neuroticism

works on the threshold of activation of the sympathetic nervous system, responsible for activating the "fight or flee" reaction when stimulated by emotions, as a primordial survival instinct (Goleman 1995). Clearly, it is noted in the answers and emotions chosen to represent how consumers of news feel, trends to the negative neurotic behavior.

On the other hand, we relating the moral foundations of the MFT with the answers collected about the moral motivations of keep consuming news during the pandemic, we realize that the foundations with the highest load form the binomial *damage and loyalty* (which together reach 54.2% of the answers), while the attributes with the second highest load refer to the binomial *purity and justice* (23.8% if added). Within the key individuation/collectivism of the MFT, the greatest motivation to consume news during the pandemic seems to have been an exercise of personal survival and compassion to the collectivity, and as a secondary motivation the expectation for the resuming of the sanitary condition on the planet and for finding the reasons and motivators for Covid-19 pandemic to have happened.

If by hypothesis we formulate a sentence with the moral characteristics of each binomial, in the first *(damage and loyalty)* we will have the respondents who continue to consume news during the pandemic, according to the individual interest of avoiding the primary damage (contracting covid-19) or secondary to the pandemic (economic, psychological aspects, etc.), while collectively constraining themselves, through a notion of belonging to a large group of affected, known through their social groups but in a national and global way by the media. The commitment to such a group may have been through the feeling of compassion for others or self-sacrifice, loyalty and vigilance against betrayal, which leads us directly to the adoption of practices to contain the spread of the virus, such as the use of masks, personal hygiene and social distancing.

In the second binomial, *purity* and *justice,* an individualizing moral sentence is revealed that values the physical and spiritual influence on healthiness and the control of desires, probably referring to how to remain physically healthy and mentally balanced during the period of quarantine and social isolation. In the case of Brazil, there was no national protocol of procedures during the pandemic, but independent decisions taken in each state of the country, which relegated the population having to decide what their contagion prevention strategy would be like while having to continue working in person, exposing itself to Covid-19. The collectivist moral attribute of this binomial, justice, concerns the global notion of rights, but also of cooperation, competition and/or cheating in society. In a context such as June 2020, when the survey was applied, there was already a lack of virus protection (PPE) resources in the country, such as masks and alcohol, mechanical respirators and hospital beds. Probably, the motivation of the interviewed readers was how the management of these scarce resources had been carried out, whether fairly or unfairly. In fact, Brazilian articles from that period reported the embezzlement of money destined for the purchase of personal protective equipment, or how there were a considerable number of citizens who did not adhere to social isolation, quarantine or mask use, which may have activated the moral foundation of (in)justice in the recipients.

References

Agamben G (2020) A medicina como religião. [Medicine like religion]. IHU Online Magazine. Retrieved 08 Oct 2020 from http://www.ihu.unisinos.br/78-noticias/598583-a-medicina-como-religiao-artigo-de-giorgio-agamben
AVAAZ (2020) Brasil está sofrendo uma infodemia de Covid-19. [Brazil is suffering an infodemia of Covid-19]. Retrieved 08 Oct 2020, from: https://secure.avaaz.org/campaign/po/brasil_infodemia_coronavirus/
Bates JE (1989) Applications of temperament concepts. In: Beatty MJ, McCroskey JC, Valencic KM (2001) The biology of communication: a communibiological perspective. Hampton Press
Bauman Z (2007) Fragmented Life: essays on postmodern morals. Relógio D'água, Lisbon
Beatty MJ, McCroskey JC, Valencic KM (2001) The biology of communication: a communibiological perspective. Hampton Press
Beatty MJ (2008) Communibiology. Wiley Library. Retrieved October 08, 2020, from: https://onlinelibrary.wiley.com/doi/abs/https://doi.org/10.1002/9781405186407.wbiecc056.10.1002/9781405186407.wbiecc056
Bakir V, McStay A (2018) Fake news and the economy of emotions: problems, causes, solutions. Digit J 6(2):154–175. https://doi.org/10.1080/21670811.2017.1345645
Beckett C, Deuze M (2016) On the role of emotion in the future of journalism. Soc Media + Soc 2(3):2056305116662395
Berger PL, Luckmann T (2004) The social construction of reality: a treatise on sociology of knowledge. Dinalivro, Lisbon
Bourdieu P (2001) O Poder simbólico [The symbolic power]. Difel, Lisboa
Capoano E, Costa PR (2020) Survey Perception of the information generated during Covid-19. Retrieved 08 Oct 2020, from: https://forms.gle/yPwerVMtE1DxYrD1A
Costa B, Viegas DJ, Moreira TA, Abreu PA (2020) The anti-vaccine movement on YouTube inpost-truth times: health education or misinformation? Rev Mídia e Cotidiano 14(1):220–239. https://doi.org/10.22409/rmc.v14i1.38210
Damásio A (2010) O livro da consciência—A construção do cérebro consciente. [The book of consciousness—The construction of the conscious brain]. Círculo de Leitores, Lisboa
Delgado MR, Frank RH, Phelps EA (2005) Perceptions of moral character modulate the neural systems of reward during the trust game. Nat Neurosci 8(11):1611–1618. https://doi.org/10.1038/nn1575
Dimock M (2019) Defining generations: where millennials end and generation Z begins. Pew Res Cent 17:1–7. Retrieved 08 Oct 2020, from: http://tony-silva.com/eslefl/miscstudent/downloadpagearticles/defgenerations-pew.pdf
Goleman D (1995) Emotional intelligence. New York, NY, England
Graham J, Haidt J, Nosek BA (2009) Liberals and conservatives rely on different sets of moral foundations. J Pers Soc Psychol 96(5):1029–1046. https://doi.org/10.1037/a0015141
Graham J, Nosek BA, Haidt J, Iyer R, Koleva S, Ditto PH (2011) Mapping the moral domain. J Pers Soc Psychol 101(2):366
Graham J, Haidt J, Koleva S, Motyl M, Iyer R, Wojcik SP, Ditto PH (2013) Moral foundations theory: the pragmatic validity of moral pluralism. Adv Exp Soc Psychol 47:55–130. Academic Press
Haidt J (2009) Moral psychology and the misunderstanding of religion. In: Schloss J, Murray M (eds) The believing primate: Scientific, philosophical, and theological reflections on the origin of religion. Oxford, New York, pp 278–291
Lévy BH (2020) Este virus que nos enlouquece. [This virus that drives us crazy]. Guerra e Paz, Lisboa
Mangus JM, Adams A, Weber R (2015) Media neuroscience. Emerging trends in the social and behavioral sciences. Interdisc Searchable Linkable Resour 1–14. https://doi.org/10.1002/9781118900772.etrds0213.cCroskey; Beatty JCMJ, Valencic KM (2001) The biology of communication: a communibiological perspective. Hampton Press (NJ)s

PAHO (2020) Understand the infodemia and misinformation in the fight against Covid-19. Department of evidence and intelligence para action in health. Retrieved 08 Oct 2020, from: https://iris.paho.org/bitstream/handle/10665.2/52054/Factsheet-Infodemic_por.pdf?sequence=14

Pennycook G, Rand DG (2019) Lazy, not biased: susceptibility to partisan fake news is better explained by lack of reasoning than by motivated reasoning. Cognition 188:39–50. https://doi.org/10.1016/j.cognition.2018.06.011

Siqueira OS, Bronsztein KP (2015) Social games and advertising: about reflecting the four levels of engagement of digital consumer. Culturas Midiáticas 8(1):138–154

Stets JE, Turner JH (2008) The sociology of emotions. In: Handbook of emotions. The Guilford Press, New York, pp 32–46. ISBN 978-1-59385-650-2

Tandoc EC Jr, Lim ZW, Ling R (2018) Defining "fake news". A typology of scholarly definitions. Digit J 6(2):137–153. https://doi.org/10.1080/21670811.2017.1360143

Weber, R. (2019), in Tassof, H. (2019). An Impassioned Speech. The Current. UC Santa Barbara. Avaiable at https://www.news.ucsb.edu/2019/019645/impassioned-speech

Tiedens LZ (2001) Anger and advancement versus sadness and subjugation: the effect of negative emotion expressions on social status conferral. J Pers Soc Psychol 80(1):86–94

Zahn R, de Olivera-Souza R, Moll J (2011) The neuroanatomical basis of moral cognition and emotion. From DNA Soc Cogn, pp. 123–138. https://doi.org/10.1002/9781118101803

Chapter 21
Social Media Creating Resilient Communities During COVID-19: India, Bangladesh & Pakistan

Ali Saha

Abstract Unlike the general norm that social media use leads to negative impacts, this chapter reflects on how these platforms helped create resilient communities during the pandemic. Using survey method along with Katz et al.'s (1973) Uses and gratification theory, I conceptualize the gratification sought through creating, watching, and sharing media contents in developing countries-India, Pakistan, and Bangladesh. Precisely, the data collected demonstrates that social media helped people deal with the impacts of COVID-19 and its unprecedented threat to physical and mental health. Moreover, the increasing use of media (apart from the urge to gather more information, lockdown, and resistance to physical mobility) was to fulfill the need to release stress and escape from the reality of high uncertainty and crisis. This chapter ends with demonstrating that despite the cultural uniqueness among different countries, media as a common platform helped create resilience among the population.

Keywords Pandemic · COVID-19 · Social media · Resilience · Uses and gratification

21.1 Introduction

In November 2019, a cluster of atypical cases of pneumonia was reported in Wuhan, China. This was later designated as COVID-19, the acronym for Coronavirus disease 2019 by the World Health Organization (WHO) was designated on 11 Feb 2020 (Anand et al. 2020; Jena 2020). The causative virus, SARS-CoV-2, was identified as a novel strain of coronaviruses that shares 79% genetic similarity with the SARS-CoV from the 2002 to 2003 SARS outbreak (Anand et al. 2020; Xiong et al 2020). Unlike the others previous strains, SARS-CoV-2 was identified to be a highly transmissive type of virus. Statistically, it took 67 days (~7th march) for COVID-19 to infect 100,000 people, an additional 12 days (~19th March) to infiltrate another 100,000,

---

A. Saha (✉)
Monash University, Melbourne, Australia
e-mail: ali.saha@monash.edu

D. M. Berube (ed.), *Pandemic Communication and Resilience*, Risk, Systems and Decisions, https://doi.org/10.1007/978-3-030-77344-1_21

and a third invasion that took only 4 days (~23rd March). Following the death of more than 4,291 people and 118,000 cases in 114 countries, on 11 Mar 2020, the WHO declared the outbreak a global pandemic (Anand et al. 2020; WHO 2020).

21.2 Impact of Virus on Human Lives

The onset of this virus has drastically altered people's lives, as well as multiple aspects of the global, public, and private economy. Decline in tourism, aviation, agriculture, and the finance industry owing to the COVID-19 outbreak resulted in massive reductions in both the supply and demand aspects of the economy (Nicola et al. 2020). The lack of specific treatment and the swift spread of the pandemic had triggered the governments in most of the countries to impose forceful lockdown measures to restrict the spread caused by mingling of the masses. The uncertainties and fears associated with the virus outbreak, along with social distancing, mass lockdowns and economic recession had an additional impact on lives. The situations worsened by limiting people to social connections, travel, and in certain conditions even restricted people to the four walls of the house without any access to open spaces, parks, exercise, or entertainment. Movie premieres, concerts, and festivals among many other media events were cancelled, and entertainment venues closed to prevent mass gatherings in public spaces. This had a direct impact on in-home media consumption around the world. Watson (2020a, b) suggests that 35% of total respondents read more books or listened to more audiobooks, 18% having listened to more radio due to the COVID-19 pandemic, whilst more than 40% of consumers spent longer on messaging services and social media during COVID-19 pandemic.

21.3 Communication via Media During Pandemic

While novel coronavirus impacted the population psychologically, accessing different forms of media to cope with the situation has been at the forefront of many consumers' minds since COVID-19 hit their country. On one hand, there were reported manifestations of distress, anxiety, depression, fear of contracting the infection, or guilt of being a carrier, and insomnia in general populations. Along with feelings of loneliness, and anger (Xiong et al. 2020) and a projected increase in suicide reported in the USA, Canada, Pakistan, India, France, Germany, and Italy (Hao et al. 2020; Mamun and Ullah 2020; Xiong et al. 2020; Thakur and Jain 2020; Wang et al. 2020). On the other hand, in countries all over the world, mainstream media, social media and news consumption grew as consumers balanced entertaining themselves with keeping up to date. In fact, several news channels in Italy saw audience growth of more than 100–200% and live TV viewing in the United Kingdom jumped by almost 50% (Statista Research Department 2020; Watson 2020). Of all the other media sources, the era exhibits an unexpected boost to the engagement on social media platforms such as WhatsApp, Facebook, Instagram, Tik Tok and many more.

The Harris poll conducted in the United States reflected an overall increase in the use of social media platforms during the pandemic, with people spending more time compared to the previous years (Samet 2020). L&A Social census (an agency that analyzes social media trends for business and marketing) noted a 30% increase in social media use among Australians (Page 2020). A similar increase was noted in the South Asian countries of India, Bangladesh, and Pakistan. While the uses of internet and smartphone ownership worldwide is rapidly increasing, due to the limited access to phone and internet usage guided by multitude of factors, a big section of Indian, Bangladesh, and Pakistan population do not have proper technological access, social media and hence are victims of digital divide. Nonetheless, Pew Research Center claims that social media use has been increasing in these developing countries and has shown an increase by 53% in 2017 where roughly 4-in-10 adults used social networking sites and it has grown closer to twice the rate in 2019.

Pew Research further adds that among people who use the internet worldwide especially for entertainment and communication purposes, those in developing countries often turn out to be more likely than their counterparts in advanced economies. This rate of internet usage has shown a significant increase during the pandemic and has been discussed in other sections of this chapter. Currently with over 560 million internet users, India is the second largest online market in the world, ranked only behind China. Bangladesh ranks 9th with 96.2 million users and Pakistan with 76.2 million (Clement 2020). Approximately half the numbers of the total population having access to internet use social media. In the next section, this chapter discusses the increased use of social media platforms and later conceptualise the increased usage to resilience by utilising examples from the social media use in the three coronavirus-affected and densely populated South Asian countries named above.

21.4 Social Media Penetration and Use

While the internet penetration worldwide in 2020 has shown a rapid development, the social media statistics reflect a greater rate of penetration than ever before. With the continuous development of information technology, social media has gradually penetrated all areas of human social life, such as people's daily life, education, and work. Social media is becoming the most important channel of communication through which messages can reach audiences and attention can be focused on the required discourse, especially when visual graphics (pictures and video) are involved. Social media has also been transforming the concept of public sphere, where it gives the opportunity to create new virtual communities which recognizes different arguments: and accomplishes the interest of most members of the virtual community. Chan (2013) adds that the characteristics of social media can be harnessed and used as a tool with an organisational capacity to demonstrate resilience in response to crises, by creating new avenues for collaboration to help build more resilient communities over time. Bedard-Gilligan, an Associate Professor in the Department of Psychiatry and Behavioral Sciences at the University of Washington states "Resilience is when you have the ability to break and get back together" (Princing 2020). Helsloot and Ruitenberg (2004) and Jurgens and Helsloot (2007) call citizens naturally responsive and resilient when confronted with disasters, even though they generally hardly prepare for them. In fact, they state that the media, especially social media plays an important role in creating resilience within their citizens during the time of social breakdowns. McFarlane and Norris, call disaster as a potentially traumatic event caused by natural, technological, or human that is collectively experienced, has an acute onset, and is time-delimited (2006). Following that the coronavirus, evolved naturally, started suddenly with a short or absent warning period, and has caused death or injury to many people simultaneously, in this chapter the pandemic surely qualifies as a disaster.

A common man looks for ways to gratify their needs in daily life. In the era of media and communication, media became one of the crucial modes to gratify their needs. The people are in lookout for new information and during the pandemic, when the information seeking activity was limited to isolation, the media became the major mode to communicate and exchange information. Many times, media was the mode to even validate one's ideas, thoughts, and presence. Moreover, social media gave the opportunity for initiators and participants to construct new online communities to share common interests, attitudes, and values. Therefore, the participants could independently involve in expressing their own interests in the discussions of virtual communities. While the use of social media for posting photos, commenting, liking or for entertainment was quite common even before, the pandemic gave a new meaning to the use of social media. During the pandemic, the social media was used for 'social' purposes, i.e., to interact and communicate. Overall, the use of social media was taking a new turn.

21.5 Coronavirus in India, Bangladesh and Pakistan and Media

To understand how and why the people used social media during the pandemic and how the use of social media platforms created resilience, literature analysis, survey, and interviews were conducted among the population of India, Bangladesh, and Pakistan. These countries were one of several countries with the highest rate of increase in coronavirus. With a huge population and comparatively less access to medical facilities and hospitals, COVID-19 took a drastic turn, whereby mid-February 2021, it had infected 11 million people in India alone (the second highest rate of infection in the world) and killed more than 15 thousand of the people. In India, the first confirmed case was reported in Thrissur, Kerala, on January 30, 2020, where the affected individual had a travel history to Wuhan, China. The transmission escalated during March first week, and on March 24, 2020, the prime minister of

India-Narendra Modi, called for a complete nation-wide lockdown, which lasted until August 2020 (Iqbal and Dar 2020).

Bangladesh confirmed their first COVID-19 case only on the 8th of March 2020 (IEDCR 2020) followed by a nationwide lockdown from March 26 which was extended several times until May 30, 2020. To prevent human transmission. The government deployed armed forces to facilitate social distancing but as 11 million people left Dhaka (the capital city of Bangladesh) to return to their home districts, the disease spread nationwide.

Eventually, due to economic breakdown, partial lockdown was restored on April 25, 2020, when all ready-made-garment (RMG) factories, industries, private offices, and business centres could open, and less community awareness about the disease led to an increased transmission among millions of people (Bodrud-Doza et al. 2020).

In Pakistan, the first case of COVID-19 was reported from Karachi around the same time on February 26, 2020 with provincial lockdown instituted on March 23, 2020 (Chandir et al. 2020). With estimated population of 204.65 million, despite the lockdown, the virus spread into various regions nationwide and within 45 days, on April 10, 2020, Pakistan's tally reached 4601 confirmed cases of COVID-19 with 66 fatalities (Abid et al. 2020).

These countries which have numerous informal settlements, ranging from dense, urban slums near the city centre to less dense, urban sprawl fading into peri-urban areas are susceptible to debilitating diseases, given that they live in densely packed areas with poor access to health services, information, and low rates of immunisation coverage. Moreover, the initial response of the public to the emerging threat of COVID-19 in the three countries was that of a generally reported apathy and indifference. Lack of public awareness was commonplace throughout the countries and mass prayer, events, public gatherings, continued even as alarms were set off as to how such public activities could exacerbate the spreading of the pathogen.

As discussed elsewhere, with the increase in the number of coronavirus cases and strict lockdown measures, in these three countries, the engagement with social media showed a sudden change. This change was not just 'how much' the people used social media but 'how' they juggled between using the different platforms for distinctive information and functions. Similarly, the interaction between audiences and media is constantly being reshaped during the pandemic. When compared to 2019, more people having access to the internet logged into social media platforms. The rate of logging in further doubled during the pandemic. The percentage of internet usage grew to an overwhelming number of 40.6% by June 2020 in India, 58.4% in Bangladesh and 32.4% in Pakistan (Internet world stats 2020). A survey among 110 participants in each of the three countries reflected that approximately 83% of the population having access to internet and smartphones, agreed to be spending comparatively more time on media during the pandemic than before. Interestingly out of these 83% only 61% of them agree to be spending more time on social media while the others used video streaming platforms like Netflix and Amazon Prime as the escape from reality. On an average the population agree to have been spending around 5–6 h on social media, while a few of them agreed to be spending more than 12–13 h a day on these platforms. A similar data from a survey among the population in the countries of Australia, and USA conducted by Trifonova (2020) derives that 43% of the population especially admit logging in for longer because of the outbreak. Overall, the data and literature analysis reflect that irrespective of the demography the engagement with media and communication increased due to the pandemic.

21.6 Theoretical Approach to Understanding Engagement in the Three Countries

Engagement, according to Higgins (2006) is to be occupied, involved and interested in something. Engagement, especially media engagement, can be both personal and social with subdivision based on its purposes such as (i) utilitarian, (ii) hedonic and/or (iii) social psychological (Mersey et al. 2010). Some scholars state that engagement about media consumption is the collective experiences that readers or viewers have with a media brand that create preference among other options (Mersey et al. 2010). The process of selecting information sources is achieved based on two factors: usefulness and repetition. Engagement can be understood as the utility that audiences get from a particular media. It is often the final product of satisfactory/gratifying experiences that audiences frequently discover on the same platform.

As the coronavirus affected people at various levels, physical, mental, and psychological, the media communication technologies irrespective of the speculations around it being a valid or an invalid source, became one of the most sought-after source for information and acted as the platform which the users used to gratify their needs and create resilience. The active audiences are in the lookout for more tools to intensely engage while performing more specialised scrutiny about the content with which they generate any level of involvement. Media audiences, the evolving spectators of the worlds, receive gratification in the process through both receiving of messages and by actively creating them.

McQuail (2010) cited that gratification on media platforms comes through (1) information, (2) personal identity, (3) integration and social interaction, and (4) entertainment. The gratification obtained by sharing information, supporting the other members of the society, improving social knowledge and being virtually visible in the absence of physical social interaction gave individuals the ability to "bounce back", i.e., be resilient during the pandemic. In a digital era where personal and social gratification play a lead role to generate a close relationship between transmitters and receivers, and create resilience, Katz, Blumler, and Gurevitch's User and Gratification theory provides a behavioural understanding of the social media use during pandemic.

Uses and gratifications theory (UGT) is an audience-centred approach to understanding mass communication and is used to understanding why and how people actively seek out specific media to satisfy specific needs. Advocates of the UGT believe that the reason users use a certain technology or media is to gratify their motivations or needs, and once these motivations or needs are satisfied, they will tend to regain the same experience.

21.7 Social Media Creating Resilient Communities

The survey conducted among 110 participants in the three countries reflected the stream time on each of the social media and video steaming platforms and were likely proportional to the level of gratification obtained. WhatsApp followed by Instagram and Facebook were the commonly used social media accounts. YouTube followed by Netflix and Amazon prime were most used video steaming platform. When questioned about the audience's propensity to use a platform more frequently than the others, they stated that Whats app provided the people with the opportunity to connect at a more personal level different from that of other social media accounts. Whereas the audiences resorted more to YouTube because it did not require subscription. Interestingly, 85% of the population stated that their media use during the pandemic was somewhat different from the pre COVID-19 times and reflected that the conscious/ unconscious inclination to media platforms acted as ways to fill in the extra times during the lockdown and create resilience through gratifying their needs during the pandemic. Liang et al. (2011, cited in Sun et al. 2020) indicates that "communication visibility or gratification captures the propensity of users to use technology". In a way the resilience created through gratification obtained by engaging on social media platforms were responsible for increase in the overall use of social media during the pandemic.

The data also evidence that the participants/media users juggled between the various social media platforms more often during the pandemic than before. While Facebook was the most common form of social media use in all the three counties before the pandemic, Instagram, Tik Tok, WhatsApp and LinkedIn started to be widely used during the pandemic. The participants between the age 16–28 who agreed to be 'addicted' to Instagram before the pandemic, reflected to have used Facebook more often during the pandemic. According to Katz (1974) people are active and can meet their requirements to use some media based on their needs. Their needs provoke motives that affect media use, which leads to their cognitive, affective and behaviour outcomes. Hence, the reason behind the changes in the engagement patterns on different social media platforms before and during the pandemic lies in the different gratifications obtained from using different social media platforms. In a way, it can be stated that the shift in the patterns or practise are modes to achieve maximum gratification hence resilience during the pandemic.

Resilience, as we know, is built over time as our experiences interact with our identity. It is the way the body responds to surrounding situations to balance the negative experiences.

Harvard University's Centre on the Developing Child explains resilience as a seesaw or balance scale, where negative experiences tip the scale toward bad outcomes, and positive experiences tip it toward good outcomes. During the pandemic, while the disaster due to the pandemic itself was the bad experience', the way people reshaped the way they would respond to the stress was by using the various communication technologies to relax, pass the time, escape, seek information, communicate or seek solutions to their problems lead to positive experiences.

The two core principles of UGT are: (i) media users are active in their selection of the media they consume, and (ii) they are aware of their reasons for selecting different media options. During COVID-19, resorting to various media practices some of which existed (liking, random scrolling, reading messages) before but now it was done to meet higher order needs like finding meaning and creating resilience. The survey conducted in the developing countries reflected several motivations in their consumption of social media during the pandemic. Below, the motivations are conceptualised in relation to resilience to explain how they connect.

21.7.1 Motivations to Consumption of Social Media

Inherent during a pandemic is a high degree of uncertainty and the perception of severe threat to oneself and close ones which acts as the major source of stress. The Harvard team stated that 'reducing the source of stress' as the most important factor for developing resilience. In an uncertain situation, there is a fundamental need experienced by those affected physically or emotionally by the pandemic to acquire useful information that may help restore a sense of normality. Those affected feel the need to know what has happened, what the current situation is, and how others are dealing with the situation. The social media provided an avenue for the people to share and investigate others' lifestyle where they often drew on the other's experience to prepare and deal with the physical distancing, joblessness, and shutdown.

Even if it takes a few minutes to click a selfie or imitate a post, recreate one in one's own way, or following a hashtag trend, the people learnt to practice self-care. Many times, although the practice was short lived, the experience of sharing with an appropriate tagline created a sense of gratification among the users. The participants who posted selfies during the pandemic identified their posts as being motivated by a need to convey the feelings of happiness or boredom, and to boost their self-esteem. However, half of them also agreed that the gratifications were dependent on constant validation through likes, comments, and shares. Many have been tweeting or 'instagramming' their experiences in self-isolation, in the hopes of inspiring others who may be into small apartment-induced boredom. Social media helped them develop resilience by adding positive experiences- lightening the load on the negative side of the resilience scale and eventually reducing stress.

The COVID-19 wave showed an increase in the number of hashtag trends. #Covid19 / #Covid-19 / #Coronavirus trended the most, followed by #StayHomeStaySafe, #Stayhome, #Quaantineandchill. Other hashtags that were equally influential in the three countries and produced a huge number of contents were #Dalgona coffee challenge, and #blacklivesmatter. The people took to social media to show how one was spending all their new-found time indoors and what hobby they learnt on any given day. Whether one was trying out a new recipe, sharing social media picture-video editing tips, or have figured out how to use a make a new cocktail at home, people creating, posting liking, and commenting on light-hearted contents further reduced their stress. Some researchers suggest resilience is "trait-like"—that is, hard-wired into one's personality—others say it can be learned and acquired later in life (Bellizzi 2020). Imitating others on social media platforms and using same hashtags or follow similar practices helped create resilience, among the latter ones.

While the lack of sufficient information and to communicate the requirements create stress, citizens working together for both themselves and others to overcome disaster-induced challenge creates resilience. In the social sciences, it is commonly acknowledged that individuals are often self-resilient during and in the aftermaths of crises (Helsloot and Rzitenberg 2004; Perry and Lindell 2003; Scanlon et al. 2014). With or without professional responders or other authorities' citizens can rescue themselves from (immediate) danger and provide aid and care to those in need (Helsloot and Ruitenberg 2004; Scanlon et al. 2014). When COVID-19 started growing rapidly, in the hours before India introduced curfews and imposed a complete lockdown and closure of business and public transport, many Indians took to Twitter and Instagram and used the hashtag #Lockdown now as a silent protest to urge the authorities to consider a total lockdown. Palen and Vieweg (2008) in their study claimed that collective behaviour, the self-organising activities of the public creates resilience. The way the masses converged on social media platform to provide aid to those in need, volunteered to safeguard the countrymen helped create gratification—a sense of satisfaction and resilience during the pandemic. Eventually, social media became a platform, a virtual place for people to give and receive light-hearted banter, memes and commentary on the situations around COVID-19.

The survey we conducted in the three countries led us to notice video creation and image posting on social media platforms had been accelerated by the outbreak. One of the major reasons was that creating videos and posting images were another way to gratify needs and seek resilience during the pandemic. The data analysis reflected that the tendency to create video and post photos increased twice than that of pre-lockdown era in the three countries. A similar research project conducted in US and UK by Trifonova (2020) reflected that the population between 16 and 23 are at the forefront of this trend, with almost 3 in 10 creating more videos because of the outbreak—an increase of 8% points between April and July.

While the videos were created and posted on Facebook, Instagram, and YouTube, the main facilitator was the viral video-sharing app Tik Tok, which saw a huge spike in visitors during the lockdown. Tik Tok had often been touted as the natural home for the younger age groups and has attracted diverse audiences from developing and developed countries looking for entertainment during lockdown. The survey data reflected that there had been an increasing influx of 22% parents with young children creating and uploading videos more on video-sharing sites during the outbreak in India. The percentages are 10 and 13% in Pakistan and Bangladesh. Nonetheless, percentage was closer to double 36% for India, 18% for Pakistan and 23% for Bangladesh for the parents uploading images. The elderly parents and grandparents were seen to have participated in different activities to bond with their children and lead to an era of content creation where one could encounter the growing prevalence of family material on Tik Tok. Although rising in popularity, following the conflicts and arguments with China on security, Tik Tok was banned in India during the lockdown. Followed by this, people resorted to Instagram's function named as 'reels' to seek experiences and gratification like that of Tik Tok. Overall, creating videos, on social media or bonding because of social media created a way for the users to gain gratification and create resilience during the pandemic.

21.7.2 The Positive Side

Eidi and Delam (2020) and Trifonova's study indicate that people took to the internet to fulfil social and communal needs. The Harvard team stated that resilience is created especially through responsive relationships. That is by maintaining connections with family and friends which was now possible through social media. Hwang et al. (2020) in his study on pandemic and mental stress states that keeping in touch with friends and family by phone calls or video calls during quarantine eases the distress from social isolation. The main distinguishing feature for social media during a disaster is the level of communication. Social media allows for communication that are more prominent, extensive, multi-level, and multidirectional.

More information could be exchanged with a higher number of people spread over a vaster area. It is evident, then, that social media can be highly practical in a pandemic situation, as they provide new ways for people to communicate with family and friends, to seek information and help, and to assist affected others.

Communication may thus be essential in dealing with disaster, the main elements being the exchange of information to reduce uncertainty, to deal with arising problems, and to cope with the event. Our data outlines a recent surge in messaging and video calling which was increasingly used to connect with others and maintain a sense of community. The citizens often resorted to engage in responsive interactions through messaging, audio, and video chats on the various social media platforms. Approximately 33% of the participants in the three countries said they opened more on messaging platforms like WhatsApp and 31% said they have done the same on public platforms like Facebook.

Moreover, the crisis has encouraged consumers to look to their wider communities for support, as people have felt comfortable sharing what they experience in the public sphere as with immediate friends and family. The data also suggests that the recent surge in messaging and video calling is increasingly used to connect with others and maintain a sense of community. When asked what content the population found most inspirational during the months of lockdown, content from the local community as in Indian community, Pakistani Community and Bangladeshi community was the second-most popular answer after that of close friends and family. Our research stated that 6 in 10, approximately 66% of the India, Pakistan and Bangladesh users reported sharing more personal news and updates on their social channels and this behaviour was most prominent among millennials. This has not been limited to messaging platforms or 1-to-1 conversations. In fact, during the outbreak people have opened about the struggles they have been facing on social media platforms which has helped them effectively cope with the disaster and the isolation. The study reflected that 63% of population in the three countries say social media has helped them feel less lonely during the outbreak, reconstituting normalcy, relationships, and a sense of community. Whereas, just under half believed it contributed to lessening stress and anxiety.

Previous studies on social media and gratification reflects that Facebook's usage produces maximum gratification in comparison to other platforms (Alhabash and Ma 2017; Quan-Haase and Young 2010; Cheung et al. 2011). It provides entertainment value, purposive value, helps maintain relationship, social enhancement and arenas for self-discovery leading to greater opportunities for achieving gratification and resilience (Cheung et al. 2011). Interestingly the social media usage data on Statistica and Statscounter 2020 reflects an eventual increase in Facebook use during the lockdown not only in the developing countries of India, Bangladesh and Pakistan but also in the developed countries of UK, USA and Australia. This increase in use of Facebook during COVID-19 coincides with Park et al.'s (2009) study which uncovered needs for Facebook group participation. Those needs included *socializing* by staying in touch and meeting people, *entertainment* using Facebook for amusement or leisure, seeking self-status by maintaining one's image, and seeking information is to learn about events and products. The best ways for creating resilience is by building on strengths, seeking opportunities for connection, having a positive outlook, seeking out for fun and, identifying one's talents and interests, and set meaningful goals that satisfy ones needs (Gatt 2020). Therefore, increased engagement on Facebook, is proportional to gratification and resilience during the time of the pandemic.

Park et al. (2009) on uses and gratifications of the internet state information seeking as a new gratification. Houston et al. (2015) adds that effective disaster communication can lessen the impact of a disaster, whereas ineffective disaster communication may cause a disaster or make its effects worse. Social media has eventually introduced new tools to support the information seeking and information providing behaviour of people. The latter holds particularly true when people are dissatisfied with the information spread by mainstream media. While the social media sites became the practical platform for the spreading of public health messages to audiences and tools for inducing attitude and behavioral change (Gough et al. 2017), it is also argued to be fuelling panic and fear among people affecting the mental health of social media users negatively (Ahmed and Murad 2020).

The field of infodemiology called the COVID-19 pandemic as the first social media infodemic because it has accelerated information and misinformation worldwide (Ahmad and Murad 2020). ABC News also reported the Ipos poll (aimed at understanding behaviours during COVID-19) claiming that in the age of social media, anxiety about the coronavirus spreads faster than the virus itself, resulting in public panic worldwide (Jackson et al. 2020). Nonetheless, the data analysis among the citizens of the three developing South Asian countries reflect that social media users tend to gravitate more towards social media platform for news than mainstream media. The percentage is particularly high among the ones below the age of 28. The survey results further demonstrate that reading the news on social media helped gratify the need of information while also allowing media users to express their emotions, feelings, and thoughts about the information published. Overall, the greater autonomy to work with the message that has been published on the social media site helps produce gratification and resilience compared to news and information disseminated on mainstream media platforms. Demand for up-to-date information in the early stages of the crisis was unparalleled, with social media providing quick and easy access to relevant updates. The citizens claimed that after the news on lockdown was flashed on mainstream media, many accessed Instagram, Facebook, and twitter to gather more information. By accessing the above, they could not only read breaking news, but could also read the various expert opinions, and discuss these with their friends and family. Additionally, on social media false information could easily be corrected by knowledgeable people. While Trifonova's study showed that using social media to stay up to date with news and current events is the top reason for logging in globally, at 36%. The survey reflected the rate to be much higher among the youngsters in India, Bangladesh and Pakistan and ranged up to approximately 46%. Out of which, a whopping 76% of them stated social media as the source for the most accurate and reliable information. Overall, social media played a central role in helping the citizen develop self-resilience during the pandemic by improving lines of communication.

21.7.3 Alterations in people's Approach to Social Media Use

The pandemic displays a reversal in media consumer's sentiment about the consumption and use of social media. Those most concerned about the time they spend on social media before the pandemic agree to have derived the most benefit from visiting social channels during the disaster. The survey suggested that assessing entertaining content is the third most popular reason media users log on social media during COVID-19 (43%), just behind random scrolling (39%). Previous digital detoxers are more likely than other internet users to say social media has helped them feel less lonely or anxious during the pandemic. One reason for this is that people have felt more comfortable being themselves on these channels amid the outbreak. 42% of internet users who use public platforms agree there has been less pressure to portray an unrealistic image of their life on social media and derived satisfaction in claiming in whatever mental or physical situation they were in during the pandemic.

Interestingly, there was a huge increase in the number of images and videos posted on body image and joblessness during the lockdown than before. People took to social media sites to reflect how the lockdown and pandemic has created an added layer to their already existing grievances and unfulfilling situations. Consumers eventually identified social media in a more comforting light and a platform/mode to achieve resilience. They have agreed that social media can have positive effect on mental well-being, and that social media and mental health are not at odds with each other, but rather help reduce stress.

The response to stress and adversity—like that from the COVID-19 pandemic—can be different for different individuals. UGT emphasis on individual differences led researchers to examine the way personality impacts people's motivations for using media. Weaver's (2003) study on psychological origins of TV viewing by the Virginia Polytechnic Institute and State University identified that motivations of participants with neurotic personalities included passing the time, companionship, relaxation, and stimulation. This was the reverse for participants with extraverted personalities. Moreover, while the neurotic personality types favored the companionship motive most, extraverted personality types strongly rejected this motive as a reason to watch TV. The data reflects a similar result for people's consumption patterns during COVID-19 in developing countries as well. Those who are more socially isolated, emotional, or shy, demonstrated a strong affinity for online streaming services or were more involved in random scrolling or reading news to gratify themselves than resorting to social media for human-to-human virtual interaction. Meanwhile, those that were more sociable, and outgoing found TV and video streaming as a poor substitute for real-life social interactions. Those with extraverted personalities posted photos, videos, WhatsApp status, Facebook, and Instagram stories more often and preferred to engage on social platforms than being a passive consumer. *Gratification of needs*, which refers to the needs gratified by media use is the relationship between the communication visibility and consumers use of social media., was particularly important for the ones with extravagated personalities. In both the cases the practices were gratifying for the consumers and helped balance the emotions and impacts of the pandemic and create resilience.

According to the UGT, the gratification of needs will in turn affect their usage behaviour. Alhabash and Ma (2017) indicated that when social media meets the needs of the users, it can further increase their frequency of using the media. Lastly, some scholars have found that gratification of needs can affect users' excessive use. In the absence of many opportunities to gratify the needs, the users' gratified using social media, eventually leading to the increase in intensity of use of technology to obtain further gratification. Therefore, based on the UGT, the heightened increase in the use of social media during the pandemic is related to gratification and hence developed the capability to 'bounce back', i.e., be resilient to the pandemic and its consequences.

21.8 Conclusion

Overall, the four components of self-resilience, that are information gathering, information dissemination, collaborative problem-solving, and coping are supported by social media. It allows citizens to search for reliable information to reduce their uncertainty about the situation far beyond the regular but out-of-date information provided by authorities.

Nonetheless, the information read is not just passive consumption of content, but it provides active modes of participation by commenting, liking, and sharing. Stress during COVID-19 could be alleviated. mental health promoted, and resilience developed by connecting with loved ones and indulging in activities to keep oneself busy and helping one feel as an important and relevant part of the society.

Social media enabled the media users to feel better by engaging in various activities, such as making posts, obtaining information, and communicating with others.

Also, the ability of social media to connect people through time and space enhances collaborative problem-solving and citizens' ability to make sense of the situation and cope with it. Whether in India, Bangladesh or Pakistan, the social media consumers had similar patterns of social media use during the lockdown. While scholars and psychologists argue social media content may have negative impacts on the audience, the data we generated reflected that social media helped users feel one with the society, gratify their needs, and help them develop resilience. Taken together, irrespective of the economic situation of a country and its population, social media has been playing a crucial role in helping people deal with the impacts of COVID-19 and its unprecedented threat to physical, mental health. Communication through social media platforms and virtual visibility provide a unique capability for highly visible communication that supports social cohesion, social networking, and knowledge sharing. This provides the potential to gratify the different needs of the people and enhance collaborative problem-solving and citizens' ability to make sense of the situation, cope with it, and create resilience to fight the pandemic.

References

Abid K, Bari YA, Younas M, Javaid ST, Imran A (2020) Prosgress of covid-19 epidemic in Pakistan. Asia-Pac J Public Health

Ahmad AR, Murad HR (2020) The impact of social media on panic during the COVID-19 pandemic in Iraqi Kurdistan: online questionnaire study. J Med Internet Res 22(5):e19556

Alhabash S, Ma M (2017) A tale of four platforms: Motivations and uses of Facebook, Twitter, Instagram, and Snapchat among college students?. Social Media+ Soc 3(1):2056305117691544

Anand KB, Karade S, Sen S, Gupta RM (2020) SARS-CoV-2: Camazotz's curse. Med J Armed Forces India 76(2):136

Bodrud-Doza M, Shammi M, Bahlman L, Islam ARM, Rahman M (2020) Psychosocial and socio-economic crisis in Bangladesh due to COVID-19 pandemic: a perception-based assessment. Front Public Health 8:341

Bellizzi MK (2020, July 27) Developing resilience is an important tool to help you deal with coronavirus and the surge in cases. Generation Next. https://www.generationnext.com.au/2020/07/developing-resilience-is-an-important-tool-to-help-you-deal-with-coronavirus-and-the-surge-in-cases/

Chan JC (2013) The role of Social media in crisis preparedness, response and recovery. http://citeseerx.ist.psu.edu/viewdoc/summary?doi=10.1.1.456.6289. Accessed 10 Aug 2017

Chandir S, Siddiqi DA, Setayesh H, Khan AJ (2020) Impact of COVID-19 lockdown on routine immunisation in Karachi Pakistan. Lancet Glob Health 8(9):e1118–e1120

Cheung CM, Chiu PY, Lee MK (2011) Online social networks: Why do students use facebook? Comput Hum Behav 27(4):1337–1343

Clement J (2020, June 25) Countries with the highest number of internet users as of December 2019. Statista. https://www.statista.com/statistics/262966/number-of-internet-users-in-selected-countries/. Accessed 25 Nov 2020

Eidi A, Delam H (2020) Internet addiction is likely to increase in home quarantine caused by coronavirus disease 2019 (COVID 19). J Health Sci Surveillance Syst 8(3):136–137

Gatt JM (2020) The neuroscience of well-being: part 1 (conceptual definitions). The Wiley Encyclopedia of Health Psychology, pp 325–330

Gough A, Hunter RF, Ajao O, Jurek A, McKeown G, Hong J, Kee F (2017). Tweet for behavior change: using social media for the dissemination of public health messages. JMIR Pub Health Surveillance 3(1), e14
Hao F, Tan W, Jiang L, Zhang L, Zhao X, Zou Y, Tran B (2020) Do psychiatric patients experience more psychiatric symptoms during COVID-19 pandemic and lockdown? A case-control study with service and research implications for immunopsychiatry. Brain Behav Immun
Helsloot I, Ruitenberg A (2004) Citizen response to disasters: a survey of literature and some practical implications. J Contingencies Crisis Manage 12(3):98–111
Higgins ET (2006) Value from hedonic experience and engagement. Psychol Rev 113(3):439
Houston JB, Hawthorne J, Perreault MF, Park EH, Goldstein Hode M, Halliwell MR, Griffith SA (2015) Social media and disasters: a functional framework for social media use in disaster planning, response, and research. Disasters 39(1):1–22
Hwang TJ, Rabheru, K, Peisah C, Reichman W, Ikeda M (2020) Loneliness and social isolation during the COVID-19 pandemic. Int Psychogeriatr 32(10):1217–1220
Institute of Epidemiology, Disease Control and Research (IEDCR). August, 2020. COVID-19 situation report (on 15 August 2020). https://corona.gov.bd/storage/press-releases/August2020/DNChFppaELcHMqgLhGrj.pdf (accessed August 15, 2020)
Iqbal N, Dar KA (2020) Coronavirus disease (COVID-19) pandemic: furnishing experiences from India. Psychol Trauma Theory Res Pract Policy 12(S1):S33
Jackson C, Newall M, Yi J (2020, Dec 14) Americans prioritize frontline workers for coronavirus vaccine. Ipsos. https://www.ipsos.com/en-us/news-polls/abc-news-coronavirus-poll. Accessed 24 Feb 2021
Jena PK (2020) Impact of pandemic COVID-19 on education in India. Int J Curr Res (IJCR), 12
Jurgens M, Helsloot I (2018) The effect of social media on the dynamics of (self) resilience during disasters: a literature review. J Contingencies Crisis Manage 26(1):79–88
Katz E, Blumler JG, Gurevitch M (1973) Uses and gratifications research. Public Opin Q 37(4):509–523
Katz E (1974) Utilization of mass communication by the individual. The uses of mass communications: Current perspectives on gratifications research, pp 19–32
Liang L, Tian G, Zhang X, Tian Y (2020) Help comes from understanding: the positive effect of communication visibility on employee helping behavior. Int J Environ Res Public Health 17(14):5022
Mamun MA, Ullah I (2020) COVID-19 suicides in Pakistan, dying off not COVID-19 fear but poverty?–The forthcoming economic challenges for a developing country. Brain Behav Immun
McFarlane AC, Norris FH (2006) Definitions and concepts in disaster research. Methods Disaster Mental Health Res 3–19
McQuail D (2010) McQuail's mass communication theory (6th edition). Thousand Oaks, CA: Sage Publications Inc
Mersey RD, Malthouse EC, Calder BJ (2010) Engagement with online media. J Media Bus Stud 7(2):39–56
Nicola M, Alsafi Z, Sohrabi C, Kerwan A, Al-Jabir A, Iosifidis C, Agha R (2020) The socio-economic implications of the coronavirus pandemic (COVID-19): a review. Int J Surg (london, England) 78:185
Page R (2020, Nov 5) How the coronavirus crisis brought a new meaning to social media for brands. https://Www.Cmo.Com.Au/Article/684255/How-Coronavirus-Crisis-Brought-New-Meaning-Social-Media-Brands/
Palen L, Vieweg S (2008) The Emergence of online widescale interaction in unexpected events: assistance, alliance & retreat. In: Proceedings of the 2008 ACM conference on computer supported cooperative work, pp 117–126
Park N, Kee KF, Valenzuela S (2009) Being immersed in social networking environment: Facebook groups, uses and gratifications, and social outcomes. Cyberpsychol Behav 12(6):729–733
Perry RW, Lindell MK (2003) Preparedness for emergency response: guidelines for the emergency planning process. Disasters 27(4):336–350

Princing MK (2021) What comes after a pandemic? resilience, according to psychologists. Right as rain by UW medicine. https://rightasrain.uwmedicine.org/mind/mental-health/resilience-after-pandemic
Quan-Haase A, Young AL (2010) Uses and gratifications of social media: a comparison of Facebook and instant messaging. Bull Sci Technol Soc 30(5):350–361
Samet A (2020, June 29) How the coronavirus is changing US social media usage. Emarketer. https://www.emarketer.com/content/how-coronavirus-changing-us-social-media-usage
Scanlon J, Helsloot I, Groenendaal J (2014) Putting it all together: integrating ordinary people into emergency response
Statista Research Department (2020, March 31) Percentage growth in news websites' audience during coronavirus (COVID-19) in Italy 2020. Statista. https://www.statista.com/statistics/1108388/news-websites-audience-during-coronaviruus-in-italy/
Sun Y, Liu Y, Zhang J (2020) Excessive enterprise social media use behavior at work: role of communication visibility and perspective of uses and gratifications theory. IEEE Access, 8, 190989–191004
Thakur V, Jain A (2020) COVID 2019-suicides: A global psychological pandemic. Brain Behav Immun
Trifonova V (2020, Aug 4) Chart of the week: how the outbreak has changed the way we use social media. GWI. https://blog.globalwebindex.com/chart-of-the-week/social-media-amid-the-outbreak/
Wang C, Pan R, Wan X, Tan Y, Xu L, McIntyre RS, Ho C (2020) A longitudinal study on the mental health of general population during the COVID-19 epidemic in China. Brain Behav Immuns
Watson A (2020a, June 18).In-home media consumption due to the coronavirus outbreak among internet users worldwide as of March 2020, by country. Statista. https://www.statista.com/statistics/1106498/home-media-consumption-coronavirus-worldwide-by-country/
Watson A (2020b, Apr 9) Coronavirus impact on media consumption worldwide— statistics & facts. Statista. https://www.statista.com/topics/6134/coronavirus-impact-on-media-consumption-us/
Weaver JB III (2003) Individual differences in television viewing motives. Personality Individ Differ 35(6):1427–1437
WHO Director-General's opening remarks at the media briefing on COVID19-March 2020
World Internet Users Statistics and 2021 World Population Stats. (2020) Internet World Stats. https://www.internetworldstats.com/stats.htm
Xiong J, Lipsitz O, Nasri F, Lui LM, Gill H, Phan L, McIntyre RS (2020) Impact of COVID-19 pandemic on mental health in the general population: a systematic review. J Affecti Disord.

Chapter 22
Fake News on COVID-19 in Indonesia

Valerii L. Muzykant, Munadhil Abdul Muqsith, Risky Ridho Pratomo, and Victor Barabash

Abstract The world is not only facing a pandemic, namely COVID-19, but also an infodemic. Pandemics have become a granary for hoax news to spread, especially through social media. Fake news has accelerated very quickly on social media, given their character especially when the former US President Donald Trump has returned to the White House after a three-night stay at Walter Reed military hospital because of his and his wife Melania's positive testing for COVID-19, just over a month before the 2020 presidential election. Indonesia also did not escape from hoax attacks carried out by irresponsible parties. This research uses a quantitative approach by analyzing content from the turnbackhoax.id website, which verifies fake news circulating on social media. The authors measure the trend of fake news for six months, January–June 2020. This trend is seen from the types of fake news that are often spread and the circulated and the platforms that are used. The results show that the trend of fake news over the past six months has decreased due to many actors' educational efforts. COVID-19 is the most common issue, and WhatsApp and Facebook are the most widely distributed media. The authors conclude that collaborative efforts are needed to prevent the spread of fake news so that the public is well informed. Infodemic resilience is considered as the ability of systems to recover from drastic change.

Keywords Covid-19 · Indonesia · Media platform · Fake news · Resilience · Infodemic

22.1 Introduction

The existence of the Internet has caused a lot of disruption in various sectors. Initially, mainstream media such as television, radio, newspapers, magazines, and others dominated the consumption of news. However, many new platforms emerged, known as Web 2.0, due to Internet presence. Blank and Reisdorf (2012, p. 539) defined

V. L. Muzykant (✉) · M. A. Muqsith · V. Barabash
Peoples' Friendship University of Russia (RUDN University), Moscow, Russia

R. R. Pratomo
Pembangunan National Veteran Jakarta University, Depok, Indonesia

D. M. Berube (ed.), *Pandemic Communication and Resilience*, Risk, Systems and Decisions, https://doi.org/10.1007/978-3-030-77344-1_22

web 2.0 as using the Internet to provide platforms through which "network effects can emerge". Web platforms such as Wikipedia, MySpace, Facebook, YouTube, Google, Blogger, Rapidshare, WordPress, Hi5, Flickr, Photobucket, Orkut, Skyrock, Twitter, YouPorn, PornHub, Youku, Orkut, Redtube, Friendster, Adultfriendfinder, Megavideo, Tagged, Tube8, Mediafire, Megaupload, Mixi, Livejournal, LinkedIn, Netlog, ThePirateBay, Orkut, XVideos, Metacafe, Digg, StudiVZ, etc. are said to be typical for this "transformation of the Internet" (Fuch 2011, p. 288). Besides, there are also Instagram and WhatsApp, which are part of the transformation of community communication.

The world is currently entering an era called post-truth. The phenomenon that occurs in society today tends to prioritize emotion over reason. Post-truth it was the most popular word in 2016 as an expression of concern by those who care much about the concept being under attack (Mcintyre 2018). However, the post-truth definition from Oxford Dictionary gives a full picture about our current condition regarding fake news issue: "in this era of post-truth politics, it's easy to cherry-pick data and come to whatever conclusion you desire' or 'some commentators have observed that we are living in a post-truth age" (Fuller 2018, p. 1).

This post-truth phenomenon is synonymous with the increasingly massive news when Post Truth enables Fake News. Rochlin stressed that the post-truth age no longer cares about fact:

> As we enter the post-trust era, in which facts and evidence have been replaced by personal belief and emotion, the nature of news, and what people accept as news, is also shifting towards a belief and emotion-based market. The truth of the story no longer matters. What matters is that the story falls in line with what a person wants to hear. Fake news ...rather a news that is seen to attack a person's pre-existing beliefs. This is the truth of the post-truth era (Rochlin 2017, p. 386).

The record of fake news itself goes back a long time. However, the term has become popular in recent years. The emergence of this phenomenon has a close relationship with the increasing development of technology and people's response to the information available on Web 2.0 today.

The reason why fake news distribution is so fast can be analyzed through the Web 2.0 platform's nature. Anyone who has the desire, time, and technological access can contribute to various online resources about vaccination. People can access data and information, take existing resources and build new resources from them, and collaborate with other like-minded people. In doing so, they "enhance and enrich others' participation and the available set of resources" (Witteman and Zikmund-Fisher 2012, p. 3735).

The characteristics of Web 2.0 make fake news distribution both abundant and quicker, Not infrequently, this makes fake news more and more prevalent. These properties and advantages increase the frequency of fake news circulating. Moreover, this certainly affects people's perceptions (Anderson and Sulistyani 2019), especially when facing an emergency in the form of COVID-19. For novel epidemic and pandemic diseases, the detection, assessment, and response to rumors, stigma, and conspiracy theories and their impact on public health in real-time are a challenge (Islam et al. 2020).

Fake news assume a change of outlook of the addressee, the formation of new vital installations, valuable reference points, motives and behavior models. They contain information on urgent problems, which are of burning interest to modern society and affect all spheres of life. "Such publications influence the addressee and seek to influence his point of view concerning certain problems of society" (Zheltukhina et al. 2017, p. 86).

When it comes to fake news, there are many definitions and "typologies in circulation" (Lazer et al. 2018, p. 1094) simply defines fake news, namely as news that mimics content from the media, but not with the process or organizational intentions. They also say that hoaxes overlap with misinformation and disinformation, even though it depends on the "newsmaker's purpose" (Rubin et al. 2015, p. 3) categorizing it as deceptive news. However, their point of view only provides a narrow understanding of fake news classified into six categories: news satire, news parody, news fabrication, photo manipulation, advertising and public relations, and propaganda (Tandoc et al. 2017).

Research has shown how massive a role the various social media platforms play in fake news. For example, Loeb et al. (2019) studied a case on urological information disseminated through social media, particularly YouTube. It revealed a significant amount of commercial, biased, and misinformative content on SoMe about common urological conditions. Another case involving the Zika virus investigated by Sommariva et al. (2018) examined 120 published content (coming from the top ten news stories) about the Zika virus and found fabricated content represented "the largest share (81%), followed by misleading content (16%) and false connection (3%)". From 120 articles analyzed, 27 were rumors, 92 were verified news, and one was satire. Brewer et al. (2013) studied the impact of Stephen Colbert's reality show on public opinion. They found such satirical programs can more influence audiences' accustomed to watching satire. Satire can influence public opinion, especially when being packaged uniquely and invites humor.

WHO Director-General, Tedros A. Ghebreyesus, told a security conference in Munich on February 15 that we're not just fighting an epidemic; we're fighting an infodemic (Zarocostas 2020). That statement is true. Fake news has surfaced everywhere related to COVID-19. For example, in Japan, news circulated that a passenger originating from Wuhan with a fever escaped from the Kansai airport quarantine. This was later denied directly by the airport through its official website. Indonesia was also hit by various fake news. The Communications and Information Ministry revealed on February 3, 2020, that 54 hoaxes about the novel coronavirus were first detected in Wuhan's Chinese city—disseminated on social media platforms and messaging apps from January 23 to February 3 (Post 2020).

Indonesian people themselves are active users of social media. Based on We Are Social and Hootsuite's report, social media users in Indonesia increased their use. In 2017, the number of social media users was 106 million. Then, in the following year, it increased to 130 million. In 2019, the number increased by 20 million to 150 million active social media users. The latest data shows that this year, the number of active social media users in Indonesia is 160 million.

However, the rate information flows does not met with a corresponding effort toward quality and accuracy. Because many people can make news without paying attention to good journalistic principles, information is abundant, making it difficult to know which news convey fact information. People find it difficult to tell which news is truthful, and what is wrong. Research from Dailysocial.id in 2018 on 2032 smartphone user respondents found several facts. Most of the respondents (44.19%) admitted to doubting whether they had expertise in detecting hoax news or not, and as many as 31% found it difficult to detect hoaxes; as many as 51.03% chose to be silent and did not believe it, and only 32.87% of respondents clarified the truth.

There is a culture in social media that allow hoaxes to proliferate. Indonesia's society often uses the phrase "I got this from another group" as a reason to spread information when other sources are not known. Instead of verifying the message's content, users often choose to spread these messages because it is "easier to do" (Ilahi 2018, p. 99). According to Professor Deddy Mulyana of Padjajaran University, Bandung, Indonesia's unconventional character of having different opinions, healthy democracy, and lack of literacy culture causes hoaxes to increase in Indonesia (Fakta 2017). Eventually, the Indonesian people tend to believe and even distribute news whose accuracy is questionable. The community finally fell into the "confusion of news" (Luthfi 2017, p. 210) and mutual disbelieve.

The fact is that Fake news channels are mostly sourced on social media. Based on research from the Indonesian Telematics Society, in 2017, 92.4% originated on social media (Kristiyono and Jayanti 2017). Even though there was a decrease in 2019, the spread of hoaxes on social media was still high: Facebook 92.40%; Chat applications 62.62%; and Website 34.40% (Septanto 2018).

Apart from individual actors, the group is called Saracen often creates hoax content. According to a police report, the number of accounts associated with the Saracen group amounted to 800,000. Their job is to express SARA-based hatred and often make insults. Some of its members were caught in 2017.

According to the Nusantara's Head of the Task Force, Inspector General Pol. Gatot Edy Pramono, from mid-2017 to December 2018, there were 3,884 hoaxes and hate speech circulating on social media (Hutagaol 2019). The people were shocked by the news of an aftershock in Palu. In that year, the Palu area was hit by an earthquake and caused many losses. Certain individuals were able to spread fear among the community using this condition. Of course, the Head of Public Relations of BNPB, Sutopo Purwo Nugroho, denied this (Ibrahim 2018).

Entering the political year in 2019 is the right momentum to spread lies. Based on the Ministry of Communication and Information data, from August 2018 to November 2019, 3901 hoaxes were identified (Irianti 2019). In April 2019, which is the time for the Presidential and DPR elections, 486 hoaxes, of which 209 related to political issues (Maharani 2019).

Based on these observations, we want to see how much fake news is in Indonesia. We begin by defining fake news, the next issues being misreported, and finally, what media are spreading fake news. Because COVID-19 remains ongoing, the authors must bracket the duration of data collection.

22.2 Methods

In this study, the authors used quantitative content analysis methods to see the amount of fake news spread on social media, the media used, what topics were the material for making fake news, and fake news types. Quantitative content analysis is the systematic and replicable examination of symbols of communication. It involves the assignment of numeric values according to valid measurement rules, and the analysis of relationships involving those values using statistical methods to describe the communication, draw inferences about its meaning, or infer "from the communication to its context, of both production and consumption"(Riffe et al. 2014, p. 19).

The authors acquired their data from the turnbackhoax.id, site maintained by the Indonesian Anti Hoax Society (MAFINDO) as primary data. Our data ranged from January 1, 2020, to June 30, 2020. We then re-analyzed, classified the fake news, and quantified the number of hoaxes associated with COVID-19. MAFINDO have their classifications: misleading content, mock content, fake content, false content, manipulated content, false connection, and satire or parody. However, most of the classification is quite difficult to detect and differentiate, and each party has its own interpretation.

Therefore, we use a more generalized category made by Tandoc et al. (2017) because we think that the classification is comprehensive. They divided fake news into six category: fabricated news, satire, propaganda, advertising, photo manipulation, and video manipulation. Our rationale, the reporting began in January when COVID-19 was noticed. June 2020 marked the institution of the New Normal policy in Indonesia, a policy that aims to accelerate the handling of COVID-19 in health and socio-economic aspects. The authors collected 359 data.

22.3 Results and Discussion

If we look at the data above, three conclusions can be drawn. Everyone and anyone can be an actor and a victim of fake news. The observation supports this statement that it is much easier today for people to make and distribute a story with social media regardless of its accuracy. When we accept it emotionally in a post-truth world, facts become secondary. Hoaxes cause an impact all directions and raise various "social, political, religious, and other interests" (Simarmata et al. 2019, p. 23–24). Hoaxes are seasonal. This hoax appears only in certain situations; if that period has passed, then the hoax will also "pass and reappear when new cases emerge" (Aditawarman et al. 2019, p. 2). For example, in Indonesia, when Pancasila Day or commemorating G30S/PKI, hoaxes re-emerged.

When we talk about seasonality, there is usually a motive that we want to achieve. According to Septanto (2018), five factors cause hoaxes to spread significantly in Indonesia. First is power political motives that justify any means make hoaxes the most effective way to achieve goal. Second, the spread of hoaxes and hate speech is carried out in an organized manner. Third, the public does not yet have social awareness in selecting various information obtained through social media so that all information obtained is mostly swallowed up without checking the truth. Fourth, people or figures who have many followers and influence often abuse their influence by making or spreading their personal opinion regardless of the impact on society. Fifth, the hoax has become a promising business and industrial field. In addition, an agreed indicator is that hoax is information made (by design) so certain parties can "destabilize the situation" (Silalahi and Sevilla 2020, p. 15).

Apart from the several causes above, all hoaxes have their own goals and targets, from "simple self-publicity to huge goals, such as practical politics" (Aditawarman et al. 2019, p. 4). Various attempts have been made by many parties to suppress and prosecute the perpetrators of hoax spreaders. Starting from the government, especially the Ministry of Communication. In January to July 2017, they admit, they have blocked 6.000 sites containing hoaxes and hate speech (Denata 2017). Not only from the Government, community also take part to tackle fake news. Some create platforms that are used to verify fake news circulating. One of them is the Indonesian Anti-Slander Society, which created the turnbackhoax.id website in 2016. However, despite all the effort from many parties, fake news is hard to solve.

As 2020 continues, hoax still penetrates because of the development and context of the situation. This year, the world is suffering through the pandemic of COVID-19. Illness and death has fueled uproars in many countries. All countries begin to mitigate the impact, and this is true for Indonesia. However, when it comes to Indonesia, lack of awareness, speed in handling the pandemic, and coherent mitigation efforts, different government opinions have create a sense of uncertainty in society. These, in turn enabled the related hoax to be more distributed.

22.3.1 COVID-19 Fake News Spread in Indonesia

Since the first cases appeared in Indonesia in early March 2020, victims of COVID-19 has continued to increase. For starters, the total cases of worldwide COVID-19 in March were still around 1524: see Fig. 22.1 (data processed from https://covid19.go.id/peta-sebaran). Even though this may seem minor is number demonstrates Indonesia was not immune to COVID-19. The increase in positive cases demonstrated tremendous vulnerability in the community, so fear and anxiety begin to emerge.

From April to June, cases continued to increase. In April, the number of cases reached 8590. Furthermore, in May, it nearly doubled to a figure of 16,355 positive cases. In June, cases reached 29,912. This increase occurred due to various factors. However, when the COVID-19 cases continued to emerge, hoaxes also took on a role as a separate threat to public perceptions, adding another layer of concern contributing to the government's difficulties.

As news about COVID-19 becomes the subject of public discussion, fake news-makers began to produce false information that could upset the public. The virus

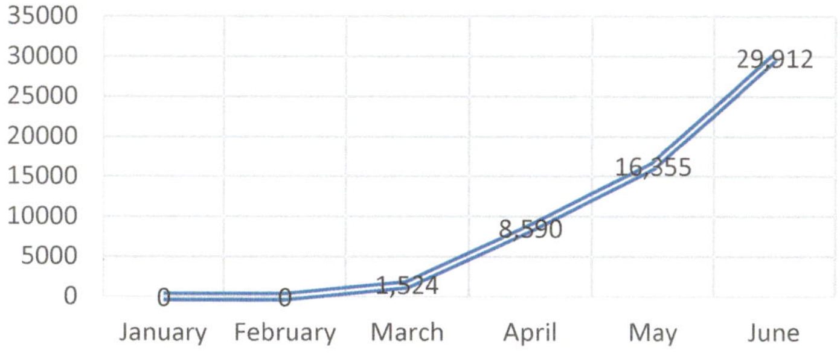

Fig. 22.1 COVID-19 cases in Indonesia

attacks made people feel afraid and anxious. In Indonesia, fake news about COVID-19 had begun to spread to the wider community. However, because COVID-19 was not yet a hot topic, there was little fake news. This was recorded in the first two months, namely January and February. There were only about 11 fake news stories in January based on our research in the turnbackhoax.id website: see Fig. 22.2 (data processed from https://turnbackhoax.id/). This is because COVID-19 has not yet reached Indonesia. Therefore, this topic is still tepid and just beginning to become a trending topic in the community.

In February, the numbers increased slightly after WHO changed the status of COVID-19 to a Public Health Emergency of International Concern (PHEIC). However, in Indonesia, the change in status seemed to be under-appreciated. The turning point occurred in March 2020. Two phenomena have made COVID-19 a topic of discussion and coverage in the community. The first phenomenon, on March 2, 2020, Indonesia reported its first case. Two people were infected because they had contact with Japanese citizens who had left Indonesia. The second phenomenon is the declaration of COVID-19 as a pandemic by WHO.

The Indonesian government was intensively taking preventive measures and providing economic stimulus to help the community. Region-specific Large-Scale Social Restrictions (PSBB) was implemented to control the spread of the virus. President Joko Widodo revealed Indonesia's first case on March 2, 2020. The number in March also increased eightfold, from 15 to 122. Although it is known that the spread

Fig. 22.2 COVID-19 fake news spread in Indonesia

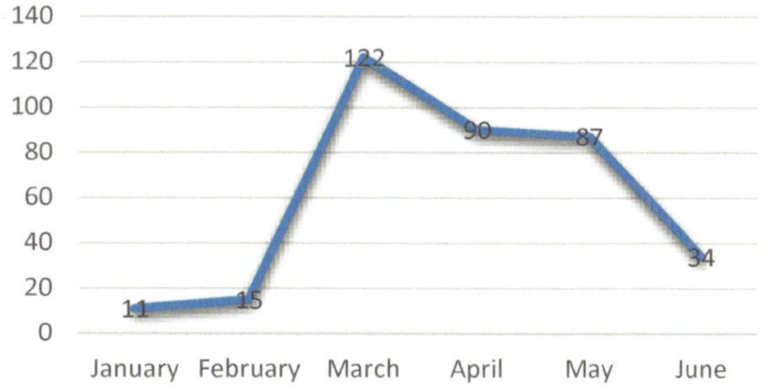

of the virus originates from saliva, it is not clear about who immune to COVID-19. Certain parties to spread fake news use this gap of information.

For example, a fake news story entitled "People with blood type O are more susceptible to catching the virus" came from toddler.com on February 11, 2020 (Salim 2020). Experts and doctors have denied this information. Epidemiology expert from the University of Indonesia, Syahrizal Syarif, said that three people are susceptible to COVID-19: people who are in close contact, people who are in social contact, and contact areas seen based on one area confirmed by COVID-19 (Gunadha and Aditya 2020). Of course, the number of people with blood type O is quite a lot, making them afraid. In Indonesia, according to the Head of Indonesia's Red Cross DKI Jakarta, dr. Salimar Salim, MARS, there are 40% population who has blood type O (Rossa and Putri 2020).

However, there was a lot of fake news that discusses COVID-19 therapeutics. There is a lot of news circulating that alcohol, bananas, and hydrochloroquine can eliminate COVID-19. Fake news given in situations like this is more dangerous because of the lack of public knowledge about COVID-19. Therefore, when people get this "valuable" information, the public's tendency to share it with other people increases. They are trying to be altruistic not mean-spirited.

An interesting phenomenon emerged when fake news about COVID-19 started decreasing. Several reasons can explain this decline. First, people are starting to be educated about the details of COVID-19. The government is one of the funnels for public information. Besides, communities, organizations, and individuals play an important role, especially in social media. For example, the star of stand-up comedy Bintang Emon makes a campaign called the #dirumahaja on Instagram. The campaign focus on raising awareness of the danger of COVID-19 and the invitation to do our activities at home.

The next explanation, the police play an active role in censoring fake news. National Police informed about 107 suspects to the perpetrators of cases of hoaxes related to COVID-19. In April, the Police handled 97 hoax cases (Antara 2020). These results came from cyber patrol work from March 19 to May 19, 2020 (Putra 2020). However, in June, the issue is more ideological rather than COVID-19 cases, because of the public holiday of Pancasila set on June 1. The community has also entered the New Normal or transitional PSBB era, where they have started to return to their normal activities with a note of tightening health protocols.

22.3.2 *Fake News and Social Media in Indonesia*

WhatsApp and Facebook become the most spreaders of hoax news in Indonesia in the last few years. The popularity of Facebook as a medium for spreading hoaxes is related to the fact that it has a user base of more than 2.2 billion (Zaenuddin 2018). Coupled with the fact that social media's advantages are the speed at which information is distributed, this has made Facebook a very substantial nest for hoaxes.

Meanwhile, WhatsApp is the second-largest media after Facebook: see Fig. 22.3 (data processed from https://turnbackhoax.id/). It has several advantages. There are five aspects of excellence (Farooq 2018, p. 109):

1. Cost: Since fake news creation does not involve much research, production costs—no reporters/journalists are sent into the field for coverage—broadcast expenses and licensing, the cost to produce fake news is much less than the real news.
2. Promotion: While news outlets might have to spend a substantial amount of money, resources, and time promoting their brand and content, fake news creators need no promotion.
3. Anonymity: The origins of fake news are difficult to trace; there is no accountability for being shared/forwarded.
4. Shelf life: Many posts shared/forwarded on messaging platforms like WhatsApp are not bound by time. Often they are historically or religiously based and tend to reinforce myths, fear, misinformation, and hate. This type of content is recycled and keeps resurfacing from time to time.
5. Impact: Fake news is different from conventional news stories as it reaches users through a network close to them. The same information might reach the user from several individuals or groups, leading to reinforcement of the information. The user has the power to edit/ manipulate the content. The sender can twist the information so that it suits the palate of the user/s and meets his/her plan.

Users may share directly without verifying the content of the information contained in the information. Users pass on fake news because they want to belong and may even feel guilty when they do not elect to share this information with their audiences.

In this pandemic, the challenge is not just to calm everyone down. Sylvie Briand, WHO Director for infectious disease management, said: "… At stake during an

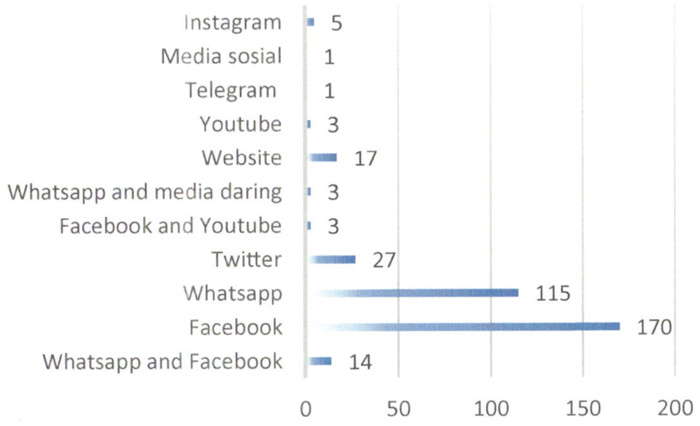

Fig. 22.3 Media platform for spreading fake news

outbreak, what is at stake in making sure people will do the right thing to control the disease or mitigate its impact. So, it is not only information to make sure people are informed; it is also making sure people are informed to act appropriately" (Hua and Shaw 2020, p. 1–12). In other words, getting everyone to do the right thing is a top priority in addition to calming the public and preventing the spread of COVID-19. Implicitly, Dr. Sylvie also said that hoaxes were one of the biggest threats to the public health sector. A group of researchers after examining 1.225 fake news stories about COVID-19 concluded that infodemic of misleading information has resulted in "false claims, half-backed conspiracy theories, and proliferation of pseudoscience health therapies regarding the diagnosis, treatment, prevention, origin, and spread of the virus. Fake news is taking over social media and putting public health at risk" (Naeem et al. 2020, p. 6).

22.3.3 Fake News Gets Easier and Easier to Spread

The Internet has lowered the cost of entry to new competitors—many of whom have rejected these norms—and undermined the traditional news source business model that has—enjoyed "high levels of public trust and credibility" (Lazer et al. 2018, p. 1094). Therefore, everyone is a newsmaker, or in journalistic terms, it is also called citizen journalism. Traditional or mainstream media must also compete with those who have the tools and speed. "It is particularly pernicious in that it is parasitic on standard news outlets, simultaneously benefiting from and undermining their credibility" (Muqsith and Muzykant 2019, p. 310).

Despite all of the Internet's side effects, McGonagle (2017) stated that technology makes it easy for criminals to quickly distribute hoax content in various formats. This also contributes to the colossal distribution of fake news. During the COVID-19 period in Indonesia and the last six months, the fabricated news level was significant, namely around 275 (76.6%) of all fake news: see Fig. 22.4 (data processed from

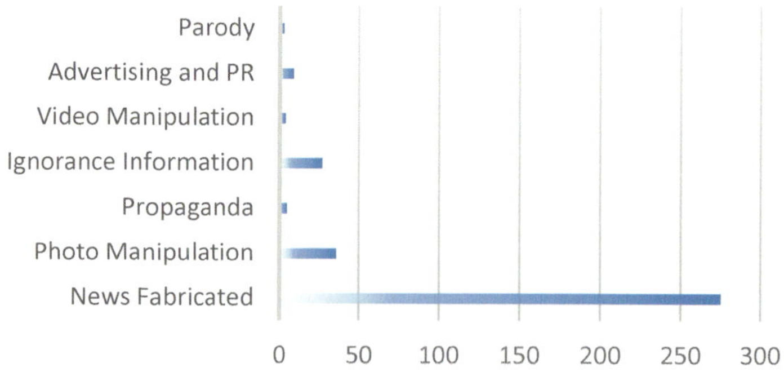

Fig. 22.4 COVID-19 fake news type in Indonesia

https://turnbackhoax.id/). This reflects that the ease and speed of technology have become counterproductive, especially during the COVID-19 pandemic.

Apart from fake news, there is one genre of fake news outside of the writer's category, namely ignorance. Ignorance news occurs when someone misunderstands content or events to associate it with factual conditions as the justification of the impact of something without conducting in-depth verification. There were quite a number of them, around 27 (7.52%), and there is a reason behind this category's emergence.

22.3.4 Ignorance Based Fake News

Rubin et al. (2015) has emphasized that filtering, checking, and verifying information online continues to be important in a library and information science (LIS), as the lines between traditional news and online information are increasingly blurred. However, this is not done for people who spread the news that is ignorance. Without going through this process, they immediately interpret it as the impact of one thing—in this case, COVID-19. As a result, this inaccurate or false information enters the public discussion.

There was a fake news report that a person who works as an online motorcycle taxi driver reported positive for COVID-19 after shaking hands with the Minister of Transportation Budi Karya Sumadi, who has been diagnosed positive. This fact was immediately denied by the parties concerned (The Jakarta Post 2020).

22.3.5 Photo Manipulation

The third-largest category is photo manipulation, around 36 (10.03%). Along with the number of photo editing applications, it becomes easier to modify photos. As a result, photos can also be used as a tool to spread lies. The remainder is propaganda with five news stories (1.4%), parody with three (8%), four in the form of video manipulation (1.1%), and nine in advertising and public relations (2.51%).

22.3.6 Issues Associated with Fake News

During the last period of research, the issue that had a lot of fake news was COVID-19 itself. Some of the numbers issues regarding the COVID-19 hoax are as follows: COVID-19 (122), policy (66), social (53), economy (32), figure (26), government (19), ethnicity, race, and between religions (17), health (9), politics (6), law (5), data (2), communism and conspiracy each one: see Fig. 22.5 (data processed from https://turnbackhoax.id/).

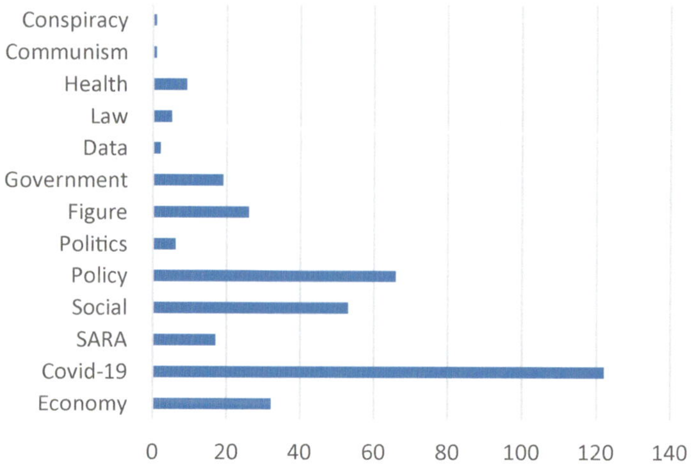

Fig. 22.5 The Hoaxes issues

Fake news about COVID-19 is the most common. The specifications are more on handling COVID-19, news on COVID-19 patients, or just false knowledge. Especially when ignorance and feelings of panic hit, people tend to trust any information. Besides, during the period, the vaccine is still in development in Indonesia, making the public more sensitive about handling COVID-19.

What is interesting is how fake newsmakers also objectify policy issues (65 stories). Although not as much as the COVID-19 issue, policy hoaxes need attention as well. Before COVID-19 hit Indonesia, the government campaigned for tourism to boost the economy. When on March 2, 2020, the government announced COVID-19 patients, there was an implicit hope that the government would focus on this direction. However, Indonesia was classified as very late in responding, government officials are not in accord with effective policies, resulting in confusing information about the policy itself.

As a result, this has become a fertile field for fake news creators to spread lies. For example, Indonesia's President Joko Widodo imposed a limited partial quarantine, even though the President had never issued such a decision, but instead referred to Large-Scale Social Restrictions. News about the regions Cirebon and Situbondo (West and East Java) imposing lockdowns was widely heard. This is due to divisions between the regions and the central government regarding whether the lockdown is enforced or not.

The policy has implications for many sectors, especially social and economic. According to the Ministry of Manpower, 1.5 million people lost their jobs. The detail is as follows: the formal sector laid off as many as 160,067 workers from 24,225 companies and sent home as many as 1,080,765 workers from 27,340 companies, while the informal sector laid off 265,881 workers from 30,466 companies (Sembiring and Sandi 2020). Due to the need for income or assistance in material or immaterial form, fake news about social assistance has spread. For example, there is some information on Facebook on May 10, 2020, that government will distribute donations for people affected by COVID-19 of two million rupiahs (Hanifan 2020).

22.4 Conclusion

The authors confirm that the world is not only facing a pandemic but also an infodemic. Nevertheless, the authors consider infodemic resilience as the ability of systems to recover from drastic change. Awareness of social responsibility is one of the keys to the successful and harmonious development of society, to prevent its stratification and internal conflicts (Muzykant 2020). On average, the public consumed almost two fake news stories about COVID-19 during January-June. Seeing this, the government and the Indonesian people responded quickly by making various fact-verification efforts so that the information that reached the public was accurate.

Fake news as influential texts designed to exert impact on reason, feeling and will of the addressee, i.e., to convince, inspire, and manipulate. The purpose of the present media texts is to influence on the addressee; attempt to force him to change the belief or to make certain action. Although hoaxes about COVID-19 have decreased due to various factors, one thing that needs to figure out is how social media becomes the dominant platform for obtaining information.

Now online platforms, particularly social media, are becoming the main news sources for a growing number of individuals, "misinformation seems to have found a new channel" Tandoc et al. (2017, p. 2). Kim et al. (2011) urge that understanding what kind of social media channel used as information sources and evaluating the news is critical to identify challenges when selecting the news from non-traditional sources.

Fake news will always be a complicated problem in society, so it requires all elements of society to solve it. According to the authors, the government may be the appropriate by centralizing all information in the central government's hands as it is in Indonesia, especially those related to economic, social, political, and legal policies. On the other hand, many communities taking part in educating and verifying information participate in increasing public understanding. However, the public must also be aware that fake news can come from nearly anywhere and must be prepared to identify it when it crosses their paths.

Acknowledgements This paper has been supported by the RUDN University Strategic Academic Leadership Program.

References

Anderson S, Sulistyani HD (2019) Detecting and combating fake news on web 2.0 technology in the 2019 political season Indonesia. J Soc Stud 15(4): 103–116

Antara (2020, April 27) Polri Tangani 97 Kasus Hoaks Selama Pandemi COVID-19. Diambil kembali dari Media Indonesia: https://mediaindonesia.com/read/detail/308054-polri-tangani-97-kasus-hoaks-selama-pandemi-COVID-19. Accessed 28 Apr 2020

Blank G, Reisdorf BC (2012) The participatory web. a user perspective on Web 2.0. Inf Commun Soc 15(4):301–318

Brewer PR, Young DG, Morreale M (2013) The impact of real news on "Fake News": intertextual processes and political satire. Int J Pub Opin Res 25(3):323–343

Denata A (2017, August 27) Blokir 6.000 Situs Hoax, Kemkominfo: 701 Penyebaran Paling Tinggi di Januari. Diambil kembali dari 702 jitunews: https://www.jitunews.com/read/64949/blokir-6-703 000-situs-hoax-kemkominfo-penyebaran-paling-tinggi-di704 januari. Accessed 28 Apr 2020

Fakta TC (2017, February 8) Kenapa "Hoax" Mudah Tersebar di Indonesia? Diambil kembali dari Kompas: https://megapolitan.kompas.com/read/2017/02/08/21160841/kenapa.hoax.mudah.tersebar.di.indonesia. Accessed 28 Apr 2020

Farooq G (2018) Politics of fake news: how WhatsApp became a potent propaganda tool in India. Media Watch 9(1):106–117

Fuch C (2011) Web 2.0, prosumption, and surveillance. Surveillance Soc 8(3):288–309

Fuller S (2018) Post-truth: knowledge as power game. Anthem Press, London

Gunadha R, Aditya R (2020, March 5) CEK FAKTA: Benarkah Orang Golongan Darah O Rentan Terinfeksi Virus Corona? Diambil kembali dari Suara.com: https://www.suara.com/news/2020/03/05/153851/cek-fakta-benarkah-orang-golongan-darah-o-rentan-terinfeksi-virus-corona?page=all. Accessed 28 May 2020

Hanifan S (2020, May 21) CEK FAKTA: Hoaks Pemerintah Beri Dana Bantuan Dampak Covid-19 Rp2 Juta: https://www.merdeka.com/cek-fakta/cek-fakta-hoaks-pemerintah-beri-dana-bantuan-dampak-covid-19-rp2-juta.html. Accessed 28 May 2020

Hua J, Shaw R (2020) Corona Virus (COVID-19) "Infodemic" and emerging issues through a data lens: the case of China. Int J Environ Res Public Health 17(2309):1–12

Hutagaol (2019, January 15) Okezone. Diambil kembali dari Berapa Jumlah Hoaks and Ujaran Kebencian dari 2017–2018? Berikut datanya: https://nasional.okezone.com/read/2019/01/15/337/2004711/berapa-jumlah-hoaks-ujaran-kebencian-dari-2017-2018-berikut-d. Accessed 28 Feb 2019

Ibrahim GM (2018, October 1) Ada Info Gempa 8,1 M dan Tsunami Susulan di Palu, BNPB: Hoax! Diambil kembali dari Detik: https://news.detik.com/berita/d-4236081/ada-info-gempa-81-m-dan-tsunami-susulan-di-palu-bnpb-hoax. Accessed 28 Dec 2018

Ilahi HN (2018) Women and Hoax News processsing on WhatsApp. Jurnal Ilmu Sosial Dan Ilmu Politik 2(2):98–111

Irianti RP (2019, December 26) Media Indonesia. Diambil kembali dari Kominfo telah identifikasi 3.901 informasi hoaks sepanjang 2019: https://m.mediaindonesia.com/read/detail/279905-kominfo-telah-indentifikasi-3901-informasi-hoaks-sepanjang-2019. Accessed 28 Dec 2019

Islam MS, Sarkar T, Khan SH, Kamal AH, Hasan SM, Kabir A, Yeasmin D, Islam MA, Chowdhury KI, Anwar KS, Chughtai AA (2020) COVID-19–related infodemic and its impact on public health: a global social media analysis. Am J Trop Med Hyg, pp 1–9

Kim K-S, Yoo-Lee E, Sin S-CJ (2011). Social media as information source: undergraduates' use and evaluation behavior. In: Proceedings of the American society for information science and technology (hal. 1–3). American Society for Information Science and Technology

Kristiyono J, Jayanti RO (2017) Fake news (hoax) and paranoid frame of mind of social media user. In: 3rd International conference on transformation in communication (ICoTiC 2017). Adv Soc Sci Educ Human Res 150:41–44

Lazer DM, Baum MA, Benkler Y, Berinsky AJ, Greenhill KM, Menczer F, Metzger MJ, Nyhan B, Pennycook G, Rothschild D, Schudson M, Sloman SA, Thorson EA, Watts DJ, Zittrain JL (2018) The science of fake news: Addressing fake news requires a multidisciplinary effort. Science 359(6380):1094–1096

Loeb S, Taylor J, Borin JF, Mihalcea R, Perez-Rosas V, Byrne N, Chiang AL, Langford A (2019) Fake news: spread of misinformation about urological conditions on social media. Eur Urol Focus 6:437-–439

Luthfi M (2017) Kitab Suci dan Hoax: Pandangan Al-Qur-an Dalam Menyikapi Berita Bohong. Wawasan: Jurnal Ilmiah Agama dan Sosial Budaya 2(2):209–222

Maharani T (2019, May 1) Kominfo Identifikasi 486 Hoax Sepanjang April 2019, 209 Terkait Politik. Diambil kembali dari detik.com: https://news.detik.com/berita/d-4532182/kominfo-identifikasi-486-hoax-sepanjang-april-2019-209-terkait-politik. Accessed 18 May 2019

McGonagle T (2017) Fake news": False fears or real concerns? Netherlands Q Hum Rights 35(4):203–209

Mcintyre L (2018) Post-truth. MIT Press, Massachusetts

Muqsith MA, Muzykant VL (2019) Effect fake news for democracy. Jurnal Cita Hukum [online] 7(3):308–318

Muzykant V, Komarov V, Shalnova O, Grinberg T, Muqsith MA (2020) Corporate social responsibility in the digital transformation of business ecosystems. Int J Innov Creativity Change 14(5):1295–1313. https://www.ijicc.net/index.php/volume-14-2020/202-vol-14-iss-5. Accessed 28 May 2020

Naeem SB, Bhatti SB, Khan A (2020) An exploration of how 786 fake news is taking over social media and putting public 787 health at risk. Health Inf Libr J 788(12):1–7. https://doi.org/10.1111/hir.12320

Post TJ (2020, February 3) Indonesia wages war against coronavirus misinformation as hoaxes spread online. Diambil kembali dari The Jakarta Post: https://www.thejakartapost.com/news/2020/02/03/indonesia-wages-war-against-coronavirus-misinformation-as-hoaxes-spread-online.html. Accessed 10 Feb 2020

Putra NP (2020, May 20) Polisi Sudah Tetapkan 107 Tersangka Kasus Hoaks Terkait Corona. Diambil kembali dari Liputan6: https://www.liputan6.com/news/read/4259573/polisi-sudah-tetapkan-107-tersangka-kasus-hoaks-terkait-corona. Accessed 25 May 2020

Riffe D, Lacy S, Fico F (2014) Analyzing media messages: using quantitative content analysis in research, 3rd edn. Routledge, New York

Rochlin N (2017) Fake news: belief in post-truth. Library Hi-Tech 35(3):386–392

Rossa F, Putri FI (2020, April 1) Stok Darah PMI Kurang, Golongan Darah A dan AB yang Paling Sedikit. Diambil kembali dari Suara.com: https://www.suara.com/health/2020/04/01/125230/stok-darah-pmi-kurang-golongan-darah-a-dan-ab-yang-paling-sedikit?page=all#:~:text=Digolo. Accessed 12 May 2020

Rubin VL, Chen Y, Conroy NK (2015) Deception detection for news: three types of fakes. In: Proceedings of the association for information science and technology (hal. 1–4). Association for Information Science and Technology

Salim HJ (2020, March 4) Cek Fakta: Hoaks Golongan Darah O Rentan Terjangkit Virus Corona. Diambil kembali dari Liputan: https://www.liputan6.com/cek-fakta/read/4193810/cek-fakta-hoaks-golongan-darah-o-rentan-terjangkit-virus-corona. Accessed 28 May 2020

Sembiring LJ (2020, April 11) 1,5 Juta masyarakat RI Kehilangan Pekerjaan karena COVID-19. https://www.cnbcindonesia.com/news/20200411213043-4-151261/15-juta-masyarakat-ri-kehilangan-pekerjaan-karena-covid-19_ Accessed 28 May 28 2020

Septanto H (2018) Pengaruh Hoax dan Ujaran Kebencian Sebuah Cyber Crime Dengan Teknologi Sederhana di Kehidupan Sosial Masyarakat. Kalbiscentia: Jurnal Sains dan Teknologi 5(2):157–162

Silalahi RR, Sevilla V (2020) Rekonstruksi makna hoaks di tengah arus informasi digital. Jurnal Global Komunikasi 1(1):8–17

Simarmata J, Iqbal M, Hasibuan MS, Limbong T, Albra W (2019) Hoaks dan media sosial: saring sebelum sharing. Yayasan Kita Menulis, Medan

Sommariva S, Vamos C, Mantzarlis A, Đào LU-L, Tyson DM (2018) Spreading the (fake) news: exploring health messages on social media and the implications for health professionals using a case study. Am J Health Educ 49(4):246–255

Tandoc Jr, EC, Lim ZW, Ling R (2017) Defining "Fake News." Digit Journalism, pp 1–17

The Jakarta Post (2020, March 22) Transportation Minister Budi Karya on the mend after contracting COVID-19: https://www.thejakartapost.com/news/2020/03/22/transportation-minister-budi-karya-on-the-mend-after-contracting-covid-19.html. Accessed 28 May 2020

Witteman HO, Zikmund-Fisher BJ (2012) The defining characteristics of Web 2.0 and their potential influence in the online vaccination debate. Vaccine 30:3734–3740

Zaenuddin A.(2018, December 20) Mengapa Facebook Jadi Sarang Hoaks? Diambil kembali dari Tirto.id: https://tirto.id/mengapa-facebook-jadi-sarang-hoaks-dca9. Accessed 28 May 2020

Zarocostas J (2020) How to fight an infodemic. The Lancet 395(10225):676. https://doi.org/10.1016/S0140-6736(20)30461-X

Zheltukhina MR, Slyshkin GG, Muzykant VL, Ponomarenko EB, Masalimova AR (2017) Functional characteristics of the English and Russian Media Texts about the Sochi 2014 Winter Olympic Games: political and linguistic aspects. Xlinguae J 10(3):83–100. https://doi.org/10.18355/XL.2017.10.03.07

Chapter 23
Fake News About Covid 19: Communication Strategies on WhatsApp in Brazil

Eloisa J. C. Klein

Abstract This text analyzes communication operations present in false content and misinformation about the coronavirus and the Covid 19 pandemic disseminated in a group of health professionals, with data collection carried out between March and May 2020. The chapter develops a conceptual discussion about fake news and disinformation, as well as modes of circulation and production of contemporary events. We analyzed the occurrence of waves of dissemination of false content that create ephemeral circuits, with a pattern of political action, simulation of journalism and the creation of an audience outside the media. The association between the concepts of fake news and misinformation allows to glimpse connections between related circuits such as confusion, fatalism and ordering of public action.

Keywords Fake news · Coronavirus · Disinformation · WhatsApp

23.1 Introduction

The discussion about the spread of false content, or fake news, has been broadened since the discovery of Cambridge Analytica's actions in the election of Trump in the United States and with Brexit. The company built psychological profiles of users, based on pages liked on Facebook and a questionnaire, which allowed access to the data of 700 thousand people (Época Negócios 2018). According to the profile of the users, the company started to distribute content, false and true, on subjects with which people felt connected—or themes that frightened them.

Since the report of such operations by a former employee of Cambridge Analytica, several actions to contain the spread of fake news have been perpetrated by Facebook, Twitter and Google (Marques 2020). Such measures did not cover WhatsApp, which served for the large-scale circulation of false content, hate speech and propaganda in the presidential elections in Brazil and India. After that, WhatsApp limited the number of contacts that could receive messages from the same sender. However,

E. J. C. Klein (✉)
Federal University of Pampa, São Borja, Brazil

D. M. Berube (ed.), *Pandemic Communication and Resilience*, Risk, Systems and Decisions, https://doi.org/10.1007/978-3-030-77344-1_23

studies indicate that groups continued to act as content multipliers (Resende et al 2019).

Whatsapp operates as a “messaging application”, according to the company that manages it. These messages can be exchanged from person to person or, also, with several people who participate in a specific group. A Whatsapp Group is a tool of this application that allows a collective chat, with the possibility of adding up to 256 people. With the limitation in the number of members, the platform aims to prevent the formation of large audiences, maintaining the character of conversation between people.

The central aspect to the existence of a conversation group on Whatsapp is the constant exchange of information between members. With intense participation, people in these groups develop some specific logics for the exchange of messages (Klein 2020), such as the use of pictures, the use of alert icons for cell phone text, short texts with a call to action. The short videos and photos (Resende et al 2019), links and appealing headlines are particularly noteworthy. These characteristics make the content circulate very quickly through Whatsapp, which guarantees a great impulse to the spread of fake news.

The public discussion of fake news took on new vigor when, as quickly as the spread of the coronavirus, rumors about the emergence of the virus, home cures, miraculous remedies, and political conspiracies began to spread. An international debate was developed, identifying fake news as a real threat to the lives of human beings.

In this chapter, we analyze the communication logics present in false content and misinformation about the Covid 19 pandemic, through materials received in a group of health professionals in Rio Grande do Norte, Brazil. Although members of the group were trained in the health-care profession, there is an intense reverberation of false content, particularly between March and April 2020.

As we are dealing with an analysis focused on the present, we conducted a study of journalistic publications on Covid 19 and the disinformation circuits associated with it. This dimension triggers the perspective of journalism as one of the ways of knowing about reality, based on an immediacy that “is the point of arrival, the result of an entire technical and rational process that involves symbolic reproduction” (Genro Filho 1987, p. 58–59). In this process, journalism can “reveal aspects of reality that other modes of knowledge are unable to reveal”, partly because journalism “not only reproduces the knowledge produced by itself, but also reproduces the knowledge produced by other institutions” (Meditsch 1997, p. 3).

We pursued investigations of WhatsApp groups (Resende et al. 2019), focusing on the concept of fake news (Tandorc et al. 2017; Allcott et al. 2017; Fiocruz 2020) and its relations with contemporary networks and its circuits (Braga 2012; Vos 2015) of event-experimentation. Theoretical systematization was the basis for a content analysis of a set of materials received by WhatsApp. We obtained material that do not reveal people’s identity from an informant who is a member of the group. The group of health professionals has members mostly linked to the city of Natal, Rio Grande do Norte. The informant was asked to send any content without confirmed authenticity about coronavirus that he received in the groups for a period of time extending from March 15 to May 5.

23.2 Fake News and Misinformation

Although the word fake news is contested by websites that specialize in journalistic checking and international organizations that try to regulate the subject, the term has become widely used. Madeleine de Cock Buning (2019), director of the European Commission's group of experts in disinformation, analyzes that "disinformation" is a more appropriate term, since over the years, the term fake news has been appropriated by people who want to disqualify anything outside their ideological field. Despite the attempt by regulators to replace the term fake news with "disinformation", this word in Brazil refers to another meaning, which is the lack of knowledge on a topic or even ignorance—especially since the Portuguese language does not have the word "misinformation". Although the fake content chains involve confusion and lack of information, there are other aspects, such as strategic production of falsification, monetization of untrue materials, or association with opinion, belief, and political preferences. Thus, it is relevant to associate the term disinformation with the terms fake news and false content.

Social media platforms have become spaces with a wide spread of content reported as fake, while they are characterized as a place to search for information.[1] Allcott et al. (2017) analyze the fact that misperceptions and misinformation are a historical problem when considering political processes but note that the use of fake news for convincing purposes has become alarming since the 2016 United States elections. The authors compared posts considered fake on Facebook and other sites and detected a reduction in fake news after the measures taken to restrict this type of content.

Tandorc et al. (2017) elaborate a typology of academic definitions about the term fake news. The authors mention that 75% of people may confuse events with fake news and, although many do nothing with such content, some people end up taking actions, such as choosing a candidate to vote. One of the characteristics of what is conventionally called fake news has to do precisely with the intentional production of disinformation, since many times those who create and share content know that it is an unverified or manipulated subject. In addition, there is the ideological correspondence factor, which makes people promote the ideas they like and don't give credit to the ideas they dislike.

The addition of aspects of predilection and ideology as part of the classification of content about reality makes it difficult to accurately separate opinion and falsification. This is because the opinion can also be, intentionally or unintentionally, used for falsification. Kumar and Shah (2018, p. 2) analyze that "according to intent, false information can be categorized as misinformation, which is created without the intent to mislead, and disinformation, which is created with the intent of misleading and

[1] In Brazil, 70% of internet users get information from social media (OLHAR DIGITAL 2019).

deceiving the reader". The malicious type is more dangerous, since it is strategically designed and operated, with the purpose of reaching a large number of people and generating engagement of this audience to execute actions. "Based on knowledge, false information is categorized as opinion-based, where a unique ground truth does not exist ..., or as fact-based, which consists of lies about entities that have unique ground truth value". Lacking a single common truth (for example, the cause of an event), opinions and views on reality are more difficult to verify, although they are often associated with misinformation processes. Strategies for adding points of view about reality can also increase shares.

A large number of people click on eye-catching (albeit fanciful) headlines, and this means that the website that issues them receives a large number of visitors and can generate large amounts of advertising. Thus, the circulation of fake news in large quantities provides monetization, due to the display of automatic ads. Because of this, a group of activists gathered under the title Sleeping Giants began to act with the objective of pressuring companies to remove ads from sites containing fake news. Activists choose a website that regularly publishes fake content and initiate a campaign for companies to retrieve ads (Pires and El País 2020). In social media, the legitimation of a content can also happen through valorization via ranking (what is most liked, commented on or shared) and by the 'majority feeling', when many people talk about some topic (Tandorc et al. 2017).

An essential feature of the term fake news has been used for a long time for media companies, which described fake news as content that copied the news format, but with unverified information and produced outside the professional field (Tandorc et al. 2017). This characteristic evokes the notion of simulation (Vincente 2005; Ingalls 2008), which in computing and engineering is related to the generation of models to predict behaviors of a system. In the field of education and science, the simulation is performed to reproduce real events on an experimental level. In journalism, the occurrence of simulation has to do with the use of language characteristics and the textual and visual pattern (with headlines and factual photographs, for example). However, these simulations of journalism often occur without the deontological concern of comparing data and versions, observing the accuracy of documents and facts, or adding a multiplicity of voices.

The term parody is similar to the notion of simulation or copy of journalism. To illustrate this aspect, Tandorc et al. (2017) take up the case of the broadcast by Orson Welles, who wanted to narrate a fictional invasion of extraterrestrials as entertainment but adopting the form of a live radio. From this example, we can think that the shape of the product prevailed over the reasonableness of the content's veracity. It was only assumed that a live broadcast should refer to facts. The significant occurrence of materials that simulate journalism or the reporting of the real, among the false contents, evokes the notion that form is a constituent of meanings. Texts that look like headlines, that simulate objective language to deal with facts, that contain quotes in quotation marks are automatically taken as 'true'.

Social media creates the possibility that ordinary people who evidence events or facts report on such situations. Thus, some people post materials they produce as witnesses—and send them directly to their audiences, without having to go through journalism. Tandorc et al. (2017) consider that when people write in 140 characters, an expanded headline model, they also partially assume the logics of the journalistic text, an effect that is amplified when it comes to authorities or celebrities writing to their audiences. The speech of people with an authority function is taken as a referent of the real, even if these people tend to falsify. During the pandemic, Twitter and Facebook had to delete posts from the Brazilian president that had fake content several times (Marques 2020).

Chaparro (1996) analyzes that there is an assimilation of journalistic language, which is replicated by institutions and public agents—a process that expands when such sources, instead of trying to schedule journalism through events and releases, create their own communication on social media. Therefore, although we try to create parameters to define what would be fake content and its difference with the real one, there are limits placed by the daily and secular experience with the media, which links true meaning to the news format or to what is said by the witnesses of events and statements by the authorities. Thus, what is endorsed by a close person or an idol or leader, tends to be framed as true.

23.3 Disinformation Circulation During Social Crisis

Pressure from users and countries has led companies that manage social media sites to seek actions to control the spread of fake news. Twitter and Facebook have adopted the policy of deleting fake content. On WhatsApp, the control of the circulation of fake news has no clear guidelines. In an analysis of groups with an access link during the truckers' strike and the 2018 Brazilian presidential elections, Resende et al (2019) collected thousands of posts, in which they observed a predominance of visual content, which guarantees expressive shares. In the electoral period, 36.2% of the images with factual content contained misinformation, 53.2% contained misleading and inconclusive content and "only 10.6% were verified as true" (Resende et al. 2019, p. 7). To reach these conclusions, the authors compared the content with analyzes of fact-checking companies. At that time, it was observed that the images appeared first on the web and then on WhatsApp, with only 6% of the content appearing first in the application. Following the analysis, it is possible to think that controlling the circulation of fake content on the internet may result in less spread of fake news on WhatsApp.

The company responsible for WhatsApp says that the application is not intended to form audiences (Spagnuolo 2019). However, in some moments, it is identified that user strategies go beyond the limits imposed by limiting members. This is what happens when looking at the case of "mass shooting" for "user base" (Folha de S. Paulo 2018), a practice that was restricted after problems in the election of Brazil and cases of lynching in India. Another way to guarantee reach to thousands of people is the administration of group networks (Klein 2020), when an administrative nucleus guarantees the receipt and transfer of content to members of several groups at almost the same time. The practice is spontaneously experienced by communities, but it is also orchestrated by companies. The articulation of audiences ensures wide circulation, without public visibility and consequent discussion about the characteristics and ethics of the contents.

During the Covid 19 pandemic, with the increase in the number of cases, events with public concentration started to be canceled, classes were suspended and even commercial activities were interrupted in some moments (Veja Saúde 2020). Such actions were also implemented in 2009, with the H1N1 outbreak. However, there was already a vaccine for H1N1, which was made available by governments worldwide, reducing measures to contain public circulation. In the Covid 19 pandemic, there was no vaccine to be made available during the period of circulation of the virus in 2020. Without a vaccine, WHO's guideline for reducing the spread of the virus was to restrict social circulation.

Despite being one of the largest countries in territorial dimensions and number of inhabitants, Brazil has not had an unified action in its 27 federal states. Because of this, the measures taken by state and municipal governments competed with public statements by the President of the Republic, who, for opposing the restriction of circulation, attacked governors, mayors and even ministers of the Supreme Court (Chaib and Teixeira 2020). With the competition of declarations and divergent measures, the circulation of false content increased, due to the emphasis on polarization.

The historian Christiane Maria Cruz de Souza (apud Biernath 2020) analyzes that there is a similar chronology in the actions carried out during pandemics already experienced: first, the authorities deny; then, tips of alleged cures are spread, such as lemon, garlic and even alcoholic drinks; then, people practice civil disobedience of the imposed rules. The denial of the existence of the pandemic is at the basis of stigmatized speeches, such as the "Spanish flu" nomenclature for a disease whose initial focus did not arise in Spain, or the attempt to call the coronavirus as the "Chinese virus".

Based on an app for reporting false content, Fiocruz (2020) analyzed the types of fake news that circulated in Brazil from March to May. In the first phase, 20% of fake news were about "homemade methods to prevent the contagion of Covid 19, 5.7% refer to bank scams, 5% mention scams about seeking funding for research institutions and 4, 3% refer to the new coronavirus as a political strategy". In the second phase, from April 11 to May 13, among the fake news "24.6% affirm that the disease is a political strategy, 10.1% teach home methods to prevent the spread of the new coronavirus, 10.1% defend the use of chloroquine and hydroxychloroquine without scientific proof of effectiveness and 7.2% are against social distancing"(Fiocruz 2020).

The type of subject shows that the speech of political authorities has a strong impact on the reverberation of false content. In March 2020, before the increase in contamination in the Americas, the President of the United States issued a statement saying that chloroquine had produced "very encouraging previous results".[2] The Brazilian president followed the American leader and made statements defending

[2] Additional information: https://www.focus.jor.br/erca-covid-19-eua-trabalham-para-aprovar-uso-da-cloroquina-em-tempo-recorde/.

the adoption of the drug. As noted by the research data, there is a direct connection between the temporality of these speeches and the wide dissemination of falsified content about this medication.

In the contemporary communicational circulation, conditions of registration and social and digital interfaces are linked, with a "continuous flow, always ahead" (Braga 2018, p. 40), which prevents one from clearly observing points of departure and arrival, or roles as distinct as senders and receivers. The indistinction of roles and stages transforms the regimes of eventualization.

In the projection of events on the Internet, direct manifestations of people are mixed with random records of their navigation and personal preferences. In addition, the use of digital platform affordances, such as the hashtag, underpins the generation of mobilization currents. Such information is not necessarily present in the speech of individuals but is also part of the production of events and reverberations on the internet. When dealing with events that mix with the logic of social networks (with network organization, narratives elaborated by the participants, instant public discussion), Henn (2013) classifies the occurrence of "cyber events", which end up returning to journalism, due to the intense reverberation in the networks.

We note that the circuits around fake news do not necessarily need to reach a totally different form from other social events. Because they are directly linked to the individual experience, the event-driven process promoted by the intense circulation of speeches and content can take place on a micro scale and have an impact on niche audiences. Because they occur in restricted domains (such as WhatsApp groups), these niche audiences may not be immediately identified by other audiences in the same society as something that exists. This is what we see in WhatsApp groups, a platform that has become central in the dissemination of information and also false content. Therefore, there is a need to observe user actions and types of content in this domain.

23.3.1 False Content About the Pandemic in a Brazilian WhatsApp Group

The focus of the analysis is on shared content and how they correlate with the circulation of fake news and misinformation. The collection took place between March 15 and May 5, a period in which we obtained 140 materials from a source that was a member of a WhatsApp group composed of health professionals from a Brazilian state. The group gathers around 100 members (the total of members varies with entries and exits), mostly dentists and some other health professionals, coming from one of the 27 Brazilian states, in the northeastern region of the country. We reached this group through reports of systematic occurrence of false content, by complaints made by one of its members to people in another group of messages, whose members are professors at the Federal University of Rio Grande do Norte. This person offered to act as an informant, assigning to our research the false content received, in order to understand the dynamics of spreading fake news and also the strategies present in these materials.

The informant only passed on materials that had untrue, manipulated content or with inconclusive information. For this analysis, debates between members were discarded. We consider links accompanied by comments, images, "WhatsApp message", audio and video files. During data collection, it was possible to observe a correspondence between statements, interviews, lives of the president, along with rumors networks involving deputies from the government base, and the waves of content dissipation in the observed WhatsApp group.

The periods of greatest generation of content are directly related to events with great media coverage. In particular, we note the impact of the Brazilian president's speeches: (1) on March 24, with an official pronouncement calling the Covid 19 a small flu, demanding an end to the quarantine in favor of "vertical isolation" and accusing the media of causing panic, and saying poor people should go back to their jobs[3]; (2) on March 25, at the G20, when he shows a medicine box composed of Hydroxochloroquine, saying that it would bring serenity to the world[4]; (3) still in the last week of Mars, there were speaks of the president to neopentecostal church leaders, in which he says that "he does not know about crowded hospitals" and that the virus "is not all what they are painting"[5]; (4) on April 3, speaking to journalists, the Brazilian president called the religious people to "national fast" to get rid of "the evil" of the virus. He also said that he would move normally down the street and said that Covid 19 was like rain, that 70% of the population would have to get wet[6]; and (5) in an official statement to the nation, on April 8, when he said that hydroxochloroquine was effective to treat Covid 19 from the initial stage).[7] There are also videos recorded by the president and broadcast to groups of militants from his government, as when he called a protest against social isolation, at the end of March.[8]

It is important to note that the topics of these speeches reverberate not only in posted content, but also in people's actions. In the first half of April, people aligned ideologically to the Brazilian president started to enter hospitals without authorization to record videos. With the death toll rising, rumors about empty hospitals have given

[3] The transcription of the speech can be read at: https://noticias.uol.com.br/politica/ultimas-noticias/2020/03/24/leia-o-pronunciamento-do-presidente-jair-bolsonaro-na-integra.htm.

[4] https://noticias.uol.com.br/saude/ultimas-noticias/redacao/2020/03/26/bolsonaro-mostra-remedio-feito-com-hidroxicloroquina-em-reuniao-do-g20.htm.

[5] A report on the speech can be checked at: https://g1.globo.com/politica/noticia/2020/04/02/desconheco-qualquer-hospital-que-esteja-lotado-diz-bolsonaro.ghtml.

[6] A report on the speech can be read at: https://jc.ne10.uol.com.br/politica/2020/04/5604814-bolsonaro-conclama-jejum-nacional-e-diz-que-vai-andar-na-rua-de-novo-mesmo-com-pandemia-de-coronavirus.html.

[7] Excerpts of the speech can be read at: https://www1.folha.uol.com.br/poder/2020/04/em-pronunciamento-bolsonaro-defende-cloroquina-e-volta-a-responsabilizar-governadores-e-prefeitos.shtml.

[8] A report about this case can be read at: https://noticias.uol.com.br/politica/ultimas-noticias/2020/04/30/covid-19-bolsonaro-incentivou-carreata-contra-isolamento-social-em-manaus.htm.

way to rumors of fabricated deaths.[9] In late April, inhabitants of Manaus started to open coffins to check if their family members were really inside, due to rumors about empty coffins. In May, a network of rumors was spread among people and WhatsApp circles and culminated in the viral video of a woman who said there were stones in the coffins.[10] The circulation of this type of material was so intense that it reached the specific WhatsApp group analyzed in this article.

In our sample, among the links posted, only two links appear for journalistic reports: a link to the Tribuna do Norte, RN, about the state government's withdrawal from advertising on measures against the virus, and a link to a video explaining what was meant by "vertical isolation", posted by BBC Brazil, on its YouTube channel. In both cases, the selection of news converges with the political position of those who post, pro-Bolsonaro and against Fátima Bezerra, governor of RN, PT, opposition party to Bolsonaro. There is also a link that directs to *Estadão*, however in the opinion section, with text entitled "comunavirus", with a post talking about "Chinese virus" and icon of the flag of China. There is a link to an opinion article in *The New York Times*, with commentary in Portuguese on Trump's change in attitude. All other links are from channels and pages of individuals and groups not necessarily linked to journalism or public production.

There is a rise in blogs and websites that mix factual information with personal opinion and imagination. We found eight cases in our sample. Some of these texts build falsifications by projecting a person's intention or thought, not manifested by actions and speeches. Other texts treat people from political fields not aligned with the president as bad, as a mention of a former president as "highly involved in darkness". The construction of fanciful scenarios is also present in texts written next to the shared link: "the fear of the demonstration that became known as 'Foda-se's Day', has left the political enemies of the president [Bolsonaro] terrified". The end of this message mixes (a) misinformation and falsification of the positions of people with divergent thoughts (directly linked to the notion of fake news), (b) opinion that stands as truth about the world, (c) and a militant organization that acts in a convening way, with the addition of religious subjects. Such styles of text build a morally good and correct position for the speaker, in addition to mentioning facts that did not happen.

Some content appears in different formats, such as a video that contains an interview made by a TV anchor to a professional who treats the use of hydroxochloroquine in a 72-year-old patient as "hope", "good news" to be celebrated. The content is then replicated in another video, a channel that deals with the use of the drug in 500 patients by the company Prevent Senior. The video is shared with the tip: "if you feel the symptoms, send this video to your doctor and ask him to make an immediate prescription, it can save your life!" The text used to share the video also says that "adopting the medicine late does not help anymore, 50% of patients who are [admitted] to the ICU die". The affirmation of drug administration in early stages is

[9] The full story can be checked at: https://piaui.folha.uol.com.br/lupa/2020/04/23/verificamos-caixoes-vazios-amazonas/.

[10] The full story can be checked at: https://www.santanafm.com.br/fake-news-autora-de-video-que-falava-em-caixao-com-pedras-em-bh-pede-desculpas/.

present in another 11 audios and videos (attributed to doctors) that reached this same group of WhatsApp.

In the sample analyzed, the idea of text-image is present, which we observed in other information dissemination circuits on WhatsApp (Klein 2020). Three of these materials refer to cards, digital leaflets of deputies, one summoning his electoral base to the "national fast" as a solution to the virus; another speaks of the allocation of electoral funds to fight the coronavirus (such a reversal of the allocation of money between ministries is not legal). Another material says that there is an attempted coup by the "PT" (Workers Party—which opposes the current president, Jair Bolsonaro) to force a new election for president and vice president (this conjecture never happened). There are general statements about "politicians" and "parties", it is interesting to note that, although issued by politicians, they do not seem to be considered as such.

There are three impact phrase records with visual form, simulating the post of an ordinary person, with a language pattern similar to the Facebook / Instagram stories: "The tendency is for Covid 19 to kill more CNPJ [companies] than CPF [people] !! Sad reality"; "Can you call it myth-chloroquine or is it early yet?" The contents are connected to waves of events formed by the correlation of political opinions, fake news and disinformation: the closing of companies as the fault of governors who imposed quarantines, and the political recommendation of using a drug.

There are five screen shots of posts on Twitter, which have already been edited, and can also be part of a wide cycle of sharing. One of them defends President Bolsonaro's pronouncement against quarantine and says that people misinterpreted it, another refers to the use of medication. Another post also defends the president's pronouncement and negatively frames China, a position that is present in other public statements of the president of Brazil and people of his government and family[11]: "Well, as Bolsonaro said: China does not want to buy from Brazil, it wants to buy Brazil".

There is a set of materials that can be thought of as a 'text message for WhatsApp', in general they are not published on other platforms, they do not refer to external visual or audiovisual content, there are no links. The content tends to be opinionated, although there is reference to episodes experienced at the time of posting and sometimes there is data, although adapted to the bias of the post. This is what happens in relation to the criticism of the use of funds for advertising by the government of RN (opposition party to Bolsonaro), which contains official publication data, but also a misinterpretation: the advertising would be made to publicize actions to confront the Covid-19 but is treated in the message as a government promotional advertisement. It is observed that the falsification can present itself in a tenuous way, based on modalities of opinion and interpretation expressed in criticisms.

There is a greater occurrence of totally false content in this type of material. One of them outlines an alleged (non-existent) association between the Supreme Court, the president of the Chamber of Deputies, together with what he calls "heads of

[11] News on the subject can be found at: https://www.em.com.br/app/noticia/nacional/2020/03/18/interna_nacional,1130290/embaixada-china-protesta-eduardo-bolsonaro-por-culpar-pais-pandemia.shtml.

left parties" (among whom the speaker places parties that are in fact center-right in Brazil). Such an association would aim to destroy Bolsonaro's government. Another text message to WhatsApp makes an accusation that the coronavirus was sought by the left to resolve the loss of elections in the world. "After Trump won the election in 2016, George Soros went to the media and said, 'Trump is the stone in the shoe for a new world order.'" Another excerpt: "With the action in progress, a new space will be opened for the left to take control, placing a socialist or communist in front of the country, thus establishing a global plan that is in progress". George Soros' statement never existed, nor is there a formed front for the rise of communism in Brazil.

The aggressive style of the content is repeated in most of these messages. In a long text that classifies those who are applying norms of social distancing as hypocrites, the aggressiveness is converted into accusation: "#hypocrites in addition to FAILING THE COUNTRY we will all die, you idiots… without money you will be without medicine without food". Another text, also as a criticism of those who are socially distant, repeats the tone: "It irritates me that phrase 'we are in the same boat'. No we are not. Do not be ridiculous. We are in the same storm, but not in the same boat". Such texts oppose people who do home-office versus people who pick up trash for living. In other materials, text operations create opposition between poor people and actors on a TV station. The critical content of the oppositions creates an idea according to which the author is defending the most vulnerable people, who would be favored with the end of the quarantine, as mentioned in the president's speeches. In these texts, there is no claim for a government interference to repair the situation of these vulnerable people.

The militant content is also present in 'calls for action', based on nationalist messages linked to biblical passages. It is pertinent to note that they call for the construction of an audience apart from the media circuit, taken as an opponent, and the creation of suspense around supposed secret revelations. "There will be a live [streaming] on YouTube today at 8:30 pm from journalist Oswaldo Eustáquio, there will be very important revelations and it will be at exactly the time that Moro will be interviewed on Globo, we will bring down the audience and follow this live … if (you) are interested, pass it on to the largest number of people". The most curious of these texts teaches how to bypass blocking of massive postings on WhatsApp and ends with the following phrase, in capital letters: "review these instructions for everyone to learn how to get out of this censorship imposed by Whatsapp".

There are insinuations that seem to go beyond the limits of reasonableness and even legality, with accusations of electoral fraud (two years after the elections), phrases calling ministers of the Supreme Court criminals, speculation that such ministers are linked to the mafia for not wanting the president to appoint the head of the Federal Police (in fact, there was an investigation underway to comply with legal steps). The most dangerous threatens members of the judiciary: "The PT (opposition to Bolsonaro) and its associates are using their last bullet: the supreme court. For them, captain [as the president is known] has the 'silver bullet'".

In the sample, two audios were collected whose authors designate themselves as doctors. In the first, the speaker says that there are no serious cases where he works and criticizes that children are at home. "If it is known that zero to two years the chance of death is one in a thousand, these had to be on the street, to contaminate themselves and create resistance. They are not, so everything is locked up. One cannot compare it with Europe. Brazil is a big country, a hot country and what they are saying is not going to happen". This audio begins with direct reference to the presidential pronouncement of March 24, to which the audio is favorable, in addition to saying that the media only cares about Covid 19 because the disease started with the rich, that if it were with the poor, like H1N1, they would not cover it. It is interesting to note that there are no data or references, but only personal preference, which is nevertheless sustained as expertise, by the attribution of the professional title of a doctor.

The second audio, also attributed to a doctor, discourages the use of a mask, stating that mass contagion would be necessary to generate "community immunization" and that patients would be treated with known medications, referred to by their names: "if these drugs are used competently, by a trained health team and even at the onset of symptoms, we would have reductions in hospitalizations and severe forms of the disease". Such content generates a proof effect, confirming reality, although based on individual preferences and opinions expressed by the author of the audio.

Videos are the most representative material in the sample, totaling 30 contents. One of the video categories is the account of personal experience. A mark of these videos is the presentation of the profession: mechanic, driver, beach salesman, or TV host. The mechanic talks about who is going to fight for their children and the concern with "financial life, because everything is stopped". The beach vendor talks about the empty coastline and the closed restaurants. The driver talks about the empty road and the difficulty in finding food and that if the trucks do not circulate, doctors and nurses would not be able to eat either. All speeches are against the media. There is a repetition of the idea of non-partisanship. The TV host appears as a political analyst, says that he would vote for ex-president Lula if he ran (ex-president, opponent of Bolsonaro), but that he defends the speech of the current president, Bolsonaro, who would be able to save the country from the imposition of quarantine by governors.

The videos with speeches from doctors mix personal opinion and expertise. A doctor orders those who watch the video to stop watching news about Covid 19 and to "calm down" because "the world is not ending". Another health professional talks about his curriculum and says he is not involved in politics: "It is necessary to make decisions and unlock the country. Are people going to die? Yes. People in risk groups are really going to be more exposed. Truth. But how many people will also die of hunger?". The video ends with criticism of governors and deputies. In all the videos of doctors, they appear wearing lab coats, sometimes embroidered with abbreviations that refer to the word doctor, followed by name. There is a common relativization of the disease.

Another category of videos includes politicians who are presented as experts. One is a physician who served in state and federal government spheres, and lists his curriculum as a manager, to define his own place as an authority. "People who are sick have to be confined, but the rest of the population has to live as normal a life as possible. This will not change a person who will be sick, who will die". The governor of the state of Santa Catarina is presented, by those who post the video in the

WhatsApp group, as someone who attenuates, but confirms what Bolsonaro presented in the speech, reinforcing the idea of not complying with social isolation. Finally, there is a deputy who presents a project to use an electoral fund to build intensive care beds, because "most deaths are not due to the worsening of the disease, but due to the absence of ICU beds". The deputy speaks shouting, without pauses, attacks what he calls 'the politicians' and 'the parties' and does not seem to see himself as one of them, even when he uses the first person plural pronoun.

Another category of videos involves montages of media content and clippings of excerpts from television programs. The clippings operate as a way of building perspective, of political orientation. One of the clippings shows the enthusiastic conversation of a TV anchor with a professional from an elderly care company. At the time of the interview, the mother of the company owner was being treated with hydroxochloroquine and, according to the professional interviewed, showed improvement. "We begin to literally create hope there, for the country and for everyone who needs to treat these patients". The name of the remedy is repeated several times, and the anchor pretends not to know how to speak correctly, so that he can be corrected and able to speak the drug's name again.

A montage appeared three times, in different formats, with a compilation of news about H1N1, from 2009. In the middle of the news excerpts, a phrase was put forward by former President Lula saying that the H1N1 virus did not seem to be of the expected scale and that Brazil was taking measures to protect the population. Such videos circulated in the weeks following the pronouncement of the current president, who calls the Covid 19 a "gripezinha", a "little cold". There is an attempt to attenuate criticism of Bolsonaro, comparing his reference to "small cold" with Lula's speech, without contextualizing the variants between epidemics, such as the existence, at the time, of the H1N1 vaccine. Another group of montages confronts media coverage with statements by doctors who defend the end of the quarantine.

A video montage simulates the characteristics of professional journalism to create fake content, considering the virus as an attempt at communist domination, and adding the idea that China is concerned with ensuring abortion in the middle of the pandemic and that the country has taken control over the World Health Organization. The beginning of the video is with a person who takes the place of a reporter (function assigned by characters). The video is interrupted by a screen photograph of media publications about Covid (with recognized sites, such as El País and BBC), and the last slides are not from traditional media, but contain a subject out of context—abortion. The combination of different content induces the viewer of the video to err. A similar strategy is observed in another video, which mixes information extracted from personal bills with funds directed by the federal government to the states. Both pieces of information are true, but together they don't make sense.

In April 2020, there is a wave of videos making denunciations, with people going to hospitals with an open camera, entering emergency rooms and saying that "there is no one" [in the place], "there is no pandemic". The videos simulate narration and takes often found in television reports, such as: the appearance of the narrator as a reporter; explanatory text accompanying the images; use of different plans and focus on details (such as hospital names and names of people on badges). There is a simulation of journalistic products, although it is limited to individual testimony.

This dispersed audience, which receives calls for WhatsApp messages, is active in these videos. And it also proves to be active due to the constant transfer of such materials in some broader social groups, such as the WhatsApp group where the materials analyzed here were posted. The call to share the messages is answered, so that the contents quickly reach groups in different states of the country.

23.4 Conclusion

As a result of the analysis, we found that there is a way to produce false content about social events and problems that goes beyond established institutions, such as science, medicine, international organizations, and governmental institutions that follow international standards, which are typically the preferred sources of journalism. Some professional voices are used in posts as representative of an explanation about reality; for example, when a doctor's speech is shared on Whatsapp as a solution to the pandemic, despite the recommendations of organizations such as the WHO. It is also observed that there is no predilection for institutionalized journalism. There is a strategy for creating an audience outside the traditional media, generally based on texts that promise dramatic revelations or that propose militant actions. In the flow of information, there is no cohesive narrative, but constant fluctuations in the perspectives regarding events, speeches, rumors and versions regarding facts.

Different materials (videos, audios, WhatsApp messages, images) refer to common topics. Various content criticizes governors; say they are people outside of politics; mention closing trade as destroying the economy; recommend medication use; and preach collective contamination. The materials are not issued at random, there is a relationship with chains of subjects present in social media. In the case analyzed, we observed the proliferation of content linked to criticism of Brazilian governors (after the president's statement giving this guideline), the use of medicines (after videos of the presidents of Brazil and the United States talking about these drugs), the investigation of hospitals supposedly empty (after political leaders associated with the president questioned the number of deaths and the non-reopening of trade required by the Brazilian president).

These constant fluctuation challenges the logic of institutionalization. There is no repetition for long enough to create sedimentation. The circulation of content in this networks are formed and disintegrate into immediate temporalities, without clinging to institutionalizing social practices and without sewing their own narratives. Rather, they engage in fragments of great narratives, which came before them and will continue afterwards. Such fragments refer to emotions, revived from repeated behavior when dealing with such structural narratives, such as the great nation, the brave man, or the savior. However, emotion is part of the process of misinformation, as it is ephemeral. The emotional reaction takes place and then almost immediately falls apart, but in its short duration it contributes to content share. Thus, when observing the circulation of fake news on WhatsApp, we note that such content is not guided by specific parameters of informative narratives, but according to a fluctuation of fragments that stick together in several streams, simultaneously.

References

Allcott H, Gentzkow M, Yu C (2017) Trends in the diffusion of misinformation on social media. Res Polit 6(2). https://www.nber.org/papers/w25500. Accessed on Mar 2020

Braga JL (2012) Circuitos versus campus sociais. Mediação & Midiatização. Salvador, Bahia: EDUFBA, pp 31–52. http://www1.pucminas.br/imagedb/documento/DOC_DSC_NOME_ARQUI20180205111302.pdf. Accessed on Mar 2020

Chaparro MC (1996) Jornalismo na fonte. In: Dines A, Maurin M (org.) Jornalismo brasileiro: No caminho das transformaçõe. Banco do Brasil, Brasília, Distrito Federal, pp 132–154

Época Negócios (2018) Quem é Aleksandr Kogan, o professor por trás do escândalo da Cambridge Analytica. Época Negócios, 23 Apr2018. https://epocanegocios.globo.com/Tecnologia/noticia/2018/04/quem-e-aleksandr-kogan-o-professor-por-tras-doescandalo-da-cambridge-analytica.html. Accessed on Jan 2019

Fiocruz (2020) Estudo identifica principais fake news relacionadas à Covid 19. Fiocruz. 21 May 2020. https://portal.fiocruz.br/noticia/estudo-identifica-principais-fake-news-relacionadas-Covid19. Accessed on Mar 2020

Genro Filho A (1987) O segrego da pirâmide: para uma Teoria Marxista do Jornalismo. Editora Tchê, Porto Alegre, Rio Grande do Sul

Ingalls RG (2008) Introduction to simulation. In: 2008 Winter simulation conference. https://www.informs-sim.org/wsc08papers/005.pdf. Accessed on Mar 2020

Klein EJC (2020) A consolidação de lógicas interacionais informativas em grupos de aplicativos de mensagens. Compós. XXIX Encontro Anual. Universidade Federal de Mato Grosso do Sul, Compós, Campo Grande MS. https://www.compos.org.br/anais_indice_autores.php?idEncontro=MzA=. Accessed on Dec 2020

Kumar S, Shah N (2018) False information on web and social media: a survey. In: Social media analytics: advances and applications. CRC press, Boca Raton, Flórida

Marques J (2020) Depois do Twitter, Facebook e Instagram também apagam post de Bolsonaro. Folha de S. Paulo, 30 de março de 2020. https://www1.folha.uol.com.br/poder/2020/03/depois-do-twitter-facebook-tambem-apaga-post-de-bolsonaro.shtml. Accessed on Mar 2020

Meditsch E (1997) O jornalismo como forma de conhecimento? Conferência. Cursos de verão da Arrábida. *BOCC*. Disponível em http://www.bocc.ubi.pt/pag/meditscheduardo-jornalismo-conhecimento.pdf. Accessed on Mar 2020

Resende G et al (2019) (Mis)Information dissemination in whatsapp: gathering, analyzing and countermeasures. In: International world wide web conference committee 19. 13–17 May 2019, San Francisco, CA, USA. Disponível em: https://people.mpi-sws.org/~johnme/pdf/resende. Accessed on Oct 2020

Tandorc Jr EC, Lim ZW, Ling R (2017) Defining "fake news. A typology of scholarly definitions. Digit Journalism. https://www.researchgate.net/publication/319383049_Defining_Fake_News_A_typology_of_scholarly_definitions. Londres, Reino Unido: Routledge, Taylor & Francis group. Accessed on Oct 2020

Vicente P (2005) O uso de simulação como metodologia de pesquisa em ciências sociais. Cadernos EBAPE.BR. vol 3, no 1. Rio de Janeiro, RJ

Vos T (2015) Revisiting gatekeeping theory during a time of transition. In: Vos TP, Heinderyckx F (eds) Gatekeeping in transition. Taylor & Francis, New York

Epilogue

This team of authors worked through an international pandemic in order to pull together a book on pandemic communication and resiliency in six months. One of us turned in a chapter with a newborn in tow. We had physical injuries including a broken wrist, some bad reactions to vaccines, and a host of other events that affected our schedules.

We are mostly academics though we had professionals in health services and public relations contribute as well. One of the reasons communication studies is so interesting is the host of people with eclectic perspectives who come together to analyze and offer recommendations to deal with some of the most problematic issues of the day.

After half a decade working on Zika, it surprised me that we approached the outbreak in South America in a highly disorganized fashion. While Puerto Rico, Brazil, and Colombia shouldered the worst of the epidemic, the world was fortunate it never spread into a pandemic like SARS-CoV-2 has this year. Still, we seemed to have learned so little from Zika. When it passed into history, either due to herd immunity or having evolved into a much more benign variant, it was forgotten and we seemed to go about our merry ways until Covid-19 hit us. We acted as if this was an "act of God" and maybe it was. Nevertheless, pandemic experts seem to warn every time we encounter a pandemic that we are woefully underprepared to act. As more pandemic histories passed across my desk, they told horrific stories about human vulnerability to outright extinction from pandemic events and warned over and over again that we need to find some way to approach pandemics with protocols and tools in hand.

The world cannot wait until the World Health Organization declares we have a pandemic; we need to prepare. Pandemics are not unpredictable. They are the natural result of humankind encroaching into ecosystems that sustain wild animal life carrying killer viruses. As climate change occurs and weather shifts, we may find more vectors and reservoirs for viral agents. The world needs protein and people find fresh protein in wet markets and some cultures retain a closer relationship with

D. M. Berube (ed.), *Pandemic Communication and Resilience*, Risk, Systems
and Decisions, https://doi.org/10.1007/978-3-030-77344-1

species vectors as do many of us in the West. On occasion we take our chances though when we participate in the rare animal pet trade. And we assuredly cannot have the occasion where a sitting President of the USA fires the Department of State's expert pandemic team.

For most of us during Covid-19 we were teaching online and or involved in mixed online/face-to-face classes as well as shepherding some more advanced students through their examinations, theses, and dissertations. For those of us with children, we found ourselves coordinating K-12 online learning where that was possible. We served as amateur psychologists and career advisors. Our friends and partners were out of work and some of us watched as our friends and colleagues got sick some of whom succumbed. For most of us the Covid-19 pandemic passed and we were left untouched. We were fortunate.

When recruiting colleagues and some new friends for chapters for this book, there was a high level of interest. There were many different chapters offered, some near completion and others designed from scratch. Over the months that it took to pull this edited volume together, we learned so much more about this pandemic.

Our political and health care systems were stretched to the breaking points. While some of us had strong and prescient leaders, others of us had leaders who were ill-prepared or under-prepared or maybe simply unwilling to rise to the challenge. The pandemic spread (120 million worldwide) and nearly 68 million recovered, but many died, over a half million here in the USA and over 2.65 million across our planet.

There are plenty of other communication and resiliency related subjects that remain to be discussed such as how and when vaccines will be shared internationally, the effectiveness of online meeting platforms that dominated many of our daily lives, efforts to reduce misinformation on social media by government and industry and its overall effectiveness in reducing fake news, what did the term "infodemic" mean and how did it affect messaging, how conspiracies became information sources for too many people and what could be done to set the record straight, how terms like "the China" or "the Wuhan" virus may have been associated with rising violent attacks on Asians in the West, how we could reach out to people of color and people living in rural areas who were most at risk yet least willing to be vaccinated, how we could maintain mental health outreach efforts associated with lockdown and social isolation as well as the depression on first responders and health service providers, the communication shortcomings of Trump's "Operation Warp Speed" and the effects FDA first emergency use authorization status on vaccine hesitancy, how we can maintain public health messages in such a politically polarized country, like the USA, how to communicate with failed businesses and find ways to reduce unemployment among the service industries. how to improve the role of medical experts as storytellers, what levels of public engagement are called for and how can we maximize outreach to stakeholders, and many, many more.

Nonetheless, we produced a book that samples approaches to building resilience among populations of stakeholders during pandemics using the theories and models developed by communication scholars, especially those in health, crisis, and risk communication. There can be no doubt that professional conferences hosted in health promotion and health communication over the next few years will feature panels of experts examining SARS CoV-2 and the effect it has had on all of us. While these activities have intrinsic and extrinsic value it is more important that we approach the difficult issues of public understanding. The public needs to take pandemics seriously and understand that activities like mask wearing, hand washing, panic-buying avoidance, working from home and self-isolating and quarantining, and keeping away from enclosed and poorly ventilated areas are important public health determinants. Maybe more importantly, we will need to determine how to reduce vaccine hesitancy and removing vaccination from the clutches of politics and misinformation.

This book was a wonderful experience for me. I met some great people and helped to mold this edited volume with a lot of input from colleagues and friends. Each of the chapters went through peer review. I wish we could have published all of the submissions but things are what they are. If you have complaints or are interested in working on similar projects with me such as moving this debate forward in other venues or even another publication, please contact me dmberube@ncsu.edu.

Index

D. M. Berube (ed.), *Pandemic Communication and Resilience*, Risk, Systems and Decisions, https://doi.org/10.1007/978-3-030-77344-1

W

Z

GPSR Compliance

The European Union's (EU) General Product Safety Regulation (GPSR) is a set of rules that requires consumer products to be safe and our obligations to ensure this.

If you have any concerns about our products, you can contact us on ProductSafety@springernature.com

In case Publisher is established outside the EU, the EU authorized representative is:

Springer Nature Customer Service Center GmbH
Europaplatz 3
69115 Heidelberg, Germany

Batch number: 10756376

Printed by Printforce, the Netherlands